Marine Polysaccharides
Volume 1

Special Issue Editor
Paola Laurienzo

MDPI • Basel • Beijing • Wuhan • Barcelona • Belgrade

MDPI

Special Issue Editor
Paola Laurienzo
Institute for Polymers, Composites and Biomaterials, CNR
Italy

Editorial Office
MDPI AG
St. Alban-Anlage 66
Basel, Switzerland

This edition is a reprint of the Special Issue published online in the open access journal *Marine Drugs* (ISSN 1660-3397) from 2010–2017 (available at: http://www.mdpi.com/journal/marinedrugs/special_issues/polysaccharides-2010).

For citation purposes, cite each article independently as indicated on the article page online and as indicated below:

Lastname, F.M.; Lastname, F.M. Article title. *Journal Name* **Year**, *Article number*, page range.

First Edition 2018

Volume 1
ISBN 978-3-03842-897-8 (Pbk)
ISBN 978-3-03842-898-5 (PDF)

Volume 1–3
ISBN 978-3-03842-743-8 (Pbk)
ISBN 978-3-03842-744-5 (PDF)

Table of Contents

About the Special Issue Editor

Paola Laurienzo grew up in Naples, Italy. In 1983, she graduated in Chemistry at "Federico II" University of Naples. This was followed by a Post-doc position at Italian Research National Council. She was appointed as Researcher at the Institute for Polymers, Composites and Biomaterials (IPCB) of CNR in Pozzuoli (Naples, Italy) in 1986. During the first 15 years, her research activity was mainly devoted to polymer and copolymer synthesis with standard and innovative strategies; chemical modification of synthetic polymers; design and chemical–physical characterization of blends; the study of the structure–properties correlations of multiphase polymeric materials. Innovative films for food packaging, new polymers for applications as components in electro-optical devices, and recycling of plastics from waste through reactive blending were developed technologies. Two national patents were obtained in these years. From the year 2000 onwards, her interests have focused on biodegradable polyesters and natural polysaccharides for applications in tissue engineering, drug delivery, and hydrogels for wound healing. Her experience in the synthesis and chemical modification of polymers has now been extended to the realization of novel amphiphilic copolymers for the design of active targeted polymeric micelles for drug delivery, with a focus on cancer therapy.

Preface to "Marine Polysaccharides"

Volume 1: Advancements in the Discovery of Novel Marine Polysaccharides

The field of marine polysaccharides is constantly evolving, due to progress in the discovery and production of new marine polysaccharides. Seaweed remains the most abundant source of polysaccharides, but recent advances in biotechnology have allowed the production of large quantities of polysaccharides from a variety of micro-algae, by controlling growth conditions and tailoring the production of bioactive compounds in a bioreactor. Of particular interest are polysaccharides produced by micro-organisms from extreme marine environments, due to their recognized different biochemistry. Extracellular polysaccharides (EPSs) with unique properties produced by a number of micro-algae are known. The first volume is a collection of papers concerning the identification and characterization of novel marine polysaccharides. It is divided into three chapters; the first two are dedicated to polysaccharides from different marine sources (algae, micro-algae, animals), while the third one gathers information on the isolation, characterization and bioactivity of new EPSs.

Volume 2: Identification of the Methabolic Pathways Involved in the Biological Activity of Marine Polysaccharides

In the second volume, papers reporting on the elucidation of the mechanisms that underlie the biological activity of some marine polysaccharides are collected. The understanding of the underlying mechanisms is an important feature to give a rigorous scientific support to the potential use of many marine polysaccharides as natural drugs in a wide range of therapies. This volume is divided into three chapters, each of them devoted to a specific class of polysaccharides.

Volume 3: Biomedical and Pharmaceutical Applications of Marine Polysaccharides

Recently-developed technology for production of polysaccharides from marine sources makes their potential use as additives in pharmacological formulations, food supplements, and support material for biomedical implants a real possibility. Although development of low-cost and eco-friendly methods remains a challenge, many companies have developed methodologies for extraction and purification of high quantities of polysaccharides from a variety of natural sources, as confirmed by the high number of trademarks that have been registered to date. Moreover, refinements of technological approaches enable further exploitation of available resources. This volume is a collection of papers focusing on the concrete application of polysaccharides in the biomedical field. In the first chapter, review articles illustrating all the potential applications of polysaccharides are presented. The second chapter includes articles on new methodologies for extraction and purification of polysaccharides of different origins, with particular attention on the evaluation of potential toxicity strictly related to the production process. Finally, in the last chapter, papers dealing with specific examples of biomedical applications are reported. The proposals contained within this collection cover a wide range, including food supplements and services in aquaculture, among others.

Paola Laurienzo
Special Issue Editor

marine drugs

MDPI

Article

The Identification of a SIRT6 Activator from Brown Algae *Fucus distichus*

Minna K. Rahnasto-Rilla [1,2], **Padraig McLoughlin** [3], **Tomasz Kulikowicz** [1], **Maire Doyle** [1], **Vilhelm A. Bohr** [1], **Maija Lahtela-Kakkonen** [2], **Luigi Ferrucci** [1], **Maria Hayes** [3] and **Ruin Moaddel** [1,*]

[1] Biomedical Research Center, National Institute on Aging, National Institutes of Health, 251 Bayview Boulevard, Baltimore, MD 21224, USA; minna.rahnasto@uef.fi (M.K.R.-R.); kulikowiczt@grc.nia.nih.gov (T.K.); maire.doyle@nih.gov (M.D.); BohrV@grc.nia.nih.gov (V.A.B.); FerrucciLu@grc.nia.nih.gov (L.F.)
[2] School of Pharmacy, University of Eastern Finland, Kuopio FI-70210, Finland; maija.lahtela-kakkonen@uef.fi
[3] Food Biosciences Department, Teagasc Food Research Centre, Ashtown, Dublin 15, Ireland; Padraig.McLoughlin@teagasc.ie (P.M.); Maria.Hayes@teagasc.ie (M.H.)
* Correspondence: moaddelru@mail.nih.gov; Tel.: +1-410-558-8294

Received: 22 May 2017; Accepted: 16 June 2017; Published: 21 June 2017

Abstract: Brown seaweeds contain many bioactive compounds, including polyphenols, polysaccharides, fucosterol, and fucoxantin. These compounds have several biological activities, including anti-inflammatory, hepatoprotective, anti-tumor, anti-hypertensive, and anti-diabetic activity, although in most cases their mechanisms of action are not understood. In this study, extracts generated from five brown algae (*Fucus dichitus*, *Fucus vesiculosus* (Linnaeus), *Cytoseira tamariscofolia*, *Cytoseira nodacaulis*, *Alaria esculenta*) were tested for their ability to activate SIRT6 resulting in H3K9 deacetylation. Three of the five macroalgal extracts caused a significant increase of H3K9 deacetylation, and the effect was most pronounced for *F. dichitus*. The compound responsible for this in vitro activity was identified by mass spectrometry as fucoidan.

Keywords: sirtuin activators; fucoidan; brown seaweed

1. Introduction

SIRT6 is an NAD$^+$-dependent histone deacetylase (HDACs EC 3.5.1.98) that functions as a regulator of many cellular processes, is evolutionarily conserved, and exists in a variety of organisms from eukaryotes to humans [1,2]. SIRT6 controls healthy ageing by regulating genomic stability, oxidative stress, and glucose metabolism, and it is considered a promising target for age-associated diseases such as chronic inflammation, diseases associated with metabolic syndrome, obesity, and insulin resistant type-2 diabetes [3–8]. In a recent study, it was demonstrated that SIRT6 inhibition could improve glycemia in a mouse model of type 2 diabetes [9]. Also, it was shown that SIRT6-deficient mice have a premature aging phenotype with a shortened lifespan, while the overexpression of SIRT6 prolongs the lifespan in male mice and prevents diet-induced obesity [4,10]. SIRT6 activity affects the development of several cancer subtypes, but it is still unclear whether it is a tumor suppressor or promoter, or both [2,11].

Some compounds that enhance SIRT6 activity have been identified, including fatty acids and endogenous fatty acid ethanolamides [12,13]. Phenolic compounds, such as quercetin and luteolin, also enhance SIRT6 deacetylation activity, albeit at very high concentrations [13].

Brown macroalgal species [14–16], specifically *Fucaceae* and *Cystoseira*, are rich in molecules that exert a large range of biological activities including phenolic compounds such as phlorotannins, and polysaccharides, such as laminarans and fucoidans. Phlorotannins protect cells against Ultraviolet

(UV)-B-induced DNA modifications by inducing the nucleotide excision repair (NER) pathway of DNA repair [17]. Fucoidans are a complex heterogeneous group of sulfated polysaccharides composed of L-fucose and sulphate ester groups with minor amounts of monosaccharides that have robust anti-inflammatory and anti-proliferative effects [18,19].

Considering the elicited physiological actions of brown seaweeds and their overlap with the reported bioactivity of SIRT6, the aim of this work was to screen brown seaweed species for novel SIRT6 modulators as potential candidates that can be used in the prevention of age-associated diseases and metabolic syndrome associated disorders, including cancer, obesity, and insulin-resistant diabetes.

We used an accelerated solvent extraction (ASE®) method with acetone:water (70:30 *v/v*) [15,19] as an extraction solvent to generate phlorotannin- and fucoidan-rich extracts from five species of brown macroalgae, namely *Fucus distichus*, *Fucus vesiculous*, *Cytoseira tamariscofolia*, *Cytoseira nodacaulis*, and *Alaria esculenta*. The generated extracts were tested for their ability to deacetylate H3K9, a proxy measure of SIRT6 activity. Three of the five macroalgal extracts significantly enhanced SIRT6 activity, and the effect was most pronounced for *F. distichus*. Herein, we identify fucoidan as the compound responsible for SIRT6 activation from *F. distichus* extract using liquid chromatography and mass spectrometry.

2. Results

2.1. Screening of Brown Algae

In this study, we used a previously developed HPLC deacetylation assay that estimates SIRT6 activity by measuring changes in the level of deacetylated peptide H3K9, over the substrate (H3K9Ac) [13,20], to determine SIRT6 activity in complex matrices. Five species of brown algae were tested for SIRT6 modulating activity at two concentrations (Figure 1). Of these, the *A. esculenta* ASE® extract had no activity, while *C. nodacaulis* displayed SIRT6 stimulating activity, with a ~five-fold increase at 1 mg/mL compared to control. The ASE® extracts from *F. distichus*, *F. vesiculosus* (Linnaeus), and *C. tamariscofolia* all displayed an approximate ~35-fold increase in SIRT6 activity when assayed at a concentration of 1 mg/mL. While the stimulation of SIRT6 activity was dose-dependent for all species tested, *F. distichus* displayed the strongest activity at 0.5 mg/mL with a ~10-fold increase. As a result, *F. distichus* was studied further.

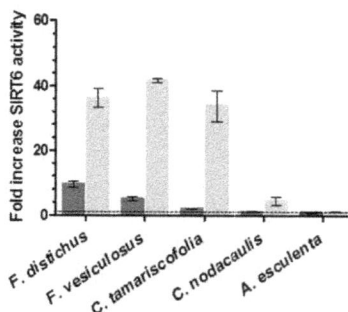

Figure 1. SIRT6 deacetylation activity in the presence of five species of brown algae extracts. The change of SIRT6 deacetylation in the presence of 0.5 mg/mL (grey) and 1.0 mg/mL (light grey) extracts is compared to controls with 500 μM NAD$^+$ and 40 μM H3K9Ac with 30 min of incubation time. The data are presented as means ± SD, *n* = 3.

2.2. Separation of F. distichus

F. distichus was separated into eight sub-fractions (F1–F8) using an XDB-C18 column (Zorbax Eclipse) guided by the SIRT6 H3K9Ac deacetylation HPLC-based assay (Figure 2). Of the eight sub-fractions, moderate activity was observed for F3–F6, and F8 (Figure 3), with a ~3-fold increase in SIRT6 activity. Interestingly, F1 and F7 were the most active, with a ~70-fold and ~40-fold increase, respectively, in SIRT6 activity, at 1 mg/mL. Due to the increased activity observed in F1, the HPLC method was scaled up for the collection of more active fractions with semi-preparative HPLC-PDA (Supplementary Figure S1) using an Eclipse XDB-C8 (9.4 mm × 250 mm, 5 µm). Five different fractions (F1–F5) were collected and the resulting F1 fraction was further purified using a Zorbax Eclipse XDB-C18 column (4.6 mm × 50 mm, 1.8 µm), resulting in a single peak (Supplementary Figure S2).

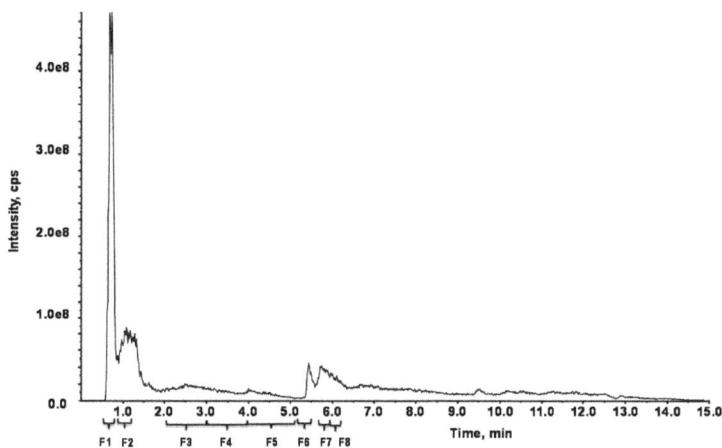

Figure 2. HPLC chromatogram of *F. distichus* and its separation into eight fractions using Zorbax Eclipse XDB-C18 column (4.6 mm × 50 mm, 1.8 µm). The collected fractions: F1 = 0.4–0.5 min; F7 = 5.4–5.5 min.

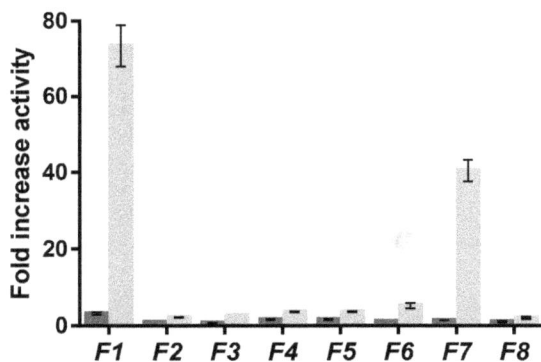

Figure 3. SIRT6 deacetylation activity (fold activity relative to control) of fractions (F1–F8) from *F. distichus* in the presence of 0.5 mg/mL (grey) and 1.0 mg/mL (light grey) fractions (F1–F8) with 500 µM NAD$^+$ and 40 µM H3K9Ac with 30 min of incubation time. The data are presented as means ± SD, n = 3.

2.3. Identification of Fucoidan

The resulting sub-fraction was characterized by mass spectrometry (Figure 4), and identified as fucoidan, a sulfated polysaccharide present in brown algae, by comparison to the reported mass spectra of isolated fucoidan from *Sargassum* genus algae [21,22]. The seaweed fucoidans are heterogenic mixtures of structurally related polysaccharides consisting of carbohydrate units

(L-fucopyranose and non-fucose ones) and non-carbohydrate substituents (mainly sulfate and acetyl groups). The precise determination of their structures with the exact location of structural elements is complex. The polysaccharide backbones of fucoidans are organized in repeating (1→3)-linked or alternatively (1→3)- and (1→4)-linked α-L-fucopyranose residues [23,24]. The backbone of the fucoidan from *F. distichus* is built up mainly of the repeating A units (Figure 5), whereas fucoidans from *F. vesiculosus* are formed mainly of B units [23–25].

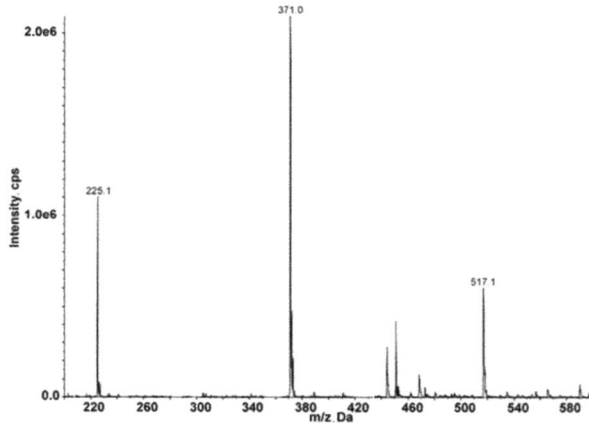

Figure 4. HPLC–MS analysis of subfraction F1 in negative ionization mode with a scan range of *m/z* 150–600.

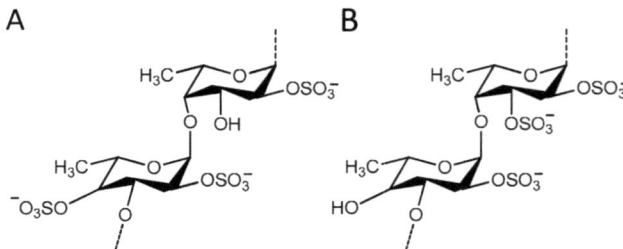

Figure 5. Reported structural elements for fucoidan isolated from the brown seaweeds (**A**) *F. distichus*, (**B**) *F. vesiculosus*. Modified from [24,25].

While the employed extraction process aimed to isolate phlorotannins, the presence of fucoidan in the extracts is not unexpected, as it has been previously reported that between 0.26% and 7.0% dry weight of the algal biomass of *F. distichus* consists of fucoidan [24]. Furthermore, Béress [26] previously reported the extraction of both polyphenols and polymers including fucoidan using water-based solvent systems. It is difficult to separate polyphenols and carbohydrates based on the differential solubility of these algal components. Pantankar previously reported that fucoidan is soluble in acetone:water [27], and this is also a well-known extraction solvent for phlorotannins. A dose-response curve of a sub-fraction of F1 was carried out, and the results demonstrated a significant increase in SIRT6 activation with a ~140-fold increase observed at a 100 µg/mL concentration (Figure 6A). However, due to the limitations in the amount of *F. distichus* available, we were unable to obtain a full dose-response curve for the sub-fraction of F1. Fucoidan (>95% pure) was purchased (isolated from *F. vesiculosus*) and a full dose-response curve was carried out. The dose-response curve obtained was very similar (Figure 6B) to that obtained for F1, with a ~355-fold increase in the activity observed at 100 µg/mL.

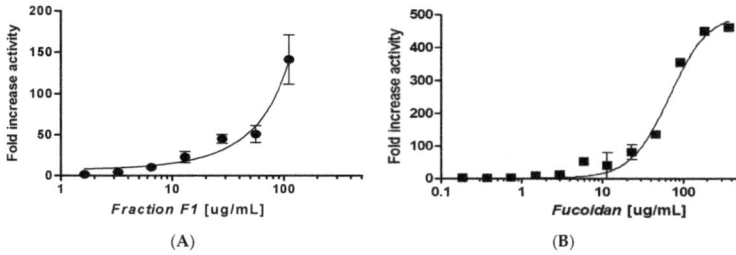

(A)   (B)

Figure 6. (**A**) Dose-response curve of the extract sub-fraction of F1 (●) in the presence of 500 µM NAD$^+$ and 40 µM H3K9Ac with 30 min of incubation time. The data are presented as means ± SD, $n = 3$; (**B**) Dose-response curve of fucoidan (■) on SIRT6 deacetylation activity in the presence of 500 µM NAD$^+$ and 40 µM H3K9Ac with 30 min of incubation time. The data are presented as means ± SD, $n = 3$.

2.4. Western Blot Analysis

The in vitro deacetylation activity was also determined by Western blot analysis, where 1 to 16 µg/mL of fucoidan was incubated with the core histones and the remaining levels of histone H3 acetylated on lysine 9 were determined. Using this technique, it was found that fucoidan activated SIRT6 deacetylation activity in a dose-dependent manner (Figure 7). Both methods demonstrate that fucoidan is an activator of SIRT6. Interestingly, it was demonstrated that higher concentrations of fucoidan (>16 µg/mL in the Western blot analysis method) resulted in the reduced activation of SIRT6 activity (data not shown). A similar observation was made in the HPLC-based assay at higher concentrations as well, indicating that fucoidan may have a dual role in SIRT6, similar to what was observed for quercetin and luteolin [13].

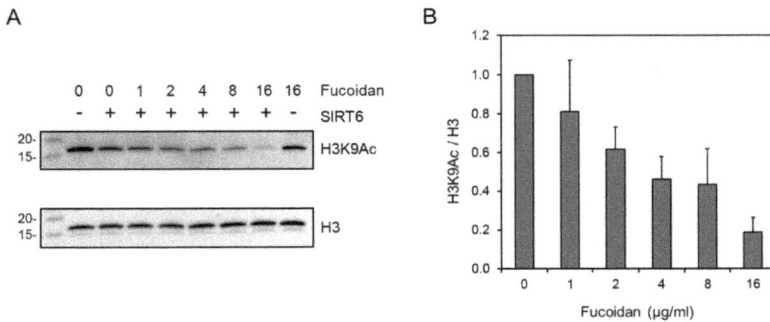

Figure 7. Western blot method for the in vitro SIRT6 deacetylation assay. Serially diluted concentrations of a SIRT6 stimulator (1–16 µg/mL) were incubated for 30 min at 37 °C in the presence of 1 µg/well of a purified recombinant GST-SIRT6 protein, 2 µg purified whole chicken core histones with 500 µM NAD$^+$ in 25 mM Tris-HCl, pH 8.0. (**A**) Acetylation level was detected with anti-H3K9Ac antibody and normalized to total H3 histone. Values indicate final fucoidan concentration in µg/mL. Molecular weight markers in kDa. (**B**) Quantification of H3K9 deacetylation. Values represent the averages of three experiments; error bars indicate standard deviation.

2.5. Selectivity for SIRT6

In order to determine whether fucoidan was selective for SIRT6, in vitro enzymatic assays were carried out against SIRT1, SIRT2, and SIRT3. Neither 10 μg/mL nor 100 μg/mL of fucoidan resulted in any change in the deacetylation activity of SIRT1 or SIRT3. There was a 20% reduction in SIRT2 activity at 100 μg/mL (Table 1). These results suggest that fucoidan activation of sirtuins is specific to SIRT6.

Table 1. Fucoidan against SIRT1-SIRT3 deacetylation activities. The data are presented as means ± SD, $n = 3$.

Fucoidan (μg/mL)	Fold Increase in Activity ± SD		
	SIRT1	SIRT2	SIRT3
10	0.96 ± 0.04	0.87 ± 0.06	0.99 ± 0.01
100	0.94 ± 0.02	0.82 ± 0.02	0.98 ± 0.02

3. Discussion

From five species of brown algae tested against SIRT6 modulating activity, *F. distichus* displayed the most robust increase of SIRT6 deacetylation activity. As a result, the active component from *F. distichus* was identified using a guided SIRT6 H3K9Ac deacetylation HPLC-based assay as fucoidan. Due to the limitations of the starting material, commercially available fucoidan (isolated from *F. vesiculosus*) was purchased and tested for SIRT6 activity. A significant increase in deacetylation activity was observed in a dose-dependent manner using the HPLC-guided deacetylation-based assay as well Western blot analysis. The commercial fucoidan, from *F. vesiculosus*, has been reported to contain fucose (>50%), galactose (6%), glucose (20%), mannose, xylose (15%), uronic acid, glucosamine, and sulfate [28]. None of these monosaccharides introduced any SIRT6 activity at a concentration up to 300 μM (data not shown), indicating that the activity is most likely due to the sulfated fucose. Percival and Ross reported that fucoidan from *F. vesiculosus* contained 31.7% sulfation [29], while the sulfate content of commercial fucoidan from *F. vesiculosus* was estimated to be 31.2% [30,31]. Previous studies revealed that the degree of sulfation significantly influences the level of anti-angiogenic activity of fucoidans in human umbilical vein endothelial cells. For example, oversulfated fucoidan from *F. vesiculosus* with sulfate contents at 52.4% were significantly stronger at inhibiting angiogenesis than natural fucoidan from *F. vesiculosus* with sulfate contents of 31.2% [30,32]. The modulation of sirtuin activity by negatively charged sulfated polysaccharides, while novel, is not completely unexpected. For example, Tong et al. [33] reported the activation of SIRT7 deacetylation activity with negatively charged DNA. In addition, heparin and heparan sulfate proteoglycans have been reported to be potent inhibitors of HAT (Histone acetyltransferases) activity [34]. Further, fucoidan has been indicated to play a role in apoptosis [28,35,36]. In one study, it induced the cleavage of PARP (poly ADP ribose polymerase) to the 89 kDa polypeptide, suggesting that caspases were involved in the fucoidan-mediated apoptosis [37]. Furthermore, in a study in rabbits investigated fucoidan injected intramuscular (i.m.)-induced apoptosis in isolated lymphoma cell lines in vitro [28]. Similarly, SIRT6 overexpression has been demonstrated to induce apoptosis in cancer cells and not in normal cells [8]. In cancer cells, the activity was mediated via the activation of both p53 and p73 signaling cascades. Min et al. [38] reported that the treatment of HepG2 cells with fucoidan (250 and 500 μg/mL) increased the upregulation of p53 and p14, which are involved in the regulation of apoptosis, by up to two- and two and half-fold, respectively, thus inhibiting the viability of HepG2 cells. Zhang et al. (2014) [39] reported a direct interaction between SIRT6 and p53, and the activation of SIRT6 expression by intact p53, which in turn leads to an elevated association of SIRT6 with FoxO1 and the subsequent inhibition of gluconeogenesis. In addition to anti-cancer properties, fucoidans have revealed numerous other health-promoting effects, including anti-oxidative and anti-inflammatory effects. In addition to observations with SIRT6, it was demonstrated in a mouse model that low molecular weight fucoidan inhibited oxidative stress and mitochondrial dysfunction through the upregulation of the expression of SIRT3 after traumatic brain injury [40]. While these studies suggest that fucoidan could also be activating other sirtuins, the results of this study demonstrate that fucoidan, isolated from *F. distichus* and *F. vesiculosus*, is a strong stimulator of SIRT6. To our knowledge, this is the first report to identify a polysaccharide which stimulates SIRT6 deacetylation activity.

4. Materials and Methods

4.1. Materials

Acetylated histone H3 (K9) peptide (residues 1-21) (H3K9Ac) was purchased from AnaSpec Incorporation (Fremont, CA, USA). Nicotinamide adenine dinucleotide (NAD), formic acid, fucoidan from *F. vesiculosus* (F5631), and anti-rabbit HRP-conjugated secondary antibody (A0545) were ordered from Sigma Aldrich (St. Louis, MO, USA). Core histones proteins (13-107) and rabbit anti-acetyl H3K9 antibody (06-942) were ordered from Merck, EMD Millpore (Temecula, CA, USA). Rabbit anti-histone H3 antibody (9715S) was purchased from Cell Signaling Technology (Danvers, MA, USA). Novex™ WedgeWell™ 10–20% Tris-Glycine Mini Gels (12-well) (XP10202BOX) and Novex® Tris-Glycine SDS Running Buffer (10×) (LC2675) were ordered from ThermoFisher Scientifics (Waltham, MA, USA).

Expression and Purification of GST-Tagged SIRT6 Protein. The human SIRT6 expression vector hSIRT6-pGEX-6P3 was kindly provided by Prof. Katrin Chua (Stanford, CA, USA). The recombinant GST (Glutathione S-transferases)-tagged SIRT6 was produced by fermentation in *Escherichia coli* BL21 (DE3)-pRARE. The production was done at +16 °C with 0.1 mM IPTG (Isopropyl β-D-1-thiogalactopyranoside), for 20 h and the soluble overexpressed protein was affinity purified on glutathione agarose (Sigma, St. Louis, MO, USA).

4.2. Plant Material and Extraction

4.2.1. Method of Preparation

Accelerated solvent extraction (ASE®) was used to generate macroalgal extracts using the Dionex PLE system (ASE 200, Dionex, ThermoFisher Scientifics). Briefly, 2 g of each freeze-dried, de-fatted, and powdered macroalga sample was mixed with 4 g of silica (Merck grades, 60 A, Sigma Aldrich, Dublin, Ireland). Silica was used as an inert dispersant and the sample plus silica mixture was then packed into 22-mL extraction cells. The automated extraction method used was 70% acetone in water and a pressure and temperature of 50 °C, 1500 psi, respectively. The extraction time consisted of four cycles of 5 min. Samples were dried using a rota-evaporator at 37 °C and subsequently freeze-dried to remove water. Hexane was used to de-fat samples as previously described [41].

4.2.2. Species Name

3-*Fucus distichus* (Newfoundland origin); 4-*Fucus vesiculosus* (Linnaeus) ISCG 0223; 6-*Cytoseira tamariscofolia* ISCG0283; 7-*Cytoseira nodacaulis* ISCG0070; 11-*Alaria esculenta* (Newfoundland origin); Location of Harvest and Time: 3-*Fucus distichus* (Newfoundland origin) supplied by Oceans Ltd. (St. John's, NL, Canada) April 2010; 4-*Fucus vesiculosus* (Linnaeus) ISCG0223, Golf Course, Galway, 8 June 2011; 6-*Cytoseira tamariscofolia* ISCG0283, Finnavara, Co. Clare, 28 September 2011; 7-*Cytoseira nodacaulis* ISCG0070, Finnavara, Co. Clare, 30 March 2010; 11-*Alaria esculenta* (Newfoundland origin) kindly supplied by Dr. Anne Mathieu, Oceans Ltd., St. John's Newfoundland. It was supplied in a freeze-dried format. All seaweeds (Irish) were rinsed and subsequently freeze-dried prior to processing and stored at −80 °C.

4.3. SIRT6 Deacetylation Assay

Solutions of macroalgal extracts at 0.5 and 1 mg/mL concentrations were prepared in DMSO. In this study, 0.6 µL of this solution and DMSO (control) were incubated for 30 min in the presence of 3 µg/well of SIRT6, 40 µM H3K9Ac, and 500 µM NAD$^+$ in Tris Buffer (25 mM, pH 8.0) at 37 °C.

Additional controls were carried out in the absence/presence of SIRT6, with or without NAD$^+$ and with and without fucoidan. During the reaction, the final solvent concentration of all samples was 1% DMSO. The samples were terminated by adding 6 uL cold 10% formic acid and subsequently centrifuging for 15 min at 13.4 rpm. Dose-response effects (0.05 mg/mL to 1 mg/mL) were carried out for sub-fraction F1 from *F. distichus* and commercial fucoidan from *F. vesiculosus*.

4.4. HPLC Analysis

The chromatographic separation of H3K9 and acetylated H3K9 was achieved on a Zorbax Eclipse XDB-C18 column (4.6 mm × 50 mm, 1.8 μm; Agilent Technologies, Santa Clara, CA, USA) at room temperature using a Shimadzu prominence system (Shimadzu Technology, Kyoto, Japan) consisting of a CBM-20A, LC-20 AB binary pumps, an SIL-20AC-HT auto-sampler, and a DGU-20A3 degassing unit. The mobile phase consisted of water with 0.02% formic acid (elute A) and acetonitrile with 0.02% formic acid (elute B). The gradient eluent at a flow rate of 0.9 mL/min was programmed as follows: 0–2.0 min, 0% B; 2.0–10 min, 0–8% B; 10–10.10 min, 8–80% B; 10.10–12 min, 80%; 12–15 min 80–0% B; 15 min, 0% B. The total run time was 15 min and the injection volume per sample was 20 μL. The HPLC system was coupled to a 5500 QTRAP from Applied Biosystems/MDS Sciex equipped with Turbo V electrospray ionization source (TIS)® (Applied Biosystems, Foster City, CA, USA). The data were acquired and analyzed using Analyst version 1.5.1 (Applied Biosystems). Positive electrospray ionization data were acquired using multiple reactions monitoring (MRM). The TIS instrumental source settings for temperature, curtain gas, ion source gas 1 (nebulizer), ion source gas 2 (turbo ion spray), entrance potential, and ion spray voltage were 550 °C, 20 psi, 60 psi, 50 psi, 10 V, and 5500 V, respectively. The TIS compound parameter settings for de-clustering potential, collision energy, and collision cell exit potential were 231, 45, and 12 V, respectively, for H3K9Ac; and were 36, 43, and 12 V, respectively, for H3K9. The standards were characterized using the following MRM ion transitions: H3K9Ac (*m/z* 766.339→760.690) and H3K9 (*m/z* 752.198→746.717).

4.5. HPLC Fingerprint

The chromatographic separation of *F. distichus* was achieved on a Zorbax Eclipse XDB-C18 column (4.6 mm × 50 mm, 1.8 μm; Agilent Technologies, Santa Clara, CA, USA) at room temperature using a Shimadzu prominence system (Shimadzu Technology, Kyoto, Japan) consisting of a CBM-20A, LC-20 AB binary pumps, an SIL-20AC-HT auto-sampler, and a DGU-20A3 degassing unit. The mobile phase consisted of water with 0.1% formic acid (elute A) and acetonitrile with 0.02% formic acid (elute B). The gradient eluent at a flow rate of 0.6 mL/min was programmed as follows: 0 min, 0%B; 3.0 min, 0% B; 6.0 min, 83% B; 11.0 min, 83%; 11.1 min 0% B; 15 min, 0% B. The total run time was 15 min and the injection volume per sample was 10 μl (7.5 mg/mL in 100 mM NaOH). Fractions were collected (Figure 2) between 0.4–0.5 min (F1), 1.0–1.3 min (F2), 2.0–3.0 min (F3), 3.0–4.0 min (F4), 4.0–5.0 min (F5), 5.1–5.3 min (F6), 5.4–5.5 min (F7), or 5.5–6.2 min (F8).

The HPLC system was coupled to a 5500 QTRAP from Applied Biosystems/MDS Sciex equipped with Turbo V electrospray ionization source (TIS)® (Applied Biosystems, Foster City, CA, USA). The data were acquired and analyzed using Analyst version 1.5.1 (Applied Biosystems). Negative electrospray ionization data were acquired using enhanced MS (EMS) from 200 to 600 *m/z*. The TIS instrumental source settings for temperature, curtain gas, ion source gas 1 (nebulizer), ion source gas 2 (turbo ion spray), entrance potential, and ion spray voltage were 500 °C, 20 psi, 50 psi, 60 psi, −10 V, and −4500 V, respectively. The TIS compound parameter settings for de-clustering potential, collision energy, and collision cell exit potential were −75, −35, and 12 V, respectively.

4.6. SemiPREP MS

The collection of the F1 sub-fraction of *F. distichus* was achieved on an Eclipse XDB-C8 column (9.4 mm × 250 mm, 5 μm; Agilent Technologies, Santa Clara, CA, USA) at room temperature using a Shimadzu prominence system (Shimadzu Technology, Kyoto, Japan) consisting of a CBM-20A,

LC-20 ADXR binary pumps, an SIL-20AC-HT auto-sampler, and a DGU-20A3R degassing unit, PDA (Photodiode array detector) SPD-M20A. The mobile phase consisted of water with 0.1% formic acid (elute A) and acetonitrile with 0.02% formic acid (elute B). The gradient eluent at a flow rate of 0.9 mL/min was programmed as follows: 0 min, 0%B; 3.0 min, 0% B; 11.0 min, 83% B; 11.1 min, 0%; 15 min, 0% B. The total run time was 15 min and the injection volume per sample was 50 μL (10 mg/mL in 100 mM NaOH). Fractions were collected (Figure 2) between 0.1–1.0 min (F1), 1.0–1.3 min (F2), 8.0–8.5 min (F3), 10.0–10.5 min (F4), or 11.0–11.4 min (F5).

The collected fractions were evaporated under nitrogen gas and dissolved two times with methanol to remove impurities. Fractions were analyzed using HPLC–MS negative ionization mode with a scan range of m/z 150–600.

4.7. H3K9 Western Blot Method

Fucoidan stock solution (10 mg/mL) was prepared and serially diluted in 25 mM Tris-HCl, pH 8.0. Subsequently, 1 μL of fucoidan solution or buffer control were incubated in 20 μL reaction for 30 min in the presence of 1 μg of a purified recombinant GST-SIRT6, 2 μg purified chicken core histones (Millipore, Billerica, MA, USA), and 500 μM NAD$^+$ in 25 mM Tris-HCl, pH 8.0 at 37 °C. The reactions were stopped with Laemmli sample buffer and separated by SDS-PAGE using 4–15% gradient gels (Bio-Rad, Hercules, CA, USA) and transferred onto polyvinylidene difluoride (PVDF) membranes. H3K9 acetylation was detected with rabbit anti-acetyl H3K9 antibody (Millipore) followed by anti-rabbit HRP-conjugated secondary antibody. Membranes were stripped and re-probed with rabbit anti-histone H3 antibody. Chemiluminescent signal detection and image acquisition were done using SuperSignal West Femto Substrate (Thermo) and ChemiDoc XRS+ with Image Lab software (Bio-Rad).

4.8. In Vitro Enzymatic Assays (SIRT1-SIRT3)

The Fluor de Lys fluorescence assays were based on the method described in the BioMol product sheet (Enzo Life Sciences, Ann Arbor, MI, USA) using the BioMol KI177 substrate for SIRT1 and the KI179 substrate for SIRT2 and SIRT3. GST-SIRT1 and GST-SIRT2 were produced as described previously [42,43]. His-SIRT3 (BML-SE270) was purchased from Enzo Life Sciences. DMSO (D2650), SIRT assay buffer (HDAC assay buffer, KI143, supplemented with 1 mg/mL BSA, A3803), and NAD$^+$ (N6522) were from Sigma. Fluor de Lys developer (KI176) and nicotinamide (BKI 283) were ordered form BioMol.

Briefly, the reaction mixture including acetylated peptide substrate (0.7 Km: 58 μM for SIRT1 [44], 198 μM for SIRT2 [44], and 32 μM for SIRT3), NAD$^+$ (0.9 Km: 558 μM for SIRT1, 547 μM for SIRT2, and 2 mM for SIRT3), and DMSO/compounds in DMSO (2.5 μL in 50 μL total reaction volume) were preincubated for 5 min at room temperature. The reaction was started by adding the enzyme following incubation for 1 h at 37 °C. After that, the developer and nicotinamide (2 mM in HDAC assay buffer giving total volume of 50 μL) were added and the incubation was continued for 45 min at 37 °C. Fluorescence readings were obtained using a VictorTM 1420 Multilabel Counter (PerkinElmer Inc., Waltham, MA, USA) with an excitation wavelength of 355 nm and an emission of 460 nm, or EnVision 2104 Multilabel Reader (PerkinElmer, Waltham, MA, USA) with an excitation wavelength of 370 nm and an emission of 460 nm.

Supplementary Materials: The following are available online at www.mdpi.com/1660-3397/15/6/190/s1, Figure S1: Fractionation (F1–F5) of *F. distichus* on the LC-MS/MS API-5500 and on the HPLC-DAD were collected using an Eclipse XDB-C8 column (9.4 mm × 250 mm, 5 μm). Fractions were collected between 0.1–1.0 min (F1), 1.0–1.3 min (F2), 8.0–8.5 min (F3), 10.0–10.5 min (F4), 11.0–11.4 min (F5). Figure S2: HPLC-MS chromatograph of subfraction F1 (10 μL). The flow rate was 0.6 mL/min a nd the injection volume was 10 μL of 7.5 mg/mL.

Acknowledgments: This work was supported by funds from the NIA Intramural Research Program (RM) and Academy of Finland (Grant no. 269341), Orion-Farmos Research Foundation (2014, 2015), Saastamoinen Foundation. We thank Sari Ukkonen for her skillful assistance.

Author Contributions: Minna K. Rahnasto-Rilla and Ruin Moaddel conceived and designed the experiments. Minna K. Rahnasto-Rilla, Padraig Mcloughlin, Tomasz Kulikowicz, Maria Hayes performed the experiments; Minna K. Rahnasto-Rilla, Padraig Mcloughlin, Maria Hayes, Ruin Moaddel analyzed the data; Minna K. Rahnasto-Rilla, Padraig Mcloughlin, Maria Hayes, Maire Doyle, Maija Lahtela-Kakkonen and Ruin Moaddel contributed reagents/materials/analysis tools; Minna K. Rahnasto-Rilla, Tomasz Kulikowicz, Vilhelm A. Bohr, Maija Lahtela-Kakkonen, Luigi Ferrucci, Maria Hayes and Ruin Moaddel contributed to the writing of the paper.

Conflicts of Interest: The authors declare no conflict of interest.

References

1. Haigis, M.C.; Sinclair, D.A. Mammalian sirtuins: Biological insights and disease relevance. *Annu. Rev. Pathol.* **2010**, *5*, 253–295. [CrossRef] [PubMed]

2. Mostoslavsky, R.; Chua, K.F.; Lombard, D.B.; Pang, W.W.; Fischer, M.R.; Gellon, L.; Liu, P.; Mostoslavsky, G.; Franco, S.; Murphy, M.M.; et al. Genomic instability and aging-like phenotype in the absence of mammalian SIRT6. *Cell* **2006**, *124*, 315–329. [CrossRef] [PubMed]

3. Gertler, A.A.; Cohen, H.Y. SIRT6, a protein with many faces. *Biogerontology* **2013**, *14*, 629–639. [CrossRef] [PubMed]

4. Kanfi, Y.; Naiman, S.; Amir, G.; Peshti, V.; Zinman, G.; Nahum, L.; Bar-Joseph, Z.; Cohen, H.Y. The sirtuin SIRT6 regulates lifespan in male mice. *Nature* **2012**, *483*, 218–221. [CrossRef] [PubMed]

5. Rodgers, J.T.; Puigserver, P. Certainly can't live without this: SIRT6. *Cell Metab.* **2006**, *3*, 77–78. [CrossRef] [PubMed]

6. Zhong, L.; D'Urso, A.; Toiber, D.; Sebastian, C.; Henry, R.E.; Vadysirisack, D.D.; Guimaraes, A.; Marinelli, B.; Wikstrom, J.D.; Nir, T.; et al. The histone deacetylase sirt6 regulates glucose homeostasis via HIF-1 alpha. *Cell* **2010**, *140*, 280–293. [CrossRef] [PubMed]

7. Dominy, J.E., Jr.; Lee, Y.; Jedrychowski, M.P.; Chim, H.; Jurczak, M.J.; Camporez, J.P.; Ruan, H.B.; Feldman, J.; Pierce, K.; Mostoslavsky, R.; et al. The deacetylase SIRT6 activates the acetyltransferase GCN5 and suppresses hepatic gluconeogenesis. *Mol. Cell* **2012**, *48*, 900–913. [CrossRef] [PubMed]

8. Van Meter, M.; Mao, Z.; Gorbunova, V.; Seluanov, A. Sirt6 overexpression induces massive apoptosis in cancer cells but not in normal cells. *Cell Cycle* **2011**, *10*, 3153–3158. [CrossRef] [PubMed]

9. Sociali, G.; Magnone, M.; Ravera, S.; Damonte, P.; Vigliarolo, T.; Von Holtey, M.; Vellone, V.; Millo, E.; Caffa, I.; Cea, M.; et al. Pharmacological sirt6 inhibition improves glucose tolerance in a type 2 diabetes mouse model. *FASEB J.* **2017**. [CrossRef] [PubMed]

10. Jiang, H.; Khan, S.; Wang, Y.; Charron, G.; He, B.; Sebastian, C.; Du, J.; Kim, R.; Ge, E.; Mostoslavsky, R.; et al. Sirt6 regulates tnf-alpha secretion through hydrolysis of long-chain fatty acyl lysine. *Nature* **2013**, *496*, 110–113. [CrossRef] [PubMed]

11. Zhang, Z.G.; Qin, C.Y. SIRT6 suppresses hepatocellular carcinoma cell growth via inhibiting the extracellular signalregulated kinase signaling pathway. *Mol. Med. Rep.* **2014**, *9*, 882–888. [PubMed]

12. Feldman, J.L.; Baeza, J.; Denu, J.M. Activation of the protein deacetylase sirt6 by long-chain fatty acids and widespread deacylation by mammalian sirtuins. *J. Biol. Chem.* **2013**, *288*, 31350–31356. [CrossRef] [PubMed]

13. Rahnasto-Rilla, M.; Kokkola, T.; Jarho, E.; Lahtela-Kakkonen, M.; Moaddel, R. *N*-acylethanolamines bind to sirt6. *Chembiochem* **2016**, *17*, 77–81. [CrossRef] [PubMed]

14. Vinayak, R.C.; Sabu, A.S.; Chatterji, A. Bio-prospecting of a few brown seaweeds for their cytotoxic and antioxidant activities. *Evid. Based Complement. Altern. Med.* **2011**, *2011*, 673083. [CrossRef] [PubMed]

15. Tierney, M.S.; Smyth, T.J.; Rai, D.K.; Soler-Vila, A.; Croft, A.K.; Brunton, N. Enrichment of polyphenol contents and antioxidant activities of irish brown macroalgae using food-friendly techniques based on polarity and molecular size. *Food Chem.* **2013**, *139*, 753–761. [CrossRef] [PubMed]

16. Wang, T.; Jonsdottir, R.; Olafsdottir, G. Total phenolic compounds, radical scavenging and metal chelation of extracts from icelandic seaweeds. *Food Chem.* **2009**, *116*, 240–248. [CrossRef]

17. Piao, M.J.; Hewage, S.R.; Han, X.; Kang, K.A.; Kang, H.K.; Lee, N.H.; Hyun, J.W. Protective effect of diphlorethohydroxycarmalol against ultraviolet b radiation-induced DNA damage by inducing the nucleotide excision repair system in hacat human keratinocytes. *Mar. Drugs* **2015**, *13*, 5629–5641. [CrossRef] [PubMed]

18. Perez, M.J.; Falque, E.; Dominguez, H. Antimicrobial action of compounds from marine seaweed. *Mar. Drugs* **2016**, *14*, 52. [CrossRef] [PubMed]

19. Tierney, M.S.; Smyth, T.J.; Hayes, M.; Soler-Vila, A.; Croft, A.K.; Brunton, N. Influence of pressurised liquid extraction and solidliquid extraction methods on the phenolic content and antioxidant activities of irish macroalgae. *Int. J. Food Sci. Technol.* **2013**, *48*, 860–869. [CrossRef]

20. Rahnasto-Rilla, M.; Lahtela-Kakkonen, M.; Moaddel, R. Sirtuin 6 (SIRT6) activity assays. *Methods Mol. Biol.* **2016**, *1436*, 259–269. [PubMed]

21. Thinh, P.D.; Menshova, R.V.; Ermakova, S.P.; Anastyuk, S.D.; Ly, B.M.; Zvyagintseva, T.N. Structural characteristics and anticancer activity of fucoidan from the brown alga sargassum mcclurei. *Mar. Drugs* **2013**, *11*, 1456–1476. [CrossRef] [PubMed]

22. Li, B.; Lu, F.; Wei, X.; Zhao, R. Fucoidan: Structure and bioactivity. *Molecules* **2008**, *13*, 1671–1695. [CrossRef] [PubMed]

23. Cumashi, A.; Ushakova, N.A.; Preobrazhenskaya, M.E.; D'Incecco, A.; Piccoli, A.; Totani, L.; Tinari, N.; Morozevich, G.E.; Berman, A.E.; Bilan, M.I.; et al. A comparative study of the anti-inflammatory, anticoagulant, antiangiogenic, and antiadhesive activities of nine different fucoidans from brown seaweeds. *Glycobiology* **2007**, *17*, 541–552. [CrossRef] [PubMed]

24. Bilan, M.I.; Grachev, A.A.; Ustuzhanina, N.E.; Shashkov, A.S.; Nifantiev, N.E.; Usov, A.I. A highly regular fraction of a fucoidan from the brown seaweed fucus *distichus* L. *Carbohydr. Res.* **2004**, *339*, 511–517. [CrossRef] [PubMed]

25. Chevolot, L.; Mulloy, B.; Ratiskol, J.; Foucault, A.; Colliec-Jouault, S. A disaccharide repeat unit is the major structure in fucoidans from two species of brown algae. *Carbohydr. Res.* **2001**, *330*, 529–535. [CrossRef]

26. Beress, A.; Wassermann, O.; Tahhan, S.; Bruhn, T.; Beress, L.; Kraiselburd, E.N.; Gonzalez, L.V.; de Motta, G.E.; Chavez, P.I. A new procedure for the isolation of anti-hiv compounds (polysaccharides and polyphenols) from the marine alga fucus vesiculosus. *J. Nat. Prod.* **1993**, *56*, 478–488. [CrossRef] [PubMed]

27. Patankar, M.S.; Oehninger, S.; Barnett, T.; Williams, R.L.; Clark, G.F. A revised structure for fucoidan may explain some of its biological activities. *J. Biol. Chem.* **1993**, *268*, 21770–21776. [PubMed]

28. Foley, S.A.; Szegezdi, E.; Mulloy, B.; Samali, A.; Tuohy, M.G. An unfractionated fucoidan from ascophyllum nodosum: Extraction, characterization, and apoptotic effects in vitro. *J. Nat. Prod.* **2011**, *74*, 1851–1861. [CrossRef] [PubMed]

29. Percival, E.G.V.; Ross, A.G. The isolation and purification of fucoidin from brown seaweeds. *J. Chem. Soc.* **1950**, 717–720. [CrossRef]

30. Soeda, S.; Kozako, T.; Iwata, K.; Shimeno, H. Oversulfated fucoidan inhibits the basic fibroblast growth factor-induced tube formation by human umbilical vein endothelial cells: Its possible mechanism of action. *Biochim. Biophys. Acta-Mol. Cell Res.* **2000**, *1497*, 127–134. [CrossRef]

31. Koyanagi, S.; Tanigawa, N.; Nakagawa, H.; Soeda, S.; Shimeno, H. Oversulfation of fucoidan enhances its anti-angiogenic and antitumor activities. *Biochemical. Pharmacol.* **2003**, *65*, 173–179. [CrossRef]

32. Soeda, S.; Shibata, Y.; Shimeno, H. Inhibitory effect of oversulfated fucoidan on tube formation by human vascular endothelial cells. *Biol. Pharm. Bull.* **1997**, *20*, 1131–1135. [CrossRef] [PubMed]

33. Tong, Z.; Wang, Y.; Zhang, X.; Kim, D.D.; Sadhukhan, S.; Hao, Q.; Lin, H. SIRT7 is activated by DNA and deacetylates histone h3 in the chromatin context. *ACS Chem. Biol.* **2016**, *11*, 742–747. [CrossRef] [PubMed]

34. Buczek-Thomas, J.A.; Hsia, E.; Rich, C.B.; Foster, J.A.; Nugent, M.A. Inhibition of histone acetyltransferase by glycosaminoglycans. *J. Cell. Biochem.* **2008**, *105*, 108–120. [CrossRef] [PubMed]

35. Vishchuk, O.S.; Ermakova, S.P.; Zvyagintseva, T.N. The effect of sulfated (1→3)-alpha-L-fucan from the brown alga saccharina cichorioides miyabe on resveratrol-induced apoptosis in colon carcinoma cells. *Mar. Drugs* **2013**, *11*, 194–212. [CrossRef] [PubMed]

36. Park, H.Y.; Kim, G.Y.; Moon, S.K.; Kim, W.J.; Yoo, Y.H.; Choi, Y.H. Fucoidan inhibits the proliferation of human urinary bladder cancer t24 cells by blocking cell cycle progression and inducing apoptosis. *Molecules* **2014**, *19*, 5981–5998. [CrossRef] [PubMed]

37. Boo, H.J.; Hyun, J.H.; Kim, S.C.; Kang, J.I.; Kim, M.K.; Kim, S.Y.; Cho, H.; Yoo, E.S.; Kang, H.K. Fucoidan from undaria pinnatifida induces apoptosis in a549 human lung carcinoma cells. *Phytother Res.* **2011**, *25*, 1082–1086. [CrossRef] [PubMed]

38. Min, E.Y.; Kim, I.H.; Lee, J.; Kim, E.Y.; Choi, Y.H.; Nam, T.J. The effects of fucodian on senescence are controlled by the p16ink4a-prb and p14arf-p53 pathways in hepatocellular carcinoma and hepatic cell lines. *Int. J. Oncol.* **2014**, *45*, 47–56. [CrossRef] [PubMed]

39. Zhang, P.; Tu, B.; Wang, H.; Cao, Z.; Tang, M.; Zhang, C.; Gu, B.; Li, Z.; Wang, L.; Yang, Y.; et al. Tumor suppressor p53 cooperates with sirt6 to regulate gluconeogenesis by promoting foxo1 nuclear exclusion. *Proc. Natl. Acad. Sci. USA* **2014**, *111*, 10684–10689. [CrossRef] [PubMed]

40. Wang, T.; Zhu, M.; He, Z.Z. Low-molecular-weight fucoidan attenuates mitochondrial dysfunction and improves neurological outcome after traumatic brain injury in aged mice: Involvement of sirt3. *Cell. Mol. Neurobiol.* **2016**, *36*, 1257–1268. [CrossRef] [PubMed]

41. Lopes, G.; Sousa, C.; Silva, L.R.; Pinto, E.; Andrade, P.B.; Bernardo, J.; Mouga, T.; Valentao, P. Can phlorotannins purified extracts constitute a novel pharmacological alternative for microbial infections with associated inflammatory conditions? *PLoS ONE* **2012**, *7*, e31145. [CrossRef] [PubMed]

42. Kiviranta, P.H.; Leppanen, J.; Rinne, V.M.; Suuronen, T.; Kyrylenko, O.; Kyrylenko, S.; Kuusisto, E.; Tervo, A.J.; Jarvinen, T.; Salminen, A.; et al. *N*-(3-(4-hydroxyphenyl)-propenoyl)-amino acid tryptamides as sirt2 inhibitors. *Bioorg. Med. Chem. Lett.* **2007**, *17*, 2448–2451. [CrossRef] [PubMed]

43. Tervo, A.J.; Kyrylenko, S.; Niskanen, P.; Salminen, A.; Leppanen, J.; Nyronen, T.H.; Jarvinen, T.; Poso, A. An in silico approach to discovering novel inhibitors of human sirtuin type 2. *J. Med. Chem.* **2004**, *47*, 6292–6298. [CrossRef] [PubMed]

44. Kiviranta, P.H.; Suuronen, T.; Wallen, E.A.; Leppanen, J.; Tervonen, J.; Kyrylenko, S.; Salminen, A.; Poso, A.; Jarho, E.M. N(epsilon)-thioacetyl-lysine-containing tri-, tetra-, and pentapeptides as SIRT1 and SIRT2 inhibitors. *J. Med. Chem.* **2009**, *52*, 2153–2156. [CrossRef] [PubMed]

marine drugs

MDPI

Article

Characterization and Comparison of the Structural Features, Immune-Modulatory and Anti-Avian Influenza Virus Activities Conferred by Three Algal Sulfated Polysaccharides

Lin Song [1,2], Xiaolin Chen [1,*], Xiaodong Liu [3], Fubo Zhang [3], Linfeng Hu [1], Yang Yue [1,2], Kecheng Li [1] and Pengcheng Li [1,*]

[1] Key Laboratory of Experimental Marine Biology, Institute of Oceanology, Chinese Academy of Sciences, No.7 Nanhai Road, Qingdao 266071, China; lylinsong@hotmail.com (L.S.), wood9818@sina.com (L.H.), yueyang12@ucas.ac.cn (Y.Y.); lkc@qdio.ac.cn (K.L.)

[2] University of Chinese Academy of Sciences, Beijing 100049, China

[3] College of Animal Science and Technology, Qingdao Agriculture University, No.700 Changcheng Road, Qingdao 266109, China; lxdau86@163.com (X.L.); tsygy@whu.edu.cn (F.Z.)

* Correspondence: chenxl@qdio.ac.cn (X.C.); pcli@qdio.ac.cn (P.L.); Tel.: +86-532-8289-8707 (X.C. & P.L.); Fax: +86-532-8289-8707 (X.C. & P.L.)

Academic Editor: Paola Laurienzo
Received: 9 October 2015; Accepted: 17 December 2015; Published: 29 December 2015

Abstract: Three marine macroalgae, *i.e.*, *Grateloupia filicina*, *Ulva pertusa* and *Sargassum qingdaoense*, were selected as the deputies of Rhodophyta, Chlorophyta and Ochrophyta for comparative analysis of the molecular structures and biological activities of sulfated polysaccharides (SP). The ratio of water-soluble polysaccharides, the monosaccharide composition and the sulfated contents of three extracted SPs were determined, and their structures were characterized by Fourier transformation infrared spectroscopy. In addition, biological activity analysis showed that all three SPs had immune-modulatory activity both *in vitro* and *in vivo*, and SPs from *S. qingdaoense* had the best effect. Further bioassays showed that three SPs could not only enhance the immunity level stimulated by inactivated avian influenza virus (AIV) *in vivo* but also significantly inhibited the activity of activated AIV (H9N2 subtype) *in vitro*. *G. filicina* SP exhibited the strongest anti-AIV activity. These results revealed the variations in structural features and bioactivities among three SPs and indicated the potential adjuvants for immune-enhancement and anti-AIV.

Keywords: marine macroalgae; sulfated polysaccharides; structure features; immune enhancement; anti-AIV

1. Introduction

Sulfated polysaccharides have attracted increasing attention due to their diversified biological and pharmacological activities, such as anti-viral, immune enhancement, anti-tumor, anti-infective and antioxidant effects (Table 1) [1–4]. Among organisms that produce sulfated polysaccharides, marine algae are regarded as the most abundant and important sources of non-animal sulfated polysaccharides [5–7]. Various biological functions of sulfated polysaccharides from marine algae have been reported in recent decades, but these analyses were mainly focused on a single species (Table 1) [8–11]. For example, the sulfated polysaccharides obtained from green alga *Enteromorpha clathrata* showed immune-enhancement activity that could stimulate TNF-α expression in serum and induce lymphocyte proliferation [9]. The sulfated polysaccharide purified from *Sargassum horneri* was reported to have antitumor activity that could inhibit the growth of human colon cancer DLD cells [12].

The *S. vulgare* polysaccharides illustrated an ability to enhance serum antibody titers and lymphocyte proliferation [10]. Furthermore, the sulfated polysaccharides extracted from the red alga *Laurencia papillosa* could inhibit breast cancer cells (MDA-MB-231) via apoptosis regulatory genes [13]. While significant attention has been paid to one or two biological and pharmacological activities of sulfated polysaccharides from a single species of marine algae, little information is available regarding the bioactivity comparison of sulfated polysaccharides from different algae and the structure-function relationship. The three main divisions of marine algae (*i.e.*, Chlorophyta, Ochrophyta and Rhodophyta) are valuable sources of structurally diverse sulfated polysaccharides. However different sulfated polysaccharides from these three algae still remain largely unknown in the comparative analysis of the molecule structures and diverse bioactivities.

Table 1. Previous studies on the biological effects of sulphated polysaccharides from seaweed.

Author	Phylum	Species	Bioactivities
Qi, *et al.*, 2005 [14]		*Ulva pertusa*	Antioxidant activity
		Ulva pertusa	
Zhang *et al.*, 2010 [2]		*Enteromorpha linza*	Antioxidant activity
		Bryopsis plumose	
Cho *et al.*, 2010 [15]	Chlorophyta	*Enteromorpha prolifera*	Antitumor and immunomodulating activities
Jiao *et al.*, 2010 [16]		*Enteromorpha intestinalis*	Antitumor and immunomodulating activities
Tabarsa *et al.*, 2012 [17]		*Ulva pertusa*	Immunomodulatory, anticancer activities
Zhang *et al.*, 2013 [18]		*Enteromorpha linza*	Immunological and antioxidant activities
Aguilar-Briseño *et al.*, 2015 [19]		*Ulva clathrata*	Antiviral activity
Zhang *et al.*, 2010 [2]		*Laminaria japonica*	Antioxidant activity
Ye *et al.*, 2008 [20]		*Sargassum pallidum*	Antitumor and antioxidant activities
Wang *et al.*, 2011 [21]		*Laminaria japonica*	Anticoagulant activity
Li *et al.*, 2012 [22]		*Sargassum pallidum*	Immune responses
Dore *et al.*, 2013 [10]		*Sargassum vulgare*	Anticoagulant, antithrombotic, antioxidant and anti-inflammatory effects
Suresh *et al.*, 2013 [23]		*Sargassum plagiophyllum*	Anticancer and antioxidant activities
Imbs *et al.*, 2014 [24]		*Fucus evanescens*	Antioxidant activity
Hwang *et al.*, 2015 [25]		*Sargassum hemiphyllum*	Anti-inflammatory
Wen *et al.*, 2014 [26]	Ochrophyta	*Sargassum horneri*	Antioxidant activity
Shao *et al.*, 2014 [27]		*Sargassum horneri*	Antioxidant and antitumor activities
Shobharani *et al.*, 2014 [28]		*Sargassum* sp.	Antioxidant and anticoagulant activities
Aguilar-Briseño *et al.*, 2015 [19]		*Cladosiphon okamuranus*	Antiviral activity
Zhang *et al.*, 2014 [29]		*Ascophyllum nodosum*	Induces Th1 and Tc1 Immune Responses
Yuan *et al.*, 2015 [30]		*Ascophyllum nodosum*	Antioxidant activity
Ammar *et al.*, 2015 [31]		*Cystoseira sedoides*, *Cystoseira compressa*, *Cystoseira crinita*	Anti-radical, anti-inflammatory and gastroprotective activities
Shao *et al.*, 2015 [32]		*Sargassum horneri*	Antioxidant and moisture-preserving activities
Athukorala *et al.*, 2005 [33]		*Grateloupia filicina*	Antioxidant activity, protecting ability for H_2O_2-induced DNA damage
Wang *et al.*, 2007 [34]		*Grateloupia longifolia* *Grateloupia filicina*	Anti-virus activity
Zhang *et al.*, 2010 [2]	Rhodophyta	*Porphyra haitanensis*	Antioxidant activity
Yu *et al.*, 2012 [35]		*Eucheuma denticulatum*	Anti-virus activity
Shi *et al.*, 2014 [36]		*Porphyra haitanensis*	Anti-allergic activity
Chen *et al.*, 2015 [37]		*Grateloupia filicina*	Anticoagulant activity
Fleita *et al.*, 2015 [38]		*Pterocladia capillacea*	Antioxidant activity

Sulfated polysaccharides from algae produce immune-modulatory activities that might have great potential for stimulating immune responses or controlling immune cell activity [8]. Karnjanapratum reported that the biological activities of water-soluble sulfated polysaccharides isolated from *Monostroma nitidum* could stimulate Raw 264.7 cells *in vitro*, and induce considerable prostaglandin-2 (PGE-2) and nitric oxide (NO) production [39]. Oral ingestion of polysaccharides isolated from *E. intestinalis* could increase the relative spleen and thymus weight of tumor-bearing animals and stimulate lymphocyte proliferation *in vitro* [16]. In addition to immune-modulatory activity, the antiviral activity of sulfated polysaccharides is also important [40]. It has been suggested that sulfated polysaccharides from algae confer activities that are anti-viral to herps simplex virus type 1 (HSV-1), herps simplex virus type 2 (HSV-2), and human immunodeficiency virus (HIV) [5].

As a lowly pathogenic avian influenza virus (AIV) group, H9N2 subtype influenza virus is considered to be the common cause of disease epidemics [41,42]. Additionally, outbreaks of H9N2 are associated with significant economic loss in the chicken industry [43,44]. More seriously, this subtype is characterized by cross-species infections and has been passed to pigs, ferrets and guinea pigs as well as to humans, in a small number of cases [45–49]. These cross-species infections indicate a potentially serious threat to human health [50,51]. The first human infection was detected in 1999, and ever since there have been several reports about the isolation of H9N2 viruses from humans and swine [52] including the latest H9N2 human case in China at the end of 2013 [53]. Sulfated polysaccharides were considered to be novel sources of natural compounds for antiviral drug discovery, but whether they could confer antiviral activity to H9N2 AIV remains elusive [54–56].

In this study, three marine algae, *i.e.*, *Grateloupia filicina*, *Ulva pertusa* and *Sargassum qingdaoense*, were selected as the deputies of Rhodophyta, Chlorophyta, and Ochrophyta for comparative analysis of the molecular structures and immune-modulatory and anti-AIV activities of sulfated polysaccharides. The variations in chemical compositions and molecular structures of three sulfated polysaccharides, including polysaccharide components and sulfate contents, might function as determinants of their bioactivities [17,34,57]. Thus, it is worth making an effort to analyze and determine the structure-function relationship of these sulfated polysaccharides. Here, we report a comprehensive analysis of structural features and immune-modulatory and anti-AIV activities of sulfated polysaccharides from three types of marine algae, *i.e.*, *S. qingdaoense*, *G. filicina*, and *U. pertusa*. These polysaccharides were characterized structurally and their biological activities were tested both *in vitro* and *in vivo*, which has not only enhanced our understanding of the characteristic of algae sulfated polysaccharides but also provided a comparison of algae from different categories and contributed further theoretical and experimental evidence for the exploration and development of polysaccharide-based immune-potentiators that are anti-AIV.

2. Results

2.1. Chemical Characterization of Three Sulfated Polysaccharides

2.1.1. Chemical Analysis

Three sulfated polysaccharides were extracted and purified from *U. pertusa*, *G. filicina*, *S. qingdaoense*, with a yield of 12.1%, 19.7% and 7.2%, respectively (Table 2). The purified sulfated polysaccharides were further characterized regarding monosaccharide composition molar ratios. Individual variations were found in the components of monosaccharides, and the most abundant components in UPP, GFP and SQP were rhamnose, galactose, and fucose, respectively. The UPP also had glucuronic acid and xylose, and a small amount of mannose, glucose, galactose and fucose was found. For GFP, except for the main component, the content of other monosaccharides, such as mannose, glucuronic acid, glucose, xylose and fucose were low. As for SQP, which mainly consists of fucose, galactose and mannose, it also contained glucuronic acid and glucose. The ratio of total saccharides to UPP (53.13%) were also shown to be much higher than the others. Regarding sulfate contents, GFP (19.89%) was higher than UPP (13.54%) and SQP (5.64%). These results suggested significant variations in chemical compositions among the three sulfated polysaccharides.

Table 2. Yield and chemical composition of three sulfated polysaccharides sample (%*w/w* of dry weight).

Sample	Yield (%)	Total Sugar (%)	Sulfate (%)	Monosaccharides Composition (Molar Ratio)						
				Man	Rha	Glc A	Glc	Gal	Xyl	Fuc
UPP	12.1	53.13	13.54	0.06	1	0.53	0.19	0.09	0.39	0.02
GFP	19.7	40.9	19.89	0.01	-	0.02	0.07	1	0.1	0.05
SQP	7.2	20.81	5.64	0.56	-	0.13	0.37	0.6	-	1

Man: mannose; Rha: rhamnose; Gal: galactose; Glc: glucose; Xyl: xylose; Fuc: fucose.

2.1.2. FT-IR Spectrometric Characterization

To further understand the structure of sulfated polysaccharides, FT-IR spectroscopy was used. The results revealed that all three sulfated polysaccharides shared several common absorption peaks (Figure 1), which were considered to correspond to the O-H stretching vibrations (3420 cm^{-1}), the S=O asymmetry stretching vibrations (1250 cm^{-1}) the C-O-H deformation vibrations (1050 cm^{-1}), and the stretching vibrations of -COO$^-$ (1650 cm^{-1} and 1420 cm^{-1}) which also indicated that the extraction was composed of acidic polysaccharides [58].

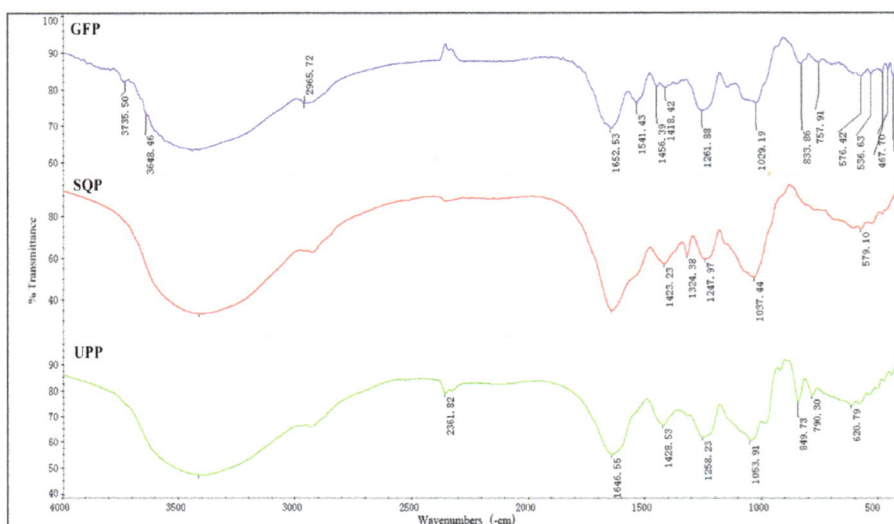

Figure 1. FT-IR spectra of three extracted sulfated polysaccharides. GFP, *Grateloupia filicina*; UPP, *Ulva Pertusa*; SQP, *Sargassum qingdaoens*.

Additionally, there exist some differences in the FT-IR spectra of the three polysaccharides. For example, the absorption peaks at 2950 cm^{-1} for C-H stretching vibrations and at approximately 830 cm^{-1} that appeared in the FT-IR spectrum of GFP indicated the presence of α-type glycosidic linkages. The band at 1030 cm^{-1} of SQP was found and corresponded to C-O-H deformation vibrations. As for UPP, the peak attributed to C-O-S symmetry stretching vibrations appeared at approximately 850 cm^{-1} [10,34,59].

2.2. Cytotoxic Activity of the Polysaccharides

MTT assays to determine the cytotoxicity of sulfated polysaccharides showed that the safe concentration for GFP was 2.5 mg/mL, whereas it was 5 mg/mL for both UPP and SQP (Table A1). The concentrations of the three polysaccharides used in our experiments were all within the safe range.

2.3. Immunologic Modulation of Three Sulfated Polysaccharides in Vitro

To understand their potential effects in immunology, these compounds were first used to test the immune response at a lymphocyte level. A lymphocyte proliferation experiment was used to evaluate the stimulation efficiency of the sulfated polysaccharides on spleen cell proliferation. As illustrated in Figure 2, the lymphocyte proliferation values for each sulfated polysaccharide treatment group were significantly higher than those of the control group (Mock; $p < 0.05$), suggesting a dramatic effect on the stimulation of spleen cell proliferation. Further analysis showed that, both UPP and GFP showed a similar effect on proliferation in response to all three treatment doses, in contrast to the fact that the stimulation efficiency of SQP was in dose-dependent manner. As a result, 500 μg/mL of SQP conferred the strongest efficiency in stimulation among all the tests.

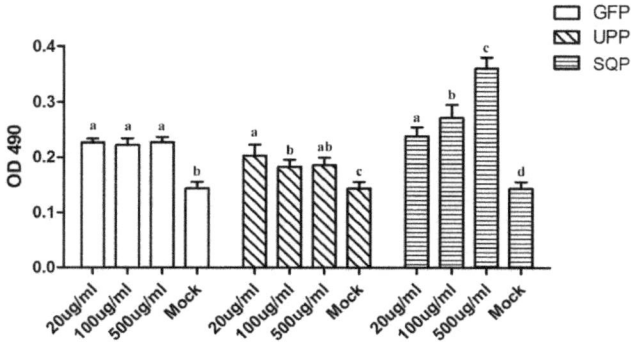

Figure 2. Mouse spleen cell proliferation effects of GFP, UPP and SQP. Mock treated with PBS instead of polysaccharides as a negative control. Values with different letters in the same column (a–d) are significantly different ($p < 0.05$) from each other. Data are shown as the Mean + SD and are fully representative of the individual experiment.

2.4. Immune-Modulation of Three Sulfated Polysaccharides in Vivo

To further verify the immune-modulation results obtained by *in vitro* analysis, *in vivo* experiments were carried out in mice. Antibody titer, cytokine production and T-cell subpopulation were tested. Additionally, H9N2-AIV was selected as the immunologic stimulant.

2.4.1. H9N2-Specific Antibody Titer

After the first injection (Prime), the antibody titer of the control group was almost undetectable, while the antibody levels of the treated group rose dramatically ($p < 0.05$; Figure 3). Moreover, after the second injection (Boost), the antibody titer of the control group still stayed at the base level, while further significant enhancement was detected relative to the prime for all the treatment doses especially for the case of 50 mg/kg of GFP and 50 mg/kg of SQP (Figure 3A,C). These results suggested that the assayed sulfated polysaccharides significantly increased H9N2-specific antibody titers.

2.4.2. Effect on Cytokine Production Stimulation

The results of cytokine production stimulation are presented in Figure 4. The levels of IFN-γ and IL-4 were significantly increased in the experimental groups compared to the control and the vaccine groups ($p < 0.05$). However, individual variations were observed among the groups treated by three sulfated polysaccharides, and the most efficient stimulations resulted from the optimal doses tested. For example, a concentration of 10 mg/kg were more efficient for stimulation of IFN-γ production compared to that of 50 mg/kg for all three sulfated polysaccharides. By contrast, 50 mg/kg was a better concentration for the IL-4 stimulation than 10 mg/kg.

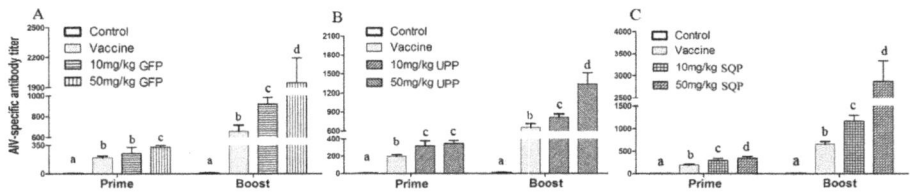

Figure 3. Avian influenza virus (AIV)-specific antibody titer detection. Kunming mice were immunized with an AIV vaccine and polysaccharides, following the prime-boost vaccination programme (days 0 and 14), respectively. (**A**) GFP; (**B**) UPP; (**C**) SQP. Values with different letters in the same column (a–d) are significantly different ($p < 0.05$) from each other. Data are shown as the Mean + SD and are fully representative of an individual experiment.

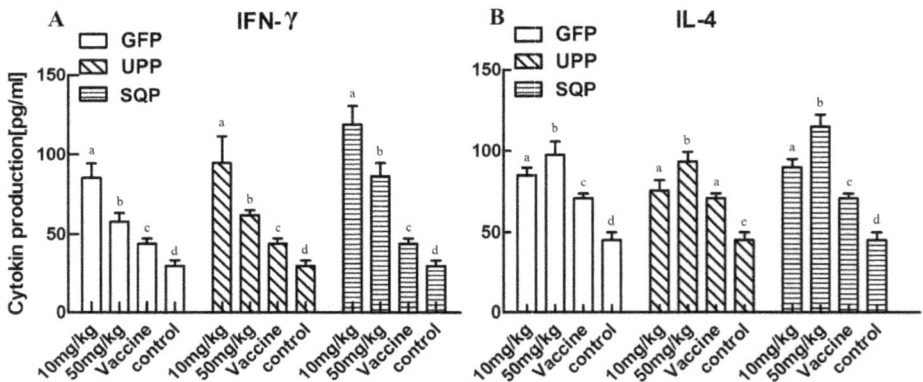

Figure 4. Cytokine production stimulating effect of GFP, UPP and SQP. Kunming mice were immunized with an AIV vaccine and polysaccharides, and sera were collected on day 28 after two immunizations to detect the cytokines IFN-γ (**A**) and IL-4 (**B**). Values with different letters in the same column (a–d) are significantly different ($p < 0.05$) from each other. Data are shown as the Mean + SD and are fully representative for the individual experiment.

2.4.3. T-Cell Subpopulation

Results presented in Figure 5 show how each of the three sulfated polysaccharides could elevate the CD3+CD4+ levels in the experimental groups significantly compared to the control and vaccine groups ($p < 0.05$), and the elevations were in a dose-dependent manner in such a way that 50 mg/kg gained a much stronger effect relative to 10 mg/kg. For stimulation of the CD3+CD8+, only the SQP stimulated groups showed a significant effect.

2.5. Anti-H9N2 Effect of Three Sulfated Polysaccharides in Vitro

Because all these sulfated polysaccharides could significantly enhance immune responses with the deactivated AIV as an immunologic stimulant, we were especially interested in whether they could show significant resistance to activated H9N2 AIV.

Based on the Hemagglutination test (HA test), treatment with 0.2 mg/mL UPP and 1 mg/mL SQP in the experimental groups decreased the virus titer significantly, although the effects of other groups were not significant (Figure 6A).

From the real-time PCR results, the expression of the H9N2 gene decreased significantly after treatment with sulfate polysaccharides. GFP was the strongest of all three test groups in terms of the inhibition of H9N2 replication, and was followed by UPP. SQP showed the weakest effect compared to the other two; however, it still had a significant effect. As for the determination of the optimal concentrations for suppressing the virus, the 20 ug/mL groups had relatively lower virus expression and that was the best suppression effect compared to the other doses tested (Figure 6B).

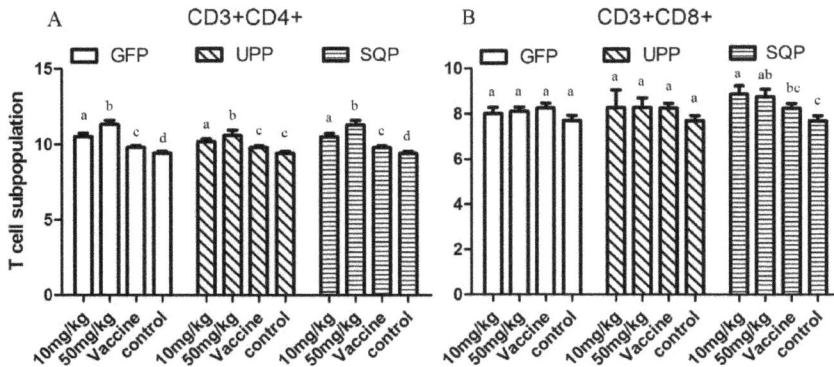

Figure 5. T-cell subpopulation tests. The blood cells of the treated mice were collected and analyzed with flow cytometry. (**A**) CD3+CD4+. (**B**) CD3+CD8+. Values with different letters in the same column (a–d) are significantly different ($p < 0.05$) from each other. Data are shown as the Mean + SD and are fully representative of the individual experiment.

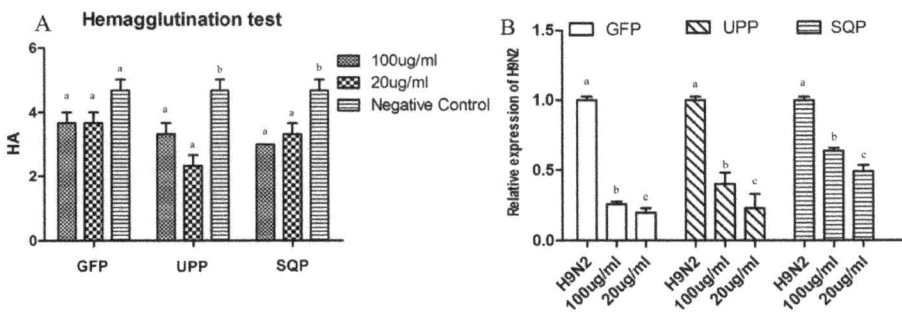

Figure 6. Haemagglutination test (HA) of the cell culture and relative expression of H9N2. Antiviral activity *in vitro* was measured with a HA test (**A**) and RT-PCR (**B**). Data from samples without polysaccharides were used as a basic control. Values with different letters in the same column (a–c) are significantly different ($p < 0.05$) from each other. Data are shown as the Mean + SD and are fully representative of the individual experiment.

3. Discussion

Sulfated polysaccharides from algae have been acknowledged to bear diversified biological and pharmacological functions [22,60–62]. Moreover, the chemical composition and structural features of polysaccharides were confirmed to function as determinants of their bioactivities [17,57]. Therefore, it is necessary to determine which group(s) in the molecule play(s) an essential roles in conferring the biological function(s), and this was one of our key focuses for the present study. As expected, significant variations in the chemical composition of monosaccharides, the ratio of total saccharides and sulfated content were detected among all three sulfate polysaccharides. The results were similar to the results of Zhang *et al.*, Suresh *et al.*, Dore *et al.*, Fleita *et al.* and Tabarsa *et al.* [10,14,17,23,38] The amount of saccharide is usually described using total saccharides, but there were some metal ions attached to the polysaccharide molecules (such as Ca^{2+}, Na^+, K^+ and Mg^{2+}), and also the SO_4^{2+}; therefore, the level of total saccharides was much lower than the real saccharide amount. Moreover, the results of FT-IR showed the variations in the structural features. For example, GFP contained C-H stretching vibrations, α-type glycosidic linkages, and the highest sulfate content compare to the other two polysaccharides. Further analyses were performed to determine the structure-function relationship.

Immune-modulatory effects are some of the most important activities of polysaccharides, and they have attracted much attention and are already a focus of research [8,63]. To verify the immune-modulatory function of sulfated polysaccharides from algae, splenic lymphocyte proliferation was tested *in vitro*. It is well known that lymphocytes play important roles in the immune response; hence, the proliferation of lymphocytes becomes the most pertinent criterion to evaluate the cellular immune functions [64,65]. Our results revealed that, all three polysaccharides tested could significantly enhance spleen cell proliferation. However, it is difficult to wholly evaluate immune responses only *in vitro*; *in vivo* tests are necessary for further verification of the effect. Thus we chose mice as the model animal and selected the AIV, which is a cross-species infection, as the immunologic stimulant to examine the immune-enhancing activity of polysaccharides *in vivo*. The antibody titer is a measure of the specific humoral immune response in animals after vaccination [66]. Cytokines are believed to function as the important mediators of immune responses, such as IFN-γ, a pleiotropic cytokine with immune-modulatory effects on different types of immune cells, while IL-4 is a cytokine essential for immune-modulation [44]. IFN-γ and IL-4 were tested in the absence or presence of the polysaccharides. CD+ T lymphocytes are primarily responsible for mediating cytotoxic effects, and CD3+, CD8+ and CD4+ are important T lymphocyte markers [65]. Collectively, the results showed that the administration of UPP, GFP and SQP could significantly enhance AIV-specific antibody production and improve the humoral immunity level, and the optimal doses that most efficiently stimulated the production of immunity level varied depended on the type of algae. It was revealed that, 50 mg/kg SQP exhibited the highest AIV-specific antibody titer and IL-4, while 10 mg/kg SQP was the best activator of IFN-γ. As for the T-cell subpopulation, 50 mg/kg of GFP and SQP elicited the best efficiency for CD3+CD4+, while for CD3+CD8+, 10 mg/kg of SQP produced higher levels compared to others. In sum, SQP might have the best immune-enhancing effects among these three algae polysaccharides, particularly at 50 mg/kg. The reason why SQP showed the best humoral immune-enhancing response might be related to the most abundant monosaccharide that SQP contained, which was fucose. This result is supported by studies that showed fucose plays an important role in immune-enhancing activity [67]. Thelen *et al.* reported that fucoidin from *F. vesiculosus* had been shown to induce macrophage activation [68]. Moreover, fucose from *F. vesiculosus* could function as an effective adjuvant as reported by Jin *et al.* [69]. Our results confirmed and inferred that the fucose group in polysaccharide molecules might contribute more to the enhancement of spleen cell proliferation and humoral immune responses than other components. These results also suggested that the polysaccharides could function as enhancers in the immune responses and SQP was recommended.

As described above, because all three sulfated polysaccharides could significantly enhance the immune response with the deactivated AIV as an immunologic stimulant, we were especially interested in whether they could resist the activated AIV. In this study, the significant suppressions of virus replication and virus gene expression were detected by both the HA test and the real-time PCR, suggesting that these sulfate polysaccharides could suppress the replication and expression of AIV *in vitro*. GFP suppressed the AIV-replication and expression more significantly compared to the other two polysaccharides, which might be related to the high sulfate content of GFP. The sulfated content was thought to be related to higher anti-virus activity [54,70]. Chen *et al.* found that the antiviral activity of sulfated polysaccharides from bush sophora roots was stronger than regular polysaccharides that were not sulfated [71]. The anti-duck hepatitis A virus (DHAV) activities of sulfated polysaccharides from *Astragalus* were stronger than those of *Astragalus* polysaccharides, both *in vitro* and *in vivo* [72]. AIV is a highly contagious disease in domestic poultry and other animals, even humans. Although a vaccination is one of the most promising measures to control AIV, the high frequency of virus antigenic variation has led to difficulties in the use of the H9N2-specific vaccines [44,64,73]. Thus, it is important to seek effective and broad-spectrum antiviral drugs for the treatment of H9N2. Based on these results, these three sulfated polysaccharides are suggested to be a potential alternative to vaccine-based prevention to reduce the breakage of AIV H9N2, among which GFP is recommended.

4. Materials and Methods

4.1. Algal Samples

Both *G. filicina*and and *U. Pertusa* were collected from the No. 2 Bathing Beach of Qingdao, China. *S. qingdaoens* was preserved at IOCAS (Institute of Oceanology, Chinese Academy of Sciences, Qingdao, China). The algae were washed with distilled water, dried at 50 °C and stored at room temperature for later use (within one year). All reagents were analytical grade and commercially available.

4.2. Extraction of Water-Soluble Sulfated Algal Polysaccharide

Water-soluble sulfated algal polysaccharides were extracted according to the protocol reported by Zhang. *et al.* protocol with improvements [18]. The water volume, temperature and extraction time for each type of seaweed were decided according to the optimized methods before. For *G. filicina*, a 50-fold volume of water was used at 100 °C for 2 h with stirring, and for *U. Pertusa*, 4000 mL H_2O/100 g was mixed at 125 °C for 4 h, while 100 g of *S. qingdaoens* was soaked with 30-fold water at 91 °C for 4 h with stirring. Then, the polysaccharides solution gained by filtration was condensed and dialyzed for salt removal. The solution was condensed again and freeze-dried to obtain the purified sulfated polysaccharides named UPP, GFP and SQP for *U. Pertusa*, *G. filicina*, and *S. qingdaoens*, respectively.

4.3. Chemical Characterization

Total carbohydrate content was analyzed with phenol–sulfuric acid method using galactose, rhamnose, and fucose as the standard for GFP, UPP, and SQP, respectively [74].

The molar ratios of monosaccharide composition were measured in reference to Zhang *et al.* [66]. Briefly, polysaccharides (10 mg/mL) were hydrolysed in trifluoroacetic acid, followed by neutralization with sodium hydroxide. Then, pre-column derivatization with 3-methyl-1-phenyl-2-pyrazolin-5-one (PMP; 99%) to neutralize the mixture was carried out and separated by HPLC on a YMC Pack ODS AQ column (4.6 mm × 250 mm) [75]. The standards for monosaccharide composition analysis and PMP were obtained from Sigma Aldrich (St Louis, MO, USA).The sulfated content was measured by the barium chloride gelatin method following Kawai *et al.* [76].

FT-IR spectra of the three types of polysaccharides were determined on a Nicolet-360 FT-IR spectrometer (36 scans, at a resolution of 6 cm^{-1}) between 400 cm^{-1} and 4000 cm^{-1}. The dried polysaccharide samples were grinded with potassium bromide (KBr) and pressed into pellets for spectrometric measurement [71].

4.4. Animals and Maintenance

Kunming mice at the age of six weeks were purchased from Qingdao Laboratory Animal Center (Qingdao, China) for this study. All animals were housed under standard environmental conditions (22 ± 0.5 °C, 55% ± 5% humidity and a 12 h light/12 h dark cycle) and maintained with free access to a standard laboratory pellet diet and water. All procedures involving animals throughout the experiments were conducted in strict accordance with the Chinese Legislation on the Use and Care of Laboratory Animals. All animal experiments were performed as per the local institutional ethic committee guidelines.

4.5. Cell Lines, Virus, and Tissue Culture

Madin-Darby canine kidney (MDCK) cells were purchased from American type culture collection (ATCC, Manassas, VA, USA). The cells were grown in Dulbecco's modified Eagle's medium (DMEM) (Corning INC., Corning, NY, USA) supplemented with 100 units/mL penicillin, 100 mg/mL streptomycin (HyClone Laboratories, Logan, Utah, USA), and 10% (*v/v*) fetal bovine serum (FBS; GIBCO BRL Life Technologies, Grand Island, NY, USA) for normal growth and 1% (*v/v*) for viral infection.

AIV H9N2 was kept in Qingdao Boite Biopharmaceutical CO., LTD Company (Qingdao, China). It was propagated on ten-day-old embryonating specific-pathogen-free (SPF) chicken eggs. Titers of the AIV H9N2 were quantified using MDCK cell monolayers by determining the 50% Tissue Culture Infective Dose (TCID50), and the 100 TCID50 of purified virus were used in the subsequent experiments.

4.6. Cytotoxic Activity Evaluation

An MTT assay was applied to determine the relative survival rate of cells during culture [77–79]. After 24 h incubation of 2.5×10^4 cells/mL MDCK cells on 96-well plates either in DMEM as a control or in sulfated polysaccharides dissolved in DMEM (the concentrations was 10 mg/mL and diluted in 2-fold steps), the cells were treated with the MTT (5 mg/mL, 30 μL/well, Beijing Solarbio Science & Technology Co. Ltd., Beijing, China) reagent. After 4 h, the supernatant was removed and 100 μL of DMSO (Beijing Solarbio Science & Technology Co. Ltd., Beijing, China) were added. The absorbance of each well was measured using a microliter enzyme-linked immunosorbent assay reader (iMark™ BIO-RAD) at a wavelength of 490 nm. All experiments were performed in triplicate. The relative survival rate of cells was calculated using the formula: living rate (%) = $(A_P/A_C) \times 100\%$, where A_C and A_P are the optical density without (A_C) and with polysaccharides (A_P), respectively. When the living rate is over 85%, it is considered that the polysaccharide does not exert toxicity on the living cells [80,81].

4.7. Immuno-Modulatory Effect

4.7.1. Mouse Splenic Lymphocyte Proliferation Assay

The spleen was harvested from a Kunming mouse aseptically, and then transferred into a petri dish with PBS. The spleen was minced and blown fully to obtain the cells suspended and filtrated through a 200 meshes steel sieve. A total of 2.5×10^4 cells per well in 96-well plates were treated with three sulfated polysaccharides at a final concentration of 0 μg/mL (control group), 20 μg/mL, 100 μg/mL, and 500 μg/mL followed by incubation for 48 h in a humid atmosphere with 5% CO_2 at 37 °C. The control group (Mock) used PBS instead of the polysaccharide solution. For each concentration, five repeats were performed. An MTT assay was used to detect the proliferation of lymphocytes. Meanwhile, the lymphocytes proliferation rate was calculated using the formula Proliferation rate (%) = $[(\overline{A} \text{ (test group)} - \overline{A} \text{ (control group)})/ \overline{A} \text{ (control group)}] \times 100\%$ [82]. $\overline{A}$ is the average absorbance of the wells with the same treatment.

4.7.2. Animals *Grouping* and *Treatment*

The mice were randomly divided into eight groups with eight mice in each group, including the experimental group, control group, and vaccine group. The experimental groups were given sulfated polysaccharides UPP, GFP and SQP, at doses of 10 mg/kg and 50 mg/kg, mixed with inactivated AIV solution by intraperitoneal injection. Mice in the control group were given the same volume of physiological saline (25 mL/kg) instead. Each mouse in the vaccine group (positive control group) was treated with inactivated AIV H9N2 solution mixed with white oil (100 mg/kg). On Day 1 and 14, the mice were separately immunized (Prime and Boost), and the sera were collected after 14 days of the prime and boost immunizations (on Day 14 and 28), respectively to be saved for the subsequent assays.

4.7.3. AIV-Specific Antibody Titer Detection

Using the sera collected in 4.7.2, the H9N2-specific antibody titers were tested with an ELISA method [83,84], and each treatment was assayed in triplicate. Briefly, a sample or standard was added in ELISA plates coated with capture antibody for 2 h at 37 °C to detect the bound cytokines using a biotinylated anti-cytokine antibody, Avidin HRP, and tetramethylbenzidine. Color development was stopped with 2 M H_2SO_4 and optical densities were read at 450 nm.

4.7.4. Cytokines Production

The sera samples collected in 4.7.2 were used to detect cytokines production. Two cytokines (*i.e.*, IL-4 and IFN-γ) in the treated mice were measured with an ELISA kit (Longtun, Shanghai, China) per the manufacturer's instructions.

4.7.5. T-Cell Subpopulation and Flow Cytometry

The blood cells of the treated Kunming mice were collected, and three-color flow cytometry analyses were performed using different mixtures of specific mAbs, CD3, CD4, and CD8 were labelled with PE-Cy5, FITC or PE, respectively. Data analysis was conducted with the FAC Scan flow cytometer (BD, FACS Aria ii, Franklin Lakes, NJ, USA) using Cell Quest software.

4.8. Anti-AIV Effect in Vitro

In the antiviral experiment, the MDCK cells treated with neither AIV nor polysaccharides served as a cell control group, and the cells infected with the virus only served as a virus control group, while the test group was treated with both the virus and polysaccharides. For all groups, MDCK cells were seeded at 2×10^5 cells/well into 24-well plates, and incubated at 37 °C in an atmosphere of 5% CO_2 until a cell monolayer had formed. The virus control group and test group were treated with 100 μL of the virus solution at a multiplicity of infection (MOI) of 0.2, while for the cell control group, 100 μL of DMEM (1% FBS) was added instead. After incubation for 1 h, the supernatant was discarded, and hereafter, 1 mL of the test polysaccharide solution (GFP, UPP, and SQP, at a concentration of 100 μg/mL and 20 μg/mL, respectively) were added to the test group, while 1 mL of DMEM was added to the virus control and cell control groups. For each group, three biological repetitions were carried out. Treated cells were then incubated at 37 °C with 5% CO_2 for 24 h.

4.8.1. Virus Titers Assay

Virus titers were determined by the HA test. A total of 25 μL of supernatant from the 24-well plate described above were serially two-fold diluted in saline on a 96-well microtiter V-plate. To each well, 25 μL of 1% red blood cell suspension was added, and the mixtures were gently mixed and incubated at 37 °C for 30 min. HA was observed and judged as positive if more than 50% of the red blood cells were agglutinated in the well [85].

4.8.2. Relative Expression of Viruses

Real-time PCR was applied for quantitative analysis of AIV replication levels *in vivo*. Briefly, total RNA was extracted from the treated MDCK cells in 4.8 section using 1 mL of RNAiso Plus Reagent (Takara BIO INC, Liaoning, China) as per the manufacturer's instructions. Reverse transcription was immediately performed with Reverse Transcriptase M-MLV (RNase H-; Takara BIO INC, Liaoning, China). Finally, first strand cDNA was used as a template for real-time PCR with a SYBR®*Premix ExTaq*™ Kit (Takara BIO INC, Liaoning, China). The primers for AIV were: Sense 5′-ACCAGTGCATGGAGACAATTC-3′ and anti-sense 5′-CAAATGTTGCATCTGCAAGAC-3′; and the primers for internal control β-actin were: Sense 5′-CTGGACTTCGAGCAGGAGATG-3′ and anti-sense 5′-CGGATGTCCACGTCACACTTC-3′. The amplification cycles were carried out as follows: 95 °C for 30 s; 95 °C for 5 s and 60 °C for 34 s (40 cycles).

β-actin was used as an internal control to normalize the relative gene expression levels calculated based on the comparative Ct method with the formula $2^{-\Delta\Delta Ct}$ [86].

4.9. Statistical Analysis

Data in the figures are expressed as the Mean + SD. A *t*-test was used to analyses the difference among groups with SPSS Software. Differences between means with $p < 0.05$ were considered statistically significant.

5. Conclusions

For understanding and comparing the molecular structures and biological activities of algae sulfated polysaccharides, three marine algae *G. filicina*, *U. pertusa* and *S. qingdaoense* were selected from Rhodophyta, Chlorophyta and Ochrophyta, respectively. We characterized the structure determination, the humoral immune responses, and cell immune responses to antiviral assays in this study. According to the results, these three types of sulfate polysaccharides shared chemical characteristics and basic structures, with some variations. Further experiments revealed that the sulfate polysaccharides not only enhanced the immunity level stimulated by inactivated AIV *in vivo* but also significantly inhibited the activity of activated AIV *in vitro*. The results indicated that these sulfated polysaccharides from algae could be a potential immune-stimulant and adjuvant against AIV. Moreover, comparative studies of different species polysaccharides were our focus with the goal of obtaining profound and refined insights into this topic. Sulfated polysaccharides from *S. qingdaoense* had the best immune enhancement bioactivity both *in vitro* and *in vivo*. *G. filicina* sulfated polysaccharides showed the best effect in anti-AIV activity. Comprehensive analyses were performed to determine whether the differences in bioactivities could be explained by the differences in structures. These comprehensive analyses enhanced our understanding of algae sulfated polysaccharides and their potential use in future research but also gave guidance for further selection of immunologic stimulant and anti-viral drugs.

Acknowledgments: This work was supported by the National Natural Science Foundation of China (21206182), the Public Science and Technology Research Funds of Ocean (No. 201305016-2), the Commonweal Item of State Oceanic Administration People's Republic of China (201405038-2), and Nantong Applied Research Projects (BK 2014068).

Author Contributions: Conceived and designed the experiments: Lin Song, Xiaodong Liu, Pengcheng Li. Performed the experiments: Lin Song, Xiaodong Liu, Fubo Zhang, Yang Yue, Linfeng Hu. Analyzed the data: Lin Song, Xiaodong Liu. Contributed reagents/materials/analysis tools: Xiaolin Chen, Pengcheng Li. Wrote the paper: Lin Song, Xiaodong Liu, Kecheng Li. Sample collection: Lin Song, Yang Yue, Linfeng Hu.

Conflicts of Interest: The author declares no conflict of interest.

Appendix

Table A1. Relative Survival Rate of the Madin-Darby canine kidney (MDCK) Cells.

Conc (mg/mL)	10	5	2.5	1.25	0.625	0.3125	0.156	0.078
UPP	0.82	1.03	1.16	1.09	1.05	1.08	1.06	1.02
GFP	0.74	0.81	0.89	0.9	0.95	0.92	1.01	1.13
SQP	0.78	0.96	1.01	1	0.97	1.05	0.99	1.03

Determination of the safe concentration by pre-experiment cytotoxicity tests. Cytotoxic activity of the sulfate polysaccharides. The relative living cell rate measured with an MTT assay reflected the cytotoxic activity of the sulfate polysaccharides.

References

1. Bohn, J.A.; Bemiller, J.N. (1→3)-β-D-Glucans as Biological Response Modifiers: A Review of Structure-Functional Activity Relationships. *Carbohydr. Polym.* **1995**, *28*, 3–14. [CrossRef]

2. Zhang, Z.; Wang, F.; Wang, X.; Liu, X.; Hou, Y.; Zhang, Q. Extraction of the Polysaccharides from Five Algae and Their Potential Antioxidant Activity *in vitro*. *Carbohydr. Polym.* **2010**, *82*, 118–121. [CrossRef]
3. Nikapitiya, C.; De Zoysa, Mahanama; Jeon, Y.-J.; Lee, J.; Jee, Y.H. Isolation of Sulfated Anticoagulant Compound from Fermented Red Seaweed *Grateloupia Filicina*. *J. World Aquac. Soc.* **2007**, *38*, 407–417. [CrossRef]
4. Genovese, G.; Faggio, C.; Gugliandolo, C.; Torre, A.; Spanò, A.; Morabito, M.; Maugeri, T.L. *In vitro* Evaluation of Antibacterial Activity of *Asparagopsis Taxiformis* from the Straits of Messina Against Pathogens Relevant in Aquaculture. *Mar. Environ. Res.* **2012**, *73*, 1–6. [CrossRef] [PubMed]
5. Bordbar, S.; Saari, F.A.N. High-value Components and Bioactives from Sea Cucumbers for Functional Foods—A Review. *Mar. Drugs* **2011**, *9*, 1761–1805. [CrossRef] [PubMed]
6. Pomin, V.H. Fucanomics and Galactanomics: Marine Distribution, Medicinal Impact, Conceptions, and Challenges. *Mar. Drugs* **2012**, *10*, 793–811. [CrossRef] [PubMed]
7. Faggio, C.; Morabito, M.; Minicante, S.A.; Piano, G.L.; Pagano, M.; Genovese, G. Potential Use of Polysaccharides from the Brown Alga *Undaria Pinnatifida* as Anticoagulants. *Braz. Arch. Biol. Technol.* **2015**, *58*. [CrossRef]
8. Wang, L.; Wang, X.; Wu, H.; Liu, R. Overview on Biological Activities and Molecular Characteristics of Sulfated Polysaccharides from Marine Green Algae in Recent Years. *Mar. Drugs* **2014**, *12*, 4984–5020. [CrossRef] [PubMed]
9. Qi, X.; Mao, W.; Gao, Y.; Chen, Y.; Chen, Y.; Zhao, C.; Li, N.; Wang, C.; Yan, M.; Lin, C. Chemical Characteristic of an Anticoagulant-Active Sulfated Polysaccharide from Enteromorpha Clathrata. *Carbohydr. Polym.* **2012**, *90*, 1804–1810. [CrossRef] [PubMed]
10. Dore, C.M.; das, C.F.A.M.G.; Will, L.S.; Costa, T.G.; Sabry, D.A.; de Souza Rego, L.A.; Accardo, C.M.; Rocha, H.A.; Filgueira, L.G.; Leite, E.L. A Sulfated Polysaccharide, Fucans, Isolated from Brown Algae *Sargassum Vulgare* with Anticoagulant, Antithrombotic, Antioxidant and Anti-Inflammatory Effects. *Carbohydr. Polym.* **2013**, *91*, 467–475. [CrossRef] [PubMed]
11. Seo, Y.; Kang, S.H.; Lee, H.J.; You, A.K.; Youn, H.J.; Lee, B.J.; Chung, H. *In vitro* Screening of Seaweed Extract on the Proliferation of Mouse Spleen and Thymus Cell. *Biotechnol. Bioprocess Eng.* **2006**, *11*, 160–163. [CrossRef]
12. Ping, S.; Jia, L.; Chen, X.; Fang, Z.; Sun, P. Structural Features and Antitumor Activity of a Purified Polysaccharide Extracted from *Sargassum Horneri*. *Int. J. Biol. Macromol.* **2015**, *73*, 124–130.
13. Murad, H.; Ghannam, A.; Al-Ktaifani, M.; Abbas, A.; Hawat, M. Algal Sulfated Carrageenan Inhibits Proliferation of Mda-Mb-231 Cells via Apoptosis Regulatory Genes. *Mol. Med. Rep.* **2015**, *11*, 2153–2158. [CrossRef] [PubMed]
14. Qi, H.; Zhang, Q.; Zhao, T.; Rong, C.; Hong, Z.; Niu, X.; Li, Z. Antioxidant Activity of Different Sulfate Content Derivatives of Polysaccharide Extracted from *Ulva Pertusa* (Chlorophyta) *in vitro*. *Int. J. Biol. Macromol.* **2005**, *37*, 195–199. [CrossRef] [PubMed]
15. Cho, M.L.; Yang, C.; Sang, M.K.; You, S.G. Molecular Characterization and Biological Activities of Water-Soluble Sulfated Polysaccharides from Enteromorpha Prolifera. *Food Sci. Biotechnol.* **2010**, *19*, 525–533. [CrossRef]
16. Jiao, L.; Xia, L.; Li, T.; Peng, J.; Zhang, L.; Wu, M.; Zhang, L. Characterization and Anti-Tumor Activity of Alkali-Extracted Polysaccharide from *Enteromorpha Intestinalis*. *Int. Immunopharmacol.* **2009**, *9*, 324–329. [CrossRef] [PubMed]
17. Tabarsa, M.; Han, J.H.; Kim, C.Y.; You, S.G. Molecular Characteristics and Immunomodulatory Activities of Water-Soluble Sulfated Polysaccharides from *Ulva Pertusa*. *J. Med. Food* **2012**, *15*, 135–144. [CrossRef] [PubMed]
18. Zhang, Z.; Wang, X.; Zhao, M.; Yu, S.; Qi, H. The Immunological and Antioxidant Activities of Polysaccharides Extracted from *Enteromorpha Linza*. *Int. J. Biol. Macromol.* **2013**, *57*, 45–49. [CrossRef] [PubMed]
19. Aguilar-Briseño, J.A.; Cruz-Suarez, L.E.; Sassi, J.-F.; Ricque-Marie, D.; Zapata-Benavides, P.; Mendoza-Gamboa, E.; Rodríguez-Padilla, C.; Trejo-Avila, L.M. Sulphated Polysaccharides from *Ulva Clathrata* and *Cladosiphon Okamuranus* Seaweeds both Inhibit Viral Attachment/Entry and Cell-Cell Fusion, in NDV Infection. *Mar. Drugs* **2015**, *13*, 697–712. [CrossRef] [PubMed]

20. Ye, H.; Wang, K.; Zhou, C.; Liu, J.; Zeng, X. Purification, Antitumor and Antioxidant Activities *in vitro* of Polysaccharides from the Brown Seaweed *Sargassum Pallidum*. *Food Chem.* **2008**, *111*, 428–432. [CrossRef] [PubMed]
21. Wang, J.; Zhang, Q.B.; Zhang, Z.S.; Hou, Y.; Zhang, H. *In-vitro* Anticoagulant Activity of Fucoidan Derivatives from Brown Seaweed Laminaria Japonica. *Chin. J. Oceanol. Limn.* **2011**, *29*, 679–685. [CrossRef]
22. Li, L.J.; Li, M.Y.; Li, Y.T.; Feng, J.J.; Hao, F.Q.; Lun, Z. Adjuvant Activity of *Sargassum Pallidum* Polysaccharides Against Combined Newcastle Disease, Infectious Bronchitis and Avian Influenza Inactivated Vaccines. *Mar. Drugs* **2012**, *10*, 2648–2660. [CrossRef] [PubMed]
23. Suresh, V.; Senthilkumar, N.; Thangam, R.; Rajkumar, M.; Anbazhagan, C.; Rengasamy, R.; Gunasekaran, P.; Kannan, S.; Palani, P. Separation, Purification and Preliminary Characterization of Sulfated Polysaccharides from *Sargassum Plagiophyllum* and its *in vitro* Anticancer and Antioxidant Activity. *Process Biochem.* **2013**, *48*, 364–373. [CrossRef]
24. Imbs, T.I.; Skriptsova, A.V.; Zvyagintseva, T.N. Antioxidant Activity of Fucose-Containing Sulfated Polysaccharides Obtained from *Fucus Evanescens* by Different Extraction Methods. *J. Appl. Phycol.* **2014**, *27*, 1–9. [CrossRef]
25. Hwang, P.A.; Hung, Y.L.; Chien, S.Y. Inhibitory Activity of *Sargassum Hemiphyllum* Sulfated Polysaccharide in Arachidonic Acid-Induced Animal Models of Inflammation. *J. Food Drug Anal.* **2015**, *23*, 49–56. [CrossRef]
26. Wen, Z.S.; Liu, L.J.; Ouyang, X.K.; Qu, Y.L.; Yin, C.; Ding, G.F. Protective Effect of Polysaccharides from *Sargassum Horneri* Against Oxidative Stress in Raw264.7 Cells. *Int. J. Biol. Macromol.* **2014**, *68*, 98–106. [CrossRef] [PubMed]
27. Ping, S.; Chen, X.; Sun, P. Chemical Characterization, Antioxidant and Antitumor Activity of Sulfated Polysaccharide from *Sargassum Horneri*. *Carbohyd. Polym.* **2014**, *105*, 260–269.
28. Shobharani, P.; Nanishankar, V.H.; Halami, P.M.; Sachindra, N.M. Antioxidant and Anticoagulant Activity of Polyphenol and Polysaccharides from Fermented *Sargassum* Sp. *Int. J. Biol. Macromol.* **2014**, *65*, 542–548. [CrossRef] [PubMed]
29. Zhang, W.; Du, J.Y.; Jiang, Z.D.; Okimura, T.; Oda, T.; Yu, Q.; Jin, J.-O. Ascophyllan Purified from *Ascophyllum Nodosum* Induces Th1 and Tc1 Immune Responses by Promoting Dendritic Cell Maturation. *Mar. Drugs* **2014**, *12*, 4148–4164.
30. Yuan, Y.; Macquarrie, D. Microwave Assisted Extraction of Sulfated Polysaccharides (fucoidan) from *Ascophyllum Nodosum* and its Antioxidant Activity. *Carbohyd. Polym.* **2015**, *129*, 101–107. [CrossRef] [PubMed]
31. Ammar, H.H.; Lajili, S.; Said, R.B.; Cerf, D.L.; Bouraoui, A.; Majdoub, H. Physico-Chemical Characterization and Pharmacological Evaluation of Sulfated Polysaccharides from Three Species of Mediterranean Brown Algae of the *Genus Cystoseira*. *Daru J. Pharm. Sci.* **2015**, *23*, 1–8. [CrossRef] [PubMed]
32. Shao, P.; Chen, X.; Sun, P. Improvement of Antioxidant and Moisture-Preserving Activities of *Sargassum Horneri* Polysaccharide Enzymatic Hydrolyzates. *Int. J. Biol. Macromol.* **2015**, *74*, 420–427. [CrossRef] [PubMed]
33. Athukorala, Y.; Lee, K.W.; Park, E.J.; Heo, M.S.; Yeo, I.K.; Lee, Y.D.; Jeon, Y.J. Reduction of Lipid Peroxidation and H_2O_2-Mediated DNA Damage by a Red Alga (*Grateloupia Filicina*) Methanolic Extract. *J. Sci. Food Agr.* **2005**, *85*, 2341–2348. [CrossRef]
34. Wang, S.C.; Bligh, S.W.; Shi, S.S.; Wang, Z.T.; Hu, Z.B.; Crowder, J.; Branford-White, C.; Vella, C. Structural Features and Anti-HIV-1 Activity of Novel Polysaccharides from Red Algae *Grateloupia Longifolia* and Grateloupia Filicina. *Int. J. Biol. Macromol.* **2007**, *41*, 369–375. [CrossRef] [PubMed]
35. Yu, G.; Li, M.; Wang, W.; Liu, X.; Zhao, X.; Lv, Y.; Li, G.; Jiao, G.; Zhao, X. Structure and Anti-Influenza A (H1N1) Virus Activity of Three Polysaccharides from Eucheuma Denticulatum. *J. Ocean U. China* **2012**, *11*, 527–532. [CrossRef]
36. Shi, C.; Pan, T.; Cao, M.J.; Liu, Q.M.; Zhang, L.J.; Liu, G.M. Suppression of Th2 Immune Responses by the Sulfated Polysaccharide from *Porphyra Haitanensis* in Tropomyosin-Sensitized Mice. *Int. Immunopharmacol.* **2015**, *24*, 211–218. [CrossRef] [PubMed]
37. Chen, X.; Yang, S.; Wang, J.; Song, L.; Xing, R.; Liu, S.; Yu, H.; Li, P. Sulfated Polysaccharides Isolated from Cloned *Grateloupia Filicina* and their Anticoagulant Activity. *Biomed Res. Int.* **2015**, *2015*, 1–5.

38. Fleita, D.; El-Sayed, M.; Rifaat, D. Evaluation of the Antioxidant Activity of Enzymatically-Hydrolyzed Sulfated Polysaccharides Extracted from Red Algae; *Pterocladia Capillacea*. *LWT—Food Sci. Technol.* **2015**, *63*, 1236–1244. [CrossRef]

39. Karnjanapratum, S.; You, S.G. Molecular Characteristics of Sulfated Polysaccharides from *Monostroma Nitidum* and their *in vitro* Anticancer and Immunomodulatory Activities. *Int. J. Biol. Macromol.* **2011**, *48*, 311–318. [CrossRef] [PubMed]

40. Komatsu, T.; Kido, N.; Sugiyama, T.; Yokochi, T. Antiviral Activity of Acidic Polysaccharides from *Coccomyxa Gloeobotrydiformi*, a Green Alga, Against An *in vitro* Human Influenza A Virus Infection. *Immunopharmacology Immunotoxicology* **2012**, *35*, 1–7. [CrossRef] [PubMed]

41. Kwon, J.S.; Lee, H.J.; Lee, D.H.; Lee, Y.J.; Mo, I.P.; Nahm, S.S.; Kim, M.J.; Lee, J.B.; Park, S.Y.; Choi, I.S. Immune Responses and Pathogenesis in Immunocompromised Chickens in Response to Infection with the H9N2 Low Pathogenic Avian Influenza Virus. *Virus Res.* **2008**, *133*, 187–194. [CrossRef] [PubMed]

42. Lee, D.H.; Park, J.K.; Lee, Y.N.; Song, J.M.; Kang, S.M.; Lee, J.B.; Park, S.Y.; Choi, I.S.; Song, C.S. H9N2 Avian Influenza Virus-Like Particle Vaccine Provides Protective Immunity and a Strategy for the Differentiation of Infected from Vaccinated Animals. *Vaccine* **2011**, *29*, 4003–4007. [CrossRef] [PubMed]

43. Yuan, J.; Xu, L.; Bao, L.; Yao, Y.; Deng, W.; Li, F.; Lv, Q.; Gu, S.; Wei, Q.; Qin, C. Characterization of an H9N2 Avian Influenza Virus from a *Fringilla Montifringilla Brambling* in Northern China. *Virology* **2015**, *476*, 289–297. [CrossRef] [PubMed]

44. Kallon, S.; Li, X.; Ji, J.; Chen, C.; Xi, Q.; Shuang, C.; Xue, C.; Ma, J.; Xie, Q.; Zhang, Y. *Astragalus* Polysaccharide Enhances Immunity and Inhibits H9N2 Avian Influenza Virus *in vitro* and *in vivo*. *J. Anim. Sci. Biotechnol.* **2013**, *4*, 325–335. [CrossRef] [PubMed]

45. Lv, J.; Wei, B.; Yang, Y.; Yao, M.; Cai, Y.; Gao, Y.; Xia, X.; Zhao, X.; Liu, Z.; Li, X.; *et al.* Experimental Transmission in Guinea Pigs of H9N2 Avian Influenza Viruses from Indoor Air of Chicken Houses. *Virus Res.* **2012**, *170*, 102–108. [CrossRef] [PubMed]

46. Lin, Y.P.; Shaw, M.; Gregory, V.; Cameron, K.; Lim, W.; Klimov, A.; Subbarao, K.; Guan, Y.; Krauss, S.; Shortridge, K. Avian-to-human Transmission of H9N2 Subtype Influenza a Viruses: Relationship Between H9N2 and H5N1 Human Isolates. *Proc. Natl. Acad. Sci.* **2000**, *97*, 9654–9658. [CrossRef] [PubMed]

47. Xu, C.; Fan, W.; Wei, R.; Zhao, H. Isolation and Identification of Swine Influenza Recombinant a/Swine/Shandong/1/2003(H9N2) Virus. *Microbes Infect.* **2004**, *6*, 919–925. [CrossRef] [PubMed]

48. Wan, H.; Sorrell, E.M.; Song, H.; Hossain, M.J.; Ramirez-Nieto, G.; Monne, I.; Stevens, J.; Cattoli, G.; Capua, I.; Chen, L.-M.; *et al.* Replication and transmission of H9N2 Influenza Viruses in Ferrets: Evaluation of Pandemic Potential. *PLoS ONE* **2008**, *3*. [CrossRef] [PubMed]

49. Richard, M.; Schrauwen, E.J.A.; de Graaf, M.; Bestebroer, T.M.; Spronken, M.I.J.; van Boheemen, S.; de Meulder, D.; Lexmond, P.; Linster, M.; Herfst, S. Limited Airborne Transmission of H7N9 Influenza a Virus Between Ferrets. *Nature* **2013**, *501*, 560–563. [CrossRef] [PubMed]

50. Dalby, A.R.; Iqbal, M. A Global Phylogenetic Analysis in Order to Determine the Host Species and Geography Dependent Features Present in the Evolution of Avian H9N2 Influenza Hemagglutinin. *PeerJ* **2014**, *2*. [CrossRef] [PubMed]

51. Lam, T.-Y.; Wang, J.; Shen, Y.Y.; Zhou, B.P.; Duan, L.; Cheung, C.-L.; Ma, C.; Lycett, S.J.; Leung, Y.H.; Chen, X.C.; *et al.* The Genesis and Source of the H7N9 Influenza Viruses Causing Human Infections in China. *Nature* **2013**, *502*, 241–244. [CrossRef] [PubMed]

52. Peiris, M.; Yuen, K.Y.; Leung, C.W.; Chan, K.H.; lp, P.L.; Lai, R.W.; Orr, W.K.; Shortridge, K.F. Human Infection with Influenza H9N2. *Lancet* **1999**, *354*, 916–917. [CrossRef]

53. Westenius, V.; Mäkelä, S.M.; Ziegler, T.; Julkunen, I.; Osterlund, P. Efficient Replication and Strong Induction of Innate Immune Responses by H9N2 Avian Influenza Virus in Human Dendritic Cells. *Virology* **2014**, *471*, 38–48. [CrossRef] [PubMed]

54. Bouhlal, R.; Haslin, C.; Chermann, J.C.; Colliec-Jouault, S.; Sinquin, C.; Simon, G.; Cerantola, S.; Riadi, H.; Bourgougnon, N. Antiviral Activities of Sulfated Polysaccharides Isolated from Sphaerococcus Coronopifolius (*Rhodophyta, Gigartinales*) and Boergeseniella Thuyoides (*Rhodophyta, Ceramiales*). *Mar. Drugs* **2011**, *9*, 1187–1209. [CrossRef] [PubMed]

55. Kim, M.; Yim, J.H.; Kim, S.Y.; Kim, H.S.; Lee, W.G.; Kim, S.J.; Kang, P.S.; Lee, C.K. *In vitro* Inhibition of Influenza a Virus Infection by Marine Microalga-Derived Sulfated Polysaccharide P-Kg03. *Antivir. Res.* **2012**, *93*, 253–259. [CrossRef] [PubMed]

56. Pujol, C.A.; Ray, S.; Ray, B.; Damonte, E.B. Antiviral Activity Against Dengue Virus of Diverse Classes of Algal Sulfated Polysaccharides. *Int. J. Biol. Macromol.* **2012**, *51*, 412–416. [CrossRef] [PubMed]
57. Fang, X.B.; Chen, X.E. Structure Elucidation and Immunological Activity of a Novel Pectic Polysaccharide from the Stems of *Avicennia Marina*. *Eur. Food Res. Technol.* **2013**, *236*, 243–248. [CrossRef]
58. Liu, C.; Chen, J.; Li, E.; Fan, Q.; Wang, D.; Li, P.; Li, X.; Chen, X.; Qiu, S.; Gao, Z.; *et al.* The Comparison of Antioxidative and Hepatoprotective Activities of *Codonopsis Pilosula* Polysaccharide (CP) and Sulfated CP. *Int. Immunopharmacol.* **2015**, *24*, 299–305. [CrossRef] [PubMed]
59. Tabarsa, M.; Lee, S.J.; You, S. Structural Analysis of Immunostimulating Sulfated Polysaccharides from *Ulva Pertusa*. *Carbohydr. Res.* **2012**, *361*, 141–147. [CrossRef] [PubMed]
60. Wang, X.; Chen, Y.; Wang, J.; Liu, Z.; Zhao, S. Antitumor Activity of a Sulfated Polysaccharide from Enteromorpha Intestinalis Targeted Against Hepatoma through Mitochondrial Pathway. *Tumor Biol.* **2013**, *35*, 1641–1647. [CrossRef] [PubMed]
61. Yu, Q.; Yan, J.; Wang, S.; Ji, L.; Ding, K.; Vella, C.; Wang, Z.; Hu, Z. Antiangiogenic Effects of GFP08, an Agaran-Type Polysaccharide Isolated from *Grateloupia Filicina*. *Glycobiology* **2012**, *22*, 1343–1352. [CrossRef] [PubMed]
62. Faggio, C.; Pagano, M.; Dottore, A.; Genovese, G.; Morabito, M. Evaluation of Anticoagulant Activity of Two Algal Polysaccharides. *Nat. Prod. Res.* **2015**. [CrossRef] [PubMed]
63. Wang, Y.; Huang, M.; Sun, R.; Pan, L. Extraction, Characterization of a Ginseng Fruits Polysaccharide and its Immune Modulating Activities in Rats with Lewis Lung Carcinoma. *Carbohydr. Polym.* **2015**, *127*, 215–221. [CrossRef] [PubMed]
64. Turan, K.; Nagata, K.; Kuru, A. Antiviral effect of *Sanicula Europaea L.* Leaves Extract on Influenza Virus-Infected Cells. *Biochem. Biophys. Res. Commun.* **1996**, *225*, 22–26. [CrossRef] [PubMed]
65. Li, D.Y.; Xue, M.Y.; Wang, C.; Wang, J.B.; Chen, P.Y. Bursopentine as a Novel Immunoadjuvant Enhances Both Humoral and Cell-Mediated Immune Responses to Inactivated H9N2 Avian Influenza Virus in Chickens. *Clin. Vaccine Immunol. CVI* **2011**, *18*, 1497–1502. [CrossRef] [PubMed]
66. Zhang, J.J.; Zhang, Q.B.; Wang, J.; Shi, X.L.; Zhang, Z.S. Analysis of the Monosaccharide Composition of Fucoidan by Precolumn Derivation HPLC. *Chin. J. Oceanol. Limnol.* **2009**, *27*, 578–582. [CrossRef]
67. Zhang, W.; Oda, T.; Yu, Q.; Jin, J.O. Fucoidan from *Macrocystis Pyrifera* has Powerful Immune-Modulatory Effects Compared to Three Other Fucoidans. *Mar. Drugs* **2015**, *13*, 1084–1104. [CrossRef] [PubMed]
68. Thelen, T.; Hao, Y.; Medeiros, A.I.; Curtis, J.L.; Serezani, C.H.; Kobzik, L.; Harris, L.H.; Aronoff, D.M. The Class A Scavenger Receptor, Macrophage Receptor with Collagenous Structure, is the Major Phagocytic Receptor for Clostridium Sordellii Expressed by Human Decidual Macrophages. *J. Immunol.* **2010**, *185*, 4328–4335. [CrossRef] [PubMed]
69. Jin, J.-O.; Zhang, W.; Du, J.-Y.; Wong, K.-W.; Oda, T.; Yu, Q. Fucoidan can Function as an Adjuvant *in vivo* to Enhance Dendritic Cell Maturation and Function and Promote Antigen-Specific T Cell Immune Responses. *PLoS ONE* **2014**, *9*. [CrossRef] [PubMed]
70. de Godoi, A. M.; Faccin-Galhardi, L.C.; Lopes, N.; Rechenchoski, D.Z.; de Almeida, R.R.; Ricardo, N.M.P.S.; Nozawa, C.; Linhares, R.E.C. Antiviral Activity of Sulfated Polysaccharide of *Adenanthera Pavonina* Against Poliovirus in Hep-2 Cells. *Evid. Based Complement. Alternat. Med.* **2014**, *2014*. [CrossRef] [PubMed]
71. Chen, Y.; Xiong, W.; Zeng, L.; Wang, D.; Liu, J.; Wu, Y.; Hu, Y. Comparison of *Bush Sophora* Root Polysaccharide and its Sulfate's Anti-Duck Hepatitis A Virus Activity and Mechanism. *Carbohydr. Polym.* **2014**, *102*, 333–340. [CrossRef] [PubMed]
72. Chen, Y.; Song, M.; Wang, Y.; Xiong, W.; Zeng, L.; Zhang, S.; Xu, M.; Du, H.; Liu, J.; Wang, D.; *et al.* The Anti-DHAV Activities of *Astragalus* Polysaccharide and its Sulfate Compared with those of BSRPs and its Sulfate. *Carbohydr. Polym.* **2015**, *117*, 339–345. [CrossRef] [PubMed]
73. Shang, R.F.; Liang, J.P.; Na, Z.Y.; Yang, H.J.; Lu, Y.; Hua, L.Y.; Guo, W.Z.; Cui, Y.; Wang, L. *In vivo* Inhibition of NAS Preparation on H9N2 Subtype Aiv. *Virol. Sin.* **2010**, *25*, 145–150. [CrossRef] [PubMed]
74. Dubois, M.; Gilles, K.A.; Hamilton, J.K.; Rebers, P.A.; Smith, F. Colorimetric Method For Determination of Sugars and Related Substances. *Anal. Chem.* **1956**, *28*, 350–356. [CrossRef]
75. Yan, W.; Niu, Y.; Lv, J.; Xie, Z.; Jin, L.; Yao, W.; Gao, X.; Yu, L.L. Characterization of a Heteropolysaccharide Isolated from Diploid *Gynostemma Pentaphyllum Makino*. *Carbohydr. Polym.* **2013**, *92*, 2111–2117. [CrossRef] [PubMed]

76. Kawai, Y.; Seno, N.; Anno, K. A Modified Method for Chondrosulfatase Assay. *Anal. Biochem.* **1969**, *32*, 314–321. [CrossRef]

77. Cardozo, F.T.; Camelini, C.M.; Cordeiro, M.N.; Mascarello, A.; Malagoli, B.G.; Larsen, I.V.; Rossi, M.J.; Nunes, R.J.; Braga, F.C.; Brandt, C.R.; *et al.* Characterization and Cytotoxic Activity of Sulfated Derivatives of Polysaccharides from *Agaricus Brasiliensis. Int. J. Biol. Macromol.* **2013**, *57*, 265–272. [CrossRef] [PubMed]

78. Miao, S.; Mao, X.; Pei, R.; Miao, S.; Xiang, C.; Lv, Y.; Yang, X.; Sun, J.; Jia, S.; Liu, Y. Antitumor Activity of Polysaccharides from *Lepista Sordida* Against Laryngocarcinoma *in vitro* and *in vivo. Int. J. Biol. Macromol.* **2013**, *60*, 235–240. [CrossRef] [PubMed]

79. Pagano, M.; Faggio, C. The Use of Erythrocyte Fragility to Assess Xenobiotic Cytotoxicity. *Cell Biochem. Funct.* **2015**, *33*, 351–355. [CrossRef] [PubMed]

80. Ma, G.X.; Yang, W.J.; Mariga, A.M.; Fang, Y.; Ma, N.; Pei, F.; Hu, Q.H. Purification, Characterization and Antitumor Activity of Polysaccharides from *Pleurotus Eryngii* Residue. *Carbohydr. Polym.* **2014**, *114*, 297–305. [CrossRef] [PubMed]

81. Seo, Y.W.; Lee, H.J.; Kim, Y.A.; Youn, H.J.; Lee, B.-J. Effects of Several Salt Marsh Plants on Mouse Spleen and Thymus Cell Proliferation Using MTT Assay. *Ocean Sci. J.* **2005**, *40*, 209–212. [CrossRef]

82. Huang, Y.; Jiang, C.; Hu, Y.; Zhao, X.; Shi, C.; Yu, Y.; Liu, C.; Tao, Y.; Pan, H.; Feng, Y.; *et al.* Immunoenhancement Effect of *Rehmannia Glutinosa* Polysaccharide on Lymphocyte Proliferation and Dendritic Cell. *Carbohydr. Polym.* **2013**, *96*, 516–521. [CrossRef] [PubMed]

83. Feng, X.; Su, X.; Wang, F.; Wei, J.; Wang, F.; Cao, R.; Zhou, B.; Mao, X.; Zheng, Q.; Chen, P. Isolation and Potential Immunological Characterization of Tpsglvy, a Novel Bursal Septpeptide Isolated from the Bursa of Fabricius. *Peptides* **2010**, *31*, 1562–1568. [CrossRef] [PubMed]

84. Liu, X.D.; Feng, X.L.; Zhou, B.; Cao, R.B.; Li, X.F.; Ma, Z.Y.; Chen, P.Y. Isolation, Modulatory Functions on Murine B Cell Development and Antigen-Specific Immune Responses of Bp11, a Novel Peptide from the Chicken Bursa of Fabricius. *Peptides* **2012**, *35*, 107–113. [CrossRef] [PubMed]

85. Sokolova, E.V.; Byankina, A.O.; Kalitnik, A.A.; Kim, Y.H.; Bogdanovich, L.N.; Solov'eva, T.F.; Yermak, I.M. Influence of Red Algal Sulfated Polysaccharides on Blood Coagulation and Platelets Activation *in vitro*. *J. Biomed. Mater. Res. Part A* **2014**, *102*, 1431–1438. [CrossRef] [PubMed]

86. Livak, K.J.; Schmittgen, T.D. Analysis of Relative Gene Expression Data Using Real-Time Quantitative PCR and the 2(-Delta Delta C(T)) Method. *Methods* **2001**, *25*, 402–408. [CrossRef] [PubMed]

marine drugs

MDPI

Article

Laminarin from Irish Brown Seaweeds *Ascophyllum nodosum* and *Laminaria hyperborea*: Ultrasound Assisted Extraction, Characterization and Bioactivity

Shekhar U. Kadam [1], Colm P. O'Donnell [1], Dilip K. Rai [2], Mohammad B. Hossain [2], Catherine M. Burgess [3], Des Walsh [3] and Brijesh K. Tiwari [2],*

[1] School of Biosystems and Food Engineering, University College Dublin, Belfield, Dublin 4, Ireland; shekhar.kadam@ucdconnect.ie (S.U.K.); colm.odonnell@ucd.ie (C.P.O.)

[2] Food Biosciences, Teagasc Food Research Centre, Dublin 15, Ireland; dilip.rai@teagasc.ie (D.K.R.); mohammad.hossain@teagasc.ie (M.B.H.)

[3] Food Safety, Teagasc Food Research Centre, Dublin 15, Ireland; Kaye.Burgess@teagasc.ie (C.M.B.); des.walsh@teagasc.ie (D.W.)

* Author to whom correspondence should be addressed; brijesh.tiwari@teagasc.ie; Tel.: +353-1-805-9785.

Academic Editor: Paola Laurienzo
Received: 14 April 2015; Accepted: 1 July 2015; Published: 10 July 2015

Abstract: Ultrasound assisted extraction (UAE), purification, characterization and antioxidant activity of laminarin from Irish brown seaweeds *Ascophyllum nodosum* and *Laminarina hyperborea* were investigated. UAE was carried out using 60% ultrasonic power amplitude and 0.1 M hydrochloric acid for 15 min. Separately, solid-liquid extraction was carried in an orbital shaker using 0.1 M hydrochloric acid at 70 °C for 2.5 h. UAE with hydrochloric acid resulted in the highest concentration of laminarin, 5.82% and 6.24% on dry weight basis from *A. nodosum* and *L. hyperborea*, respectively. Purification of all extracts was carried out using molecular weight cut off dialysis at 10 kDa. Characterization of the laminarin fraction was carried out using matrix assisted laser desorption/ionization time-of-flight mass spectrometry. Antioxidant activity of *A. nodosum* and *L. hyperborea* extracts had 2,2-diphenyl-1-picrylhydrazyl (DPPH) inhibition levels of 93.23% and 87.57%, respectively. Moreover, these extracts have shown inihibition of bacterial growth of *Staphylcoccus aureus*, *Listeria monocytogenes*, *Escherichia coli* and *Salmonella typhimurium*.

Keywords: laminarin; antioxidant; antimicrobial; bioactive; *Laminaria hyperborea*; *Ascophyllum nodosum*

1. Introduction

Half of all global biodiversity is present in the marine ecosystem. Marine organisms are an excellent source of bioactive compounds. Marine algae, one of the important components of the marine ecosystem, are a good source of carotenoids, vitamins, minerals, phenolic compounds, sulphated polysaccharides, peptides and proteins [1]. These compounds have been demonstrated to possess diverse biological activities including anti-tumor, anti-viral, anti-coagulant, anti-allergic, anti-viral, anti-tumor and antibacterial [2]. Extracts from these seaweeds have been investigated for use as ingredients in food and feed products [3]. *Ascophyllum nodosum* is a brown seaweed which belongs to the Phaeophyceae class and it is the only species in the genus *Ascophyllum*. It is good source of fucoidan, alginates, ascophyllan, laminarins and polyphenols [4]. *Laminaria hyperborea* is a large, leathery brown seaweed from the Laminariaceae species which is found on the coast of Norway, Scotland and Ireland. It is commercially exploited by the hydrocolloid industry for alginate production [5]. Crude extracts from this seaweed are rich in phenolic compounds and β-glucan polysaccharides known as laminarins.

Laminarins are composed of (1,3)-β-D-glucan [6]. They consist of (1,3)-β-D-glucopyranose residues with some 6-O-branching in the main chain and some β-(1,6)-intrachain links are also present.

The molecular weight of laminarin is approximately 5 kDa [7]. Both water soluble and insoluble forms of laminarin exist depending mainly on the level of branching [8]. In general, extraction of laminarin involves grinding, precipitation in an mild acid medium, ultrafiltration and dialysis [9]. Characterization of laminarin rich extracts can be carried out by high performance size exclusion chromatography using a refractive index detector [10], matrix assisted laser desorption/ionization time-of-flight (MALDI-TOF) with dehydroxy-benzoic acid as matrix [11] and nuclear magnetic resonance spectroscopy using the ^{1}H-^{13}C heteronuclear single quantum coherence method [12].

However, current methods of extraction of laminarin are time and energy intensive, and result in poor yields. The development of novel eco-friendly extraction techniques is required to improve extraction rate and yield. Ultrasound is a low cost, easy to use, environment friendly technology suitable for the extraction of bioactive compounds from natural sources [13]. There are no reported studies on the use of ultrasound assisted extraction for laminarin applications. Only a limited number of studies on the characterization of laminarin-rich extract from brown seaweeds are available in the literature [6,11,12]. This study investigates the efficiency of ultrasound assisted extraction of laminarin in comparison to traditional extraction methods, and characterizes the resultant laminarin rich extracts with respect to molecular weight, chemical properties in addition to antioxidant and antimicrobial activities.

2. Results and Discussion

2.1. Laminarin and Phenolics Content

In this study two seaweed species harvested from the west coast of Ireland were selected for the extraction of laminarin. Laminarin was extracted using ultrasound assisted extraction and conventional solid liquid extraction using water and 0.1 M HCl (Table 1).

Table 1. Laminarin and total phenolic content of different extracts.

Seaweed Species	Solvent Type	Extraction Method	Time (min)	Sample Code	Laminarin Assay (% db)	Total Phenolic Content (mg PGE/g$_{db}$)
Laminaria hyperborea	Water	Ultrasound	15	LHWU	5.975 ± 0.467	0.365 ± 0.039
Laminaria hyperborea	0.1 M HCl	Ultrasound	15	LHAU	6.240 ± 0.008	0.343 ± 0.003
Laminaria hyperborea	Water	Solid liquid	150	LHWS	4.362 ± 0.197	0.363 ± 0.057
Laminaria hyperborea	0.1 M HCl	Solid liquid	150	LHAS	3.254 ± 0.235	0.352 ± 0.021
Ascophyllum nodosum	Water	Ultrasound	15	ANWU	5.290 ± 0.480	0.156 ± 0.014
Ascophyllum nodosum	0.1 M HCl	Ultrasound	15	ANAU	5.822 ± 0.343	0.128 ± 0.008
Ascophyllum nodosum	Water	Solid liquid	150	ANWS	4.599 ± 0.030	0.166 ± 0.015
Ascophyllum nodosum	0.1 M HCl	Solid liquid	150	ANAS	4.304 ± 0.165	0.110 ± 0.007

The highest laminarin content was measured in the extract of *L. hyperborea* and *A. nodosum* obtained using ultrasound and 0.1 M HCl. In previous studies, 0.1 M HCl was also found to give higher extraction yields compared to water [14]. Moreover HCl at higher temperatures is more effective than at room temperature for laminarin extraction [7]. Laminarin content varies with species, with *L. hyperborea* having a higher level of laminarin compared to *A. nodosum*. Laminarin content also varies with factors such as harvesting season and geographical location. Laminarin is absent during the period of fast growth in spring, but in autumn and winter, it may represent up to 35% of the dried weight of the fronds [15]. Ultrasound was found to give higher extraction yields of laminarin. High power ultrasound treatment for 15 min achieved a higher extraction yield than conventional solid and liquid extraction for both seaweeds investigated. Laminarin yield for *L. hyperborea* extracted using ultrasound was 36.97% and 91.76% higher using water and 0.1 M HCl solvents respectively, whereas for *A. nodosum* laminarin yield was 15.02% and 35.62% higher using water and 0.1 M HCl solvents respectively. This can be attributed to the bubble cavitation phenomena generated by ultrasound waves. The implosion of cavitation bubbles generates macroturbulence, high-velocity interparticle collisions, and perturbations in microporous particles of the biomass. Cavitation near liquid-solid interfaces directs a fast-moving stream of liquid through the cavity at the surface. Impingement by these microjets results in surface peeling, erosion, and particle breakdown, facilitating the release of bioactive compounds and other components from the biological matrix. These effects increase the efficiency of extraction by increasing mass transfer by eddy and internal diffusion mechanisms [16]. Ultrasound assisted extraction (UAE) has also been successfully used for extraction of *A. nodosum* bioactive compounds including phenolic compounds, fucose and uronic acids [17,18]. The phenolic content was higher in *L. hyperborea* extracts. Water was demonstrated to be a better solvent than HCl for extraction of phenolics. This may be attributed to fact that acid solvents at a temperature of 70 °C may be detrimental to phenolic compounds leading to a lower content in acid extracts. The highest content of phenolics observed in *L. hyperborea* and *A. nodosum* was 0.365 mg PGE/g_{db} and 0.166 mg PGE/g_{db}, respectively.

2.2. Characterization of Extracts

Laminarin rich extracts were analyzed for their molecular weight distribution using Matrix Assisted Laser Desorption Ionization Quadrupole Time-of-Flight Mass Spectrometry (MALDI-Q-TOF-MS). Figure 1A,B show the mass spectra obtained in negative ion mode for a laminarin standard and an ultrasound assisted extract of *L. hyperborea*, respectively.

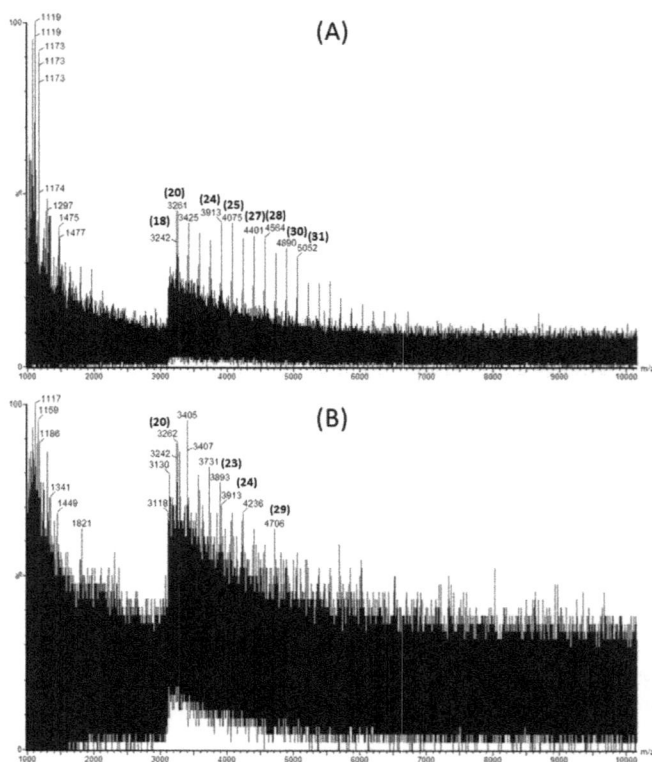

Figure 1. Matrix Assisted Laser Desorption Ionization Quadrupole Time-of-Flight Mass Spectrometry (MALDI-Q-TOF-MS) spectra showing the distribution of molecular weights of (**A**) commercial laminarin in sinapinic acid and (**B**) *Laminaria hyperborea* extract obtained by ultrasound assisted extraction using 0.1 M HCl as a solvent.

The degree of polymerization (DP) for each laminarin peak is shown in bracketed bold numbers above the corresponding m/z peaks. Four extracts were evaluated to investigate the effect of ultrasound and seaweed species on extraction of laminarin. Since M-chains and G-chains in the native laminarins differ by only 2 mass units, they cannot be distinguished from high molecular weight profiles [11]. Laminarin extracted using UAE from *L. hyperborea* had higher molecular weight laminarins ranging from 3242 to 5052 Da (corresponding to DP20 to DP31) compared to the other extracts measured. Meanwhile, *L. hyperborea* obtained with conventional extraction yielded laminarins from DP20 to DP24 only. This demonstrates that ultrasound assisted extraction is more efficient than solid liquid extraction for extracting higher molecular weight laminarins. This can be attributed to the bubble cavitation phenomena releasing high molecular weight laminarins from *L. hyperborea*. Similar results were reported for extraction of high molecular weight phlorotannins from *A. nodosum* using ultrasound assisted extraction [18].

Ultrasound treated *A. nodosum* extracts yielded laminarins from DP25 to DP30 (molecular weights in the range of 4075 to 4884 Da). *A. nodosum* extracts did not yield lower molecular weight laminarins compared to *L. hyperborea* extracts. Moreover, *L hyperborea* extracts had a wide range of degree of polymerization compared to *A. nodosum* extracts.

2.3. Bioactivities of Laminarin Rich Extract

Laminarin has been found to possess various biological activities. The antioxidant and antimicrobial activities of the crude laminarin extract are shown in Table 2. The % of 2,2-diphenyl-1-picrylhydrazyl (DPPH) inhibition of seaweed extracts was found to be highest in ultrasound treated extracts using acid solvent for *L. hyperborea* (87.58%) and *A. nodosum* (93.24%). O'Sullivan *et al.* [19] also found that *A. nodosum* was one of the most effective extracts for scavenging DPPH radicals. The extracts obtained with acid were found to have higher antioxidant activity than those obtained using water solvent. Further, authors reported that *A. nodosum* with 0.45 g/100 g (Gallic acid equivalent) of phenolic content exhibited antioxidant activity of 25.6% DPPH inhibition [19]. Balboa *et al.* [20] have extensively reviewed antioxidant activities in brown seaweeds and they have reported that phenolic compounds possess antioxidant properties.

Table 2. Antioxidant activity as %DPPH inhibition and antimicrobial activity as minimum inhibition concentration (MIC) of different extracts.

Sample	Antioxidant Activity (% DPPH Inhibition)	MIC (mg/mL)			
		E. coli	*S. typhimurium*	*S. aureus*	*L. monocytogenes*
LHWU	52.78	21.0	21.0	5.3	2.6
LHAU	87.58	13.1	13.1	6.6	3.3
LHWS	73.91	45.6	22.8	11.4	5.7
LHAS	86.04	28.2	28.2	7.0	7.0
ANWU	61.46	43.2	21.6	NI [*]	NI
ANAU	93.24	596.8	14.9	29.8	59.7
ANWS	15.13	NI	25.5	NI	NI
ANAS	87.82	NI	33.4	66.8	66.8

NI, no inhibition.

The extracts were tested for inhibition against two Gram positive (*Staphylcoccus aureus* and *Listeria monocytogenes*) and two Gram negative (*Escherichia coli* and *Salmonella typhimurium*) bacterial strains. All extracts of *L. hyperborea* were found to inhibit the growth of all the micro-organisms tested. The acid extracts had better inhibition of bacterial growth compared to water extracts. However, *A. nodosum* extracts only inhibited the growth of *S. typhimurium* completely. Only the acid extract of *A. nodosum* proved effective in inhibiting the bacterial growth of all species. The higher phenolic content and antioxidant activity of *L. hyperborea* extracts may enhance the antimicrobial efficacy of these extracts as many phenolic compounds have been shown to possess antimicrobial properties [21]. Laminarin rich extracts prepared using ultrasound and acid solvents had minimum inhibitory concentrations (MIC) of 13.1 mg/mL for *E. coli* and *S. typhimurium* and 6.6 mg/mL and 3.3 mg/mL for *S. aureus* and *L. monocytogenes*, respectively. This is the first reported study demonstrating that laminarin rich extracts possess microbial inhibitory activity. There are number of reports of seaweed extracts such as *Sargassum polyophyllum*, *Sargassum flavellum*, *Padina australis* and *Sargassum binderi* possessing antimicrobial activity [22]. However, ultrasound can be a novel alternative to high energy consuming traditional solid liquid extraction methods. These seaweed extracts have potential application in the preparation of antimicrobial products for example, a hydrogel wound dressing incorporating a seaweed *Polysiphonia lanosa* extract [23].

3. Experimental Section

3.1. Chemicals and Reagents

Laminarin produced from *Laminaria digitata*, 2,2-diphenyl-1-picrylhydrazyl (DPPH), phloroglucinol, sodium carbonate, Folin-Ciocalteu reagent (FCR), and hydrochloric acid (37%) were obtained from Sigma-Aldrich (Wicklow, Ireland). GOPOD (glucose oxidase/peroxidase) reagent and β-glucosidase were obtained from Megazyme, Wicklow, Ireland.

3.2. Seaweed Samples

Brown seaweed *A. nodosum* and *L. hyperborea* were harvested from Finavarra, Co. Clare, Ireland in May 2014. Seaweed samples were washed thoroughly with fresh water to remove epiphytes and salt. Fresh seaweed samples were freeze dried. Dried seaweed was powdered using a hammer mill. Samples were stored at 4 °C prior to extraction studies.

3.3. Ultrasound Assisted Extraction

Ten grams of *A. nodosum* and *L. hyperborea* powders were extracted using 200 mL of solvent (distilled water or 0.03 M HCl). HCl was used as the solvent for extraction based on preliminary studies and previously reported studies for the extraction of laminarin [24–26]. A 750 W ultrasonic processor (VC 750, Sonics and Materials Inc., Newtown, CT, USA) with a 13 mm diameter probe and constant frequency of 20 kHz was used. Ultrasonic energy was controlled by setting the amplitude of the sonicator probe. Ultrasound treatment was applied for 15 min at an amplitude level of 60% which corresponds to an ultrasonic intensity of 35.61 W cm^{-2}. Ultrasonic power dissipated was calculated at each amplitude level, with temperature (T) recorded as a function of time (t) under adiabatic conditions using a T-type thermocouple. From temperature *versus* time data, the initial temperature rise dT/dt was determined by polynomial curve fitting. The ultrasonic power (P) was determined using Equation (1) where dT/dt is the change in temperature over time (°C s^{-1}), C_p is the specific heat of water (4.18 kJ kg^{-1} °C^{-1}), and m is the mass (kg).

$$P = mC_p(dT/dt)_{t=0} \tag{1}$$

Ultrasonic intensity (W cm^{-2}) dissipated from an ultrasonic probe tip with diameter D (cm) is given by Equation (2)

$$UI = \frac{4P}{\pi D^2} \tag{2}$$

The traditional solid-liquid method of extraction involved stirring at 70 °C for 2.5 h using distilled water and 0.1 M HCl as solvents and no ultrasound pretreatment was employed. The extracted samples were then centrifuged at 9000 rpm for 30 min. The supernatant was separated and precipitated with ethanol overnight at 4 °C. The precipitated extract was freeze dried and stored at −20 °C for further analysis. The different methods of extraction carried out in this experiment are listed in Table 3. The extraction yield (%) was calculated by measuring the mass of freeze dried extract over the initial mass of the sample.

Table 3. Different methods of extraction and corresponding sample codes.

Seaweed Species	Solvent Type	Extraction	Time (min)	Sample Code
Laminaria hyperborea	Water	Ultrasound	15	LHWU
Laminaria hyperborea	0.1 M HCl	Ultrasound	15	LHAU
Laminaria hyperborea	Water	Solid liquid	150	LHWS
Laminaria hyperborea	0.1 M HCl	Solid liquid	150	LHAS
Ascophyllum nodosum	Water	Ultrasound	15	ANWU
Ascophyllum nodosum	0.1 M HCl	Ultrasound	15	ANAU
Ascophyllum nodosum	Water	Solid liquid	150	ANWS
Ascophyllum nodosum	0.1 M HCl	Solid liquid	150	ANAS

3.4. Laminarin Assay

Laminarin in the extract was quantified by measuring the glucose concentration released by the enzymatic hydrolysis of laminarin [27]. A 100 µL sample volume was incubated in 100 µL of β-glucosidase enzyme at 40 °C for 15 min. After incubation, 3 mL of GOPOD (glucose oxidase/peroxidase) reagent was added. This mixture was incubated at 40 °C for 20 min. Finally the absorbance of the sample was measured at 510 nm by UV-VIS spectrophotometer (UV3100PC, VWR International). Laminarin produced from *Laminaria digitata* was used as a standard.

3.5. Total Phenolic Content

Total phenolic content was determined using the method of Wang *et al.* [28]. Folin-Ciocalteau reagent was diluted with distilled water at a ratio of 1:10. An extract of 100 µL was mixed with 100 µL of diluted Folin-Ciocalteau reagent and 100 µL of sodium bicarbonate (20%, w/v) was added to the mixture and diluted to 1000 µL with distilled water. This solution was maintained at room temperature for 30 min and the absorbance was measured at 735 nm by UV-VIS spectrophotometer (UV3100PC, VWR International). Results were expressed as mg phloroglucinol equivalents (PGE)/g_{db}.

3.6. Matrix Assisted Laser Desorption Ionization Quadrupole Time-of-Flight Mass Spectrometry (MALDI-Q-TOF-MS)

Mass spectrometry of samples was performed using MALDI-Q-TOF-MS Waters Corporation, Milford, MA, USA). Prior to analysis, samples were dialyzed (molecular weight cut off of 10 kDa) in distilled water overnight. Aliquots of 5 µL of sample were mixed with 5 µL of matrix sinapinic acid. Finally 1–2 µL of sample was plated on a 96 well stainless steel MALDI plate. Samples were allowed to dry and co-crystallize with the matrix at room temperature and the plate was loaded in MALDI-Q-TOF mass spectrometer. Mass spectral data were obtained in the negative-ion mode for a mass range of *m/z* 1000 to *m/z* 10,000.

3.7. Antioxidant Activity—DPPH Method

DPPH free radical scavenging inhibition assay was used to determine the antioxidant capacity of extracted samples [29]. Extract sample of 200 µL was added to 800 µL of 60 µM DPPH in ethanol, decrease in absorbance was monitored at 517 nm by UV-VIS spectrophotometer (UV3100PC, VWR International) after 30 min incubation in dark. The readings were compared with the controls, which contained 200 µL of water instead of the seaweed extract. The percent inhibition was calculated as

$$\% \text{ Inhibition} = \frac{\text{Abs}_{control} - \text{Abs}_{sample}}{\text{Abs}_{control}} \times 100 \tag{3}$$

3.8. Antimicrobial Activity—Iodonitrotetrazolium Chloride (INT) Dye Method

3.8.1. Bacterial Strains and Culture Conditions

Seaweed extracts were tested for antimicrobial activity against the following strains of bacteria: *Staphylcoccus aureus* NCTC 8178, *Escherichia coli* DSM 1103, *Listeria monocytogenes* NCTC 11994 and *Salmonella typhimurium* SARB 65. The strains were stored on ceramic beads in glycerol at −80 °C prior to use. A bead of each strain was streaked on a nutrient agar plate and incubated for 18 h at 37 °C. A single colony was removed from each plate and inoculated into tubes containing 25 mL of sterile Mueller-Hinton Broth (MHB) and incubated for 22 h at 37 °C. Overnight cultures were vortexed and aliquots diluted appropriately in sterile MHB to produce solutions containing $\log_{10}$ 6.0 ± 0.5 cells/mL. Cell numbers were confirmed by plate counting. Antibiotic Gentamicin (0.2 mg/mL) was used as standard for negative growth.

3.8.2. Minimum Inhibitory Concentration (MIC) Assay

The MIC of each seaweed extract was carried out using a previously described microtitre method of Kenny, Smyth, Walsh, Kelleher, Hewage and Brunton [21]. Each extract (2 mg/mL) was prepared by dissolving the material in distilled water added to the first well of the plate followed by a serial dilution across the plate. Iodonitrotetrazolium chloride (INT) dye was used identify microbial growth [30]. The MIC of each extract against a bacterial strain was determined as the lowest sample concentration at which no pink color appeared. This process was repeated in triplicate for each bacterial strain to ensure reproducibility.

4. Conclusions

In this study, water and acid extracts from *L. hyperborea* and *A. nodosum* were obtained using ultrasound assisted extraction and solid liquid extraction. The extracts were purified to obtain laminarin rich extracts. The *L. hyperborea* extracts contained higher contents of laminarin. The overall laminarin content in all the extracts was low which may be attributed to seasonal and geographical factors. Ultrasound was demonstrated to be a more efficient method of extraction than solid liquid extraction based on laminarin content and molecular weight distribution observed in the extracts. The laminarin rich extracts were also studied for biological activities including anti-oxidant and anti-microbial activity. This study is the first report of laminarin rich extracts possessing anti-microbial activity. The use of laminarin as a nutraceutical ingredient should be further investigated due to its dietary fiber properties in addition to the anti-oxidant and anti-microbial activities reported in this paper.

Acknowledgments: The authors are thankful to Irish Research Council's Embark Initiative for financial support for this work. This work has also been supported by the Marine Functional Foods Research Initiative (NutraMara project) which is a programme for marine based functional food development. This project (Grant-Aid Agreement No. MFFRI/07/01) is carried out under the Sea Change Strategy with the support of the Marine Institute and the Department of Agriculture, Food and the Marine, funded under the National Development Plan 2007–2013.

Author Contributions: All authors contributed equally to the manuscript preparation. All authors approved the final version of the manuscript.

Conflicts of Interest: The authors declare no conflict of interest.

References

1. Kadam, S.U.; Prabhasankar, P. Marine foods as functional ingredients in bakery and pasta products. *Food Res. Int.* **2010**, *43*, 1975–1980. [CrossRef]
2. Gupta, S.; Abu-Ghannam, N. Bioactive potential and possible health effects of edible brown seaweeds. *Trends Food Sci. Technol.* **2011**, *22*, 315–326. [CrossRef]

3. Gupta, S.; Abu-Ghannam, N. Recent developments in the application of seaweeds or seaweed extracts as a means for enhancing the safety and quality attributes of foods. *Innov. Food Sci. Emerg. Technol.* **2011**, *12*, 600–609. [CrossRef]

4. Jiang, Z.; Okimura, T.; Yokose, T.; Yamasaki, Y.; Yamaguchi, K.; Oda, T. Effects of sulfated fucan, ascophyllan, from the brown Alga *Ascophyllum nodosum* on various cell lines: A comparative study on ascophyllan and fucoidan. *J. Biosci. Bioeng.* **2010**, *110*, 113–117. [CrossRef] [PubMed]

5. Werner, A.; Kraan, S. *Review of the Potential Mechanisation of Kelp Harvesting in Ireland*; National University of Ireland: Galway, Ireland, 2004.

6. Rioux, L.E.; Turgeon, S.L.; Beaulieu, M. Characterization of polysaccharides extracted from brown seaweeds. *Carbohydr. Polym.* **2007**, *69*, 530–537. [CrossRef]

7. Kadam, S.U.; Tiwari, B.K.; O'Donnell, C.P. Extraction, structure and biofunctional activities of laminarin from brown algae. *Int. J. Food Sci. Technol.* **2014**, *50*, 24–31. [CrossRef]

8. Miao, H.-Q.; Ishai-Michaeli, R.; Peretz, T.; Vlodavsky, I. Laminarin sulfate mimics the effects of heparin on smooth muscle cell proliferation and basic fibroblast growth factor-receptor binding and mitogenic activity. *J. Cell. Physiol.* **1995**, *164*, 482–490. [CrossRef] [PubMed]

9. Yvin, J.C.; LeVasseur, F.; Hud'Homme, F. Use of Laminarin and Oligosaccharides Derived Therefrom in Cosmetics and for Preparing a Skin Treatment Drug. U.S. Patent US5980916 A, 9 November 1999.

10. Zha, X.-Q.; Xiao, J.-J.; Zhang, H.-N.; Wang, J.-H.; Pan, L.-H.; Yang, X.-F.; Luo, J.-P. Polysaccharides in *Laminaria japonica* (LP): Extraction, physicochemical properties and their hypolipidemic activities in diet-induced mouse model of atherosclerosis. *Food Chem.* **2012**, *134*, 244–252. [CrossRef]

11. Chizhov, A.O.; Dell, A.; Morris, H.R.; Reason, A.J.; Haslam, S.M.; McDowell, R.A.; Chizhov, O.S.; Usov, A.I. Structural analysis of laminarans by MALDI and FAB mass spectrometry. *Carbohydr. Res.* **1998**, *310*, 203–210. [CrossRef]

12. Date, Y.; Sakata, K.; Kikuchi, J. Chemical profiling of complex biochemical mixtures from various seaweeds. *Polym. J.* **2012**, *44*, 888–894. [CrossRef]

13. Vilkhu, K.; Mawson, R.; Simons, L.; Bates, D. Applications and opportunities for ultrasound assisted extraction in the food industry—A review. *Innov. Food Sci. Emerg. Technol.* **2008**, *9*, 161–169. [CrossRef]

14. Deville, C.; Damas, J.; Forget, P.; Dandrifosse, G.; Peulen, O. Laminarin in the dietary fibre concept. *J. Sci. Food Agric.* **2004**, *84*, 1030–1038. [CrossRef]

15. Rinaudo, M. Seaweed polysaccharides. In *Comprehensive Glycoscience*; Johannis, P.K., Ed.; Elsevier: Oxford, UK, 2007; pp. 691–735.

16. Kadam, S.U.; Tiwari, B.K.; O'Donnell, C.P. Application of novel extraction technologies for bioactives from marine algae. *J. Agric. Food Chem.* **2013**, *61*, 4667–4675. [CrossRef] [PubMed]

17. Kadam, S.U.; Tiwari, B.K.; O'Connell, S.; O'Donnell, C.P. Effect of ultrasound pre-treatment on the extraction kinetics of bioactives from brown seaweed (*Ascophyllum nodosum*). *Sep. Sci. Technol.* **2014**, *50*, 1–6.

18. Kadam, S.U.; Tiwari, B.K.; Smyth, T.J.; O'Donnell, C.P. Optimization of ultrasound assisted extraction of bioactive components from brown seaweed *Ascophyllum nodosum* using response surface methodology. *Ultrason. Sonochem.* **2015**, *23*, 308–316. [CrossRef] [PubMed]

19. O'Sullivan, A.M.; O'Callaghan, Y.C.; O'Grady, M.N.; Queguineur, B.; Hanniffy, D.; Troy, D.J.; Kerry, J.P.; O'Brien, N.M. *In vitro* and cellular antioxidant activities of seaweed extracts prepared from five brown seaweeds harvested in spring from the west coast of Ireland. *Food Chem.* **2011**, *126*, 1064–1070. [CrossRef]

20. Balboa, E.M.; Conde, E.; Moure, A.; Falqué, E.; Domínguez, H. *In vitro* antioxidant properties of crude extracts and compounds from brown algae. *Food Chem.* **2013**, *138*, 1764–1785. [CrossRef] [PubMed]

21. Kenny, O.; Smyth, T.J.; Walsh, D.; Kelleher, C.T.; Hewage, C.M.; Brunton, N.P. Investigating the potential of under-utilised plants from the Asteraceae family as a source of natural antimicrobial and antioxidant extracts. *Food Chem.* **2014**, *161*, 79–86. [CrossRef] [PubMed]

22. Jaswir, I.; Tawakalit Tope, A.-H.; Raus, R.A.; Ademola Monsur, H.; Ramli, N. Study on anti-bacterial potentials of some Malaysian brown seaweeds. *Food Hydrocoll.* **2014**, *42*, 275–279. [CrossRef]

23. Tan, S.P.; McLoughlin, P.; O'Sullivan, L.; Prieto, M.L.; Gardiner, G.E.; Lawlor, P.G.; Hughes, H. Development of a novel antimicrobial seaweed extract-based hydrogel wound dressing. *Int. J. Pharm.* **2013**, *456*, 10–20. [CrossRef] [PubMed]

24. Black, W.A.P.; Cornhill, W.J.; Dewar, E.J.; Woodward, F.N. Manufacture of algal chemicals. III. Laboratory-scale isolation of laminarin from brown marine algae. *J. Appl. Chem.* **1951**, *1*, 505–517. [CrossRef]

25. Ermakova, S.; Men'shova, R.; Vishchuk, O.; Kim, S.-M.; Um, B.-H.; Isakov, V.; Zvyagintseva, T. Water-soluble polysaccharides from the brown alga *Eisenia bicyclis*: Structural characteristics and antitumor activity. *Algal Res.* **2013**, *2*, 51–58. [CrossRef]
26. Jin, W.; Zhang, W.; Wang, J.; Ren, S.; Song, N.; Duan, D.; Zhang, Q. Characterization of laminaran and a highly sulfated polysaccharide from *Sargassum fusiforme*. *Carbohydr. Res.* **2014**, *385*, 58–64. [CrossRef] [PubMed]
27. Devillé, C.; Gharbi, M.; Dandrifosse, G.; Peulen, O. Study on the effects of laminarin, a polysaccharide from seaweed, on gut characteristics. *J. Sci. Food Agric.* **2007**, *87*, 1717–1725. [CrossRef]
28. Wang, X.; Wu, Y.; Chen, G.; Yue, W.; Liang, Q.; Wu, Q. Optimisation of ultrasound assisted extraction of phenolic compounds from *Sparganii rhizoma* with response surface methodology. *Ultrason. Sonochemistry* **2013**, *20*, 846–854. [CrossRef] [PubMed]
29. Shetty, K.; Curtis, O.F.; Levin, R.E.; Witkowsky, R.; Ang, W. Prevention of vitrification associated with *in vitro* shoot culture of oregano (*Origanum vulgare*) by *Pseudomonas* spp. *J. Plant Physiol.* **1995**, *147*, 447–451. [CrossRef]
30. Eloff, J.N. A sensitive and quick microplate method to determine the minimal inhibitory concentration of plant extracts for bacteria. *Planta Med.* **1998**, *64*, 711–713. [CrossRef] [PubMed]

marine drugs

MDPI

Article

Degradation of Polysaccharides from *Grateloupia filicina* and Their Antiviral Activity to Avian Leucosis Virus Subgroup J

Yuhao Sun [1,2,3], Xiaolin Chen [1,3,*], Ziqiang Cheng [4], Song Liu [1,3], Huahua Yu [1,3], Xueqin Wang [1,3] and Pengcheng Li [1,3,*]

[1] Key Laboratory of Experimental Marine Biology, Institute of Oceanology, Chinese Academy of Sciences, No. 7 Nanhai Road, Qingdao 266071, China; 18669884128@163.com (Y.S.); sliu@qdio.ac.cn (S.L.); yuhuahua@qdio.ac.cn (H.Y.); xueqinwang@qdio.ac.cn (X.W.)
[2] University of Chinese Academy of Sciences, Beijing 100049, China
[3] Laboratory for Marine Drugs and Bioproducts of Qingdao National Laboratory for Marine Science and Technology, No. 7 Nanhai Road, Qingdao 266071, China
[4] College of Animal Science and Veterinary Medicine, Shandong Agricultural University, No. 61 Daizong Road, Taian 271018, China; czqsd@126.com
*   Correspondence: chenxl@qdio.ac.cn (X.C.); pcli@qdio.ac.cn (P.L.); Tel.: +86-0532-8289-8707 (X.C. & P.L.)

Received: 18 September 2017; Accepted: 1 November 2017; Published: 3 November 2017

Abstract: In this study, polysaccharides from *Grateloupia filicinia* (GFP) were extracted and several low molecular weight (Mw) *G. filicina* polysaccharides (LGFPs) were prepared by the hydrogen peroxide (H_2O_2) oxidation method. Additionally, the effect of different experimental conditions on the degradation of GFP was determined. Results showed that the GFP degradation rate was positively related to H_2O_2 concentration and temperature, and negatively related to pH. Chemical analysis and Fourier transform infrared spectra (FT-IR) of GFP and LGFPs showed that the degradation caused a slight decrease of total sugar and sulfate content. However, there was no obvious change for monosaccharide contents. Then, the anti-ALV-J activity of GFP and LGFPs were determined in vitro. Results revealed that all of the samples could significantly inhibit ALV-J and lower Mw LGFPs exhibited a stronger suppression, and that the fraction LGFP-3 with Mw 8.7 kDa had the best effect. In addition, the reaction phase assays showed that the inhibition effect was mainly because of the blocking virus adsorption to host cells. Moreover, real-time PCR, western-blot, and IFA were further applied to evaluate the blocking effects of LGFP-3. Results showed that the gene relative expression and gp85 protein for LGFPS-3 groups were all reduced. Data from IFA showed that there was less virus infected cells for 1000 and 200 µg/mL LGFPS-3 groups when compared to virus control. Therefore, lower Mw polysaccharides from *G. filicina* might supply a good choice for ALV-J prevention and treatment.

Keywords: *Grateloupia filicina*; degradation; low molecular weight polysaccharides; antiviral activity; ALV-J

1. Introduction

Avian leucosis virus subgroup J (ALV-J) is an immunosuppressive virus, which generally causes a diversity of tumors, such as hemangioma and myeloid leucosis in birds [1]. It has spread worldwide since the first time it was discovered in British broiler breeds in 1988. In China, ALV-J was first detected and officially recorded in 1999 [2], and after that the ALV-J infection appeared around the country. The virus could infect chickens through both horizontal and vertical transmission; initially it was mainly found in turkeys and broilers, but recently it also brought disease and death in egg layers and broiler breeders [3], which caused heavy losses to the poultry breeding industry. Due to the complex genetic sequence and the antigenic variability, there are still no commercial vaccines against ALV-J infection [4–6]. Currently, the only way to prevent ALV-J is by quarantining and eliminating positive chickens. Therefore, it is urgent and necessary to find effective antiviral medicine to control ALV-J spread.

On the other hand, the red seaweed genus *Grateloupia*, which belongs to Halymeniaceae, is an intertidal alga [7], which is widely distributed in the countries of the west Pacific bank such as China, Japan, and South Korea. It has enormous commercial value because it may be used as raw materials for carrageenan. In addition, the protein content of *Grateloupia turuturu* can account for 20% of the dry weight therefore being of potential in the food industry [8]. Polysaccharide is another important chemical compound of *Grateloupia*. It has attracted wide attention in recent years due to its multiple biological activities. For example, Nikapitiya et al. reported a good anticoagulant activity of a sulfated polysaccharide that is extracted from *Grateloupia filicina* [9]. In addition, Yu et al. isolated an agaran-type polysaccharide from *G. filicina* and demonstrated that the polysaccharide carried antiangiogenic effects [10]. Further, the polysaccharides that are extracted from *G. livida* were reported to have antioxidant, antibacterial, and antisachistosomal activity [11,12]. Scientists also studied the antiviral activity of *Grateloupia* and found that polysaccharides obtained from *G. indica*, *G. filicina*, and *G. longilia* had anti-HSV-1, HSV-2, HIV-1, and dengue virus activity [13,14]. However, the studies on antiviral activity of the polysaccharide from *Grateloupia* were mainly focused on human suffering virus, and few studies aimed at finding the effects against poultry viruses.

Even though the high Mw *Grateloupia* polysaccharide has various biological activities, its huge Mw causes low solubility and tough organism absorption, which greatly restricts its application. Commonly, lower Mw algal polysaccharides are obtained by the degradation of high Mw polysaccharides. When compared with original high Mw polysaccharide, low Mw polymers are characterized by having a higher water solubility and stability, and easy organism absorption [15]. In order to obtain lower Mw polysaccharide a chain of polysaccharides is usually broken, as according to the $1 \rightarrow 3$, $1 \rightarrow 4$ and $1 \rightarrow 6$ glycosidic linkages fractures, and some of their original activity might be improved in different degrees [16–18]. Additionally, in recent years, it has been found that low Mw polysaccharides from seaweeds presented some effects on the suppression H1N1 influenza virus and HIV [19–22]. However, research on the anti-ALV-J activity of algal polysaccharides still scarce. Therefore, it would be innovative and promising applying the low Mw polysaccharide from *Grateloupia* to prevent ALV-J.

In this study, polysaccharides from *G. filicina* (GFP) were extracted by hot water extraction and were degraded via H_2O_2 oxidative degradation. The effect of different conditions (including concentration of H_2O_2, pH and temperature) on the degradation of GFP was determined. Several low Mw fractions of GFP (LGFPs) were prepared and their chemical composition was tested. Finally, the anti-ALV-J activity of LGFPs was explored. The results of this study would provide a possible useful application of seaweeds polysaccharides as a veterinary medicine in the prevention of ALV-J.

2. Results

2.1. Influence of Degradation Conditions on the Mws of Polysaccharides

The yield of the GFP that was extracted by traditional hot water extraction and alcohol precipitation methods was 34.44%. Freeze-dried GFP was dissolved in distilled water and degraded by H_2O_2 oxidative method. The influence of different conditions on the degradation of GFP was determined.

2.1.1. Effect of Temperature

In order to investigate the effect of temperature on the Mw of GFP, 60 °C, 70 °C, 80 °C, 90 °C, and 100 °C were selected. As shown in Figure 1, the Mw of the products was high at a comparatively low temperature (such as 60 °C and 70 °C). At 120 min, the Mws changed to 76.9 kDa, 36 kDa and 18 kDa for 60 °C, 70 °C, and 80 °C, respectively. On the contrary, the Mws could violently decrease to

3.2 kDa and 2.2 kDa in 60 min when temperatures were 90 °C and 100 °C and the curves were tend to stay after 120 min. At last, the Mws of the polysaccharides for 60 °C, 70 °C, 80 °C, 90 °C, and 100 °C groups were 63 kDa, 16 kDa, 5.7 kDa, 2.0 kDa, and 1.8 kDa, respectively. The results showed that high temperature was more favorable for degradation and feasible to generate lower Mws polysaccharides.

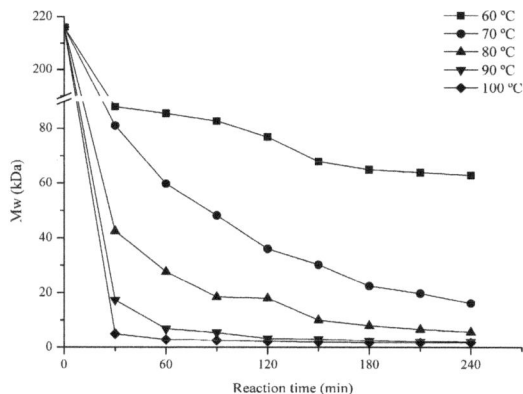

Figure 1. Effects of temperature on the molecular weight (Mw) of *Grateloupia filicinia* polysaccharide (GFP). The reaction was carried out in pH 1 and 0.15% H_2O_2 under different temperature conditions.

2.1.2. Effect of H_2O_2 Concentration

Figure 2 shows the effect of H_2O_2 concentration (0.03%, 0.15%, 0.30%, 0.6%, and 1.5%) on the Mw of the polysaccharides. In the first 60 min, Mw of 0.03% and 0.15% concentrations degradation products were similar and was bigger than the other groups, it decreased to about 9.5 kDa after 60 min. By contrast, the reaction with 0.3%, 0.6%, and 1.5% concentration groups presented no differences and were faster; the Mw of the products were all about 6.7 kDa after 60 min. However, after 90 min, the degradation results were similar for all of the groups. The curves of all H_2O_2 concentrations were closer, and the Mw of the final low molecular weight (LMw) polysaccharides were all around 2 kDa.

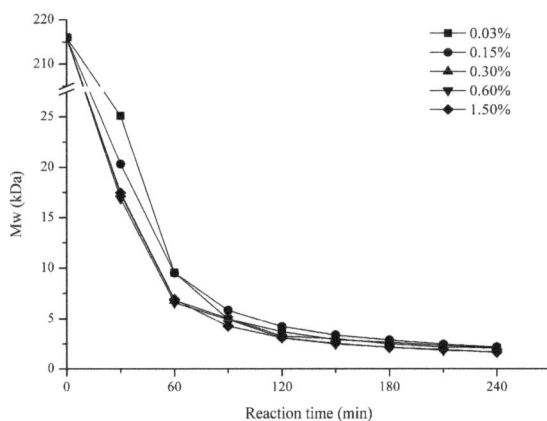

Figure 2. Effect of H_2O_2 concentration on the Mw of GFP. GFP was degraded at pH 1 and 90 °C under different H_2O_2 concentration conditions.

2.1.3. Effect of pH

Figure 3 illustrates the pH impact on the degradation of GFP. (pH 1, 2, 3, 4, and 8) were tested. As shown in Figure 3, the pH had a significant influence on GFP degradation. In 30 min, the Mw were remarkably reduced to 20 kDa and 54 kDa when under the lowest pH 1 and 2, respectively; but when pH was increased to 3, 4, or 8, the Mw of GFP were approximately 120 kDa at 30 min. After 120 min, the Mw of GFP changed slightly to a final Mw of 2.1 kDa and 3.2 kDa for pH 1 and 2, respectively. However, when the pH increased to 3, 4, or 8, the Mw decreased drastically to 4.2 kDa, 4.5 kDa and 7.6 kDa at 240 min, respectively. This suggested that the higher pH (such as 3, 4, and 8) could also degrade GFP, but a lower pH would accelerate the reaction to a shorter time.

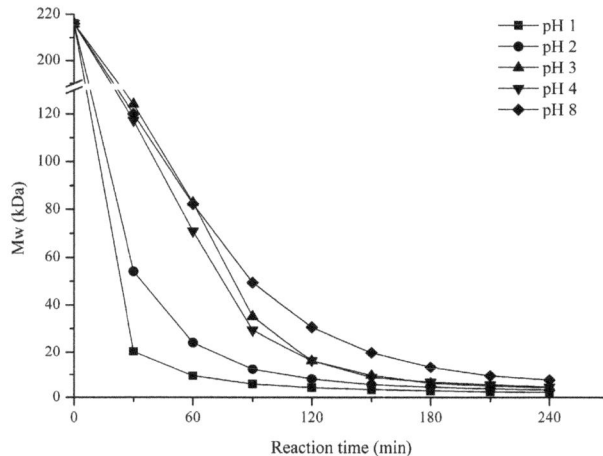

Figure 3. Effects of pH on the Mw of GFP. GFP was degraded in 0.3% H_2O_2 at 90 °C under different pH conditions.

2.2. *Preparation and Chemical Characterization of LGFP*

2.2.1. Preparation of LGFP

Depending on the influence of different conditions on the degradation of GFP, a series of conditions were screened to prepare four kinds of LGFP solutions. After being neutralized to pH 7, the solution was dialyzed against tap water for 24 h and was then changed to distilled water for 24 h. The dialysis tube for LGFP-1 to LGFP-4 were 10 kDa, 3.5 kDa, 1 kDa, and 0.5 kDa Mw cut off, respectively. After dialysis, the solutions were filtered by common filter paper, freeze dried, and then the yield was calculated. The specific preparation conditions and yield are shown in Table 1.

Table 1. Preparation conditions and yield of LGFPs.

Sample	Temperature (°C)	pH	H_2O_2 (%)	Time (min)	Yield (%)
LGFP-1	90	4	0.3	75	73.4
LGFP-2	90	4	0.3	135	46.28
LGFP-3	90	4	0.3	210	42.99
LGFP-4	90	4	0.3	240	40.38

2.2.2. Chemical Characterization

According to the conditions exhibited in Table 1, four LGFPs with different Mw were prepared. The HPLC profiles and chemical composition of the initial GFP and the LGFPs are given in Figure 4 and Table 2. The HPLC standard curve equation was $y = -2.4616x + 24.507$ with the Log Mw as the abscissa and elution time as the ordinate. The R square was 0.991. With the decrease of Mw of the polysaccharides, the total sugar content has decreased as well, but the sulfate content decreased slightly. The protein content of LGFPs also reduced. Results of monosaccharide composition showed that all samples were mainly composed of galactose.

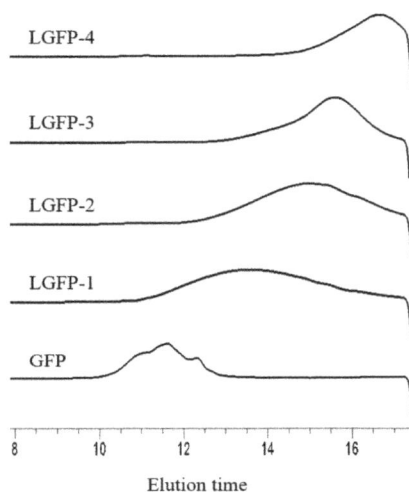

Figure 4. HPLC profiles of the GFP and LGFPs.

Table 2. Chemical composition of polysaccharides (%w/w of dry weight).

Sample	Total Sugar (%)	Sulfate (%)	Protein (%)	Mw (kDa)	Monosaccharides Composition (Molar Ratio)						
					Man	Rha	Glc A	Glc	Gal	Xyl	Fuc
GFP	55.22 ± 0.96	21.52 ± 0.04	1.45 ± 0.08	216.7 ± 1.2	0.00	0.01	0.03	0.01	1	0.03	0.03
LGFP-1	49.74 ± 1.22	20.99 ± 0.28	0.96 ± 0.01	40.2 ± 0.3	0.01	0.01	0.03	0.01	1	0.03	0.02
LGFP-2	47.21 ± 1.32	16.86 ± 0.10	0.45 ± 0.02	14.0 ± 0.3	0.01	0.01	0.05	0.01	1	0.03	0.02
LGFP-3	44.73 ± 2.43	18.33 ± 0.43	0.18 ± 0.01	8.7 ± 0.2	0.01	0.01	0.05	0.01	1	0.03	0.01
LGFP-4	41.00 ± 0.25	17.01 ± 0.44	0.29 ± 0.00	2.7 ± 0.1	0.01	0.02	0.04	0.01	1	0.02	0.02

Man: mannose; Rha: rhamnose; Glc A: glucuronic acid; Glc: glucose; Gal: galactose; Xyl: xylose; Fuc: fucose.

In order to further characterize the chemical structure of the GFP and LGFPs, the respective FT-IR spectra were examined (Figure 5). Based on previous reports [17,23,24], the O–H stretching vibration appeared at 3300 cm^{-1}, and C–H stretching vibration appeared at 2940 cm^{-1}. The absorption peaks at 1620 cm^{-1} and 1420 cm^{-1} represents the asymmetric and symmetric stretching vibration of C=O, respectively. The absorption peaks at 1220 cm^{-1} indicates S=O stretching vibration and that at 1020 cm^{-1} correspond to C–O–H deformation vibration. Feature absorption at 840 cm^{-1} reflects C–O–S symmetry stretching vibration.

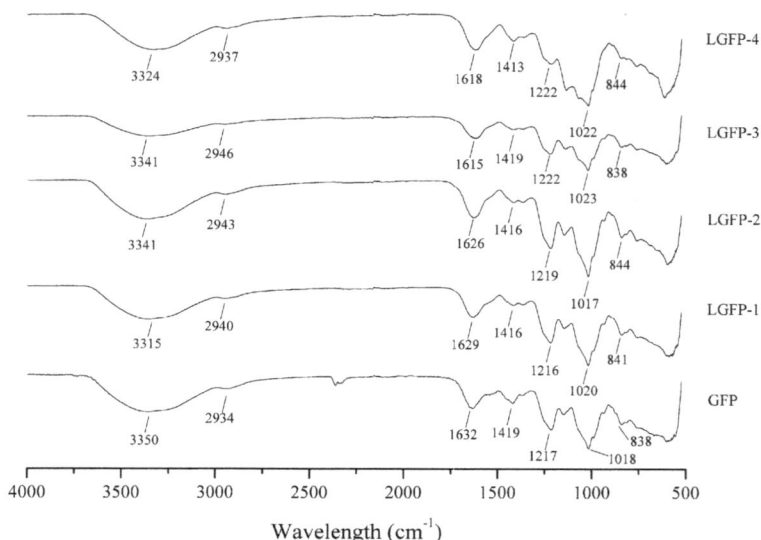

Figure 5. Fourier transform infrared (FT-IR) spectra of GFP and LGFPs in regions from 4000 to 500 cm^{-1}.

2.3. Cytotoxic Activity

The specific results of MTT (3-(4,5-dimethyl-2-thiazolyl)-2,5-diphenyl-2-H-tetrazolium bromide) assay used to detect the cytotoxicity of GFP and LGFPs on DF-1 cells were shown in Table 3. It was considered that the polysaccharides do not have cytotoxic activity to DF-1 cells as the relative survival is over 85% [25,26]. These results suggested that 2 mg/mL was still safe for all samples.

Table 3. Relative Survival Rate of DF-1 cells.

Concentration (mg/mL)	2	1	0.5	0.25	0.125	0.0625	0.03125
GFP	0.95 ± 0.02	1.00 ± 0.06	1.03 ± 0.07	1.08 ± 0.01	1.07 ± 0.04	1.05 ± 0.09	1.02 ± 0.06
LGFP-1	0.96 ± 0.02	1.00 ± 0.05	1.05 ± 0.04	1.05 ± 0.08	1.01 ± 0.04	0.99 ± 0.04	1.00 ± 0.04
LGFP-2	0.94 ± 0.05	1.06 ± 0.04	1.01 ± 0.02	1.05 ± 0.04	1.05 ± 0.05	1.01 ± 0.04	1.03 ± 0.07
LGFP-3	0.95 ± 0.02	1.02 ± 0.05	1.04 ± 0.10	0.99 ± 0.03	1.03 ± 0.03	0.99 ± 0.05	1.03 ± 0.04
LGFP-4	0.96 ± 0.10	0.98 ± 0.03	1.03 ± 0.06	1.01 ± 0.05	1.01 ± 0.03	1.04 ± 0.07	1.09 ± 0.03

The safe concentration of GFP and LGFPs was tested using MTT assay. All treatments were performed in triplicate. The relative survival rate reflected the cytotoxity of GFP and LGFPs.

2.4. Anti-ALV-J Activity In Vitro

2.4.1. ALV-Specific Antigen Detection

ELISA method was applied to preliminary verify the anti-ALV-J activity of polysaccharides preliminary. S/P value represented the relative expression of p27 antigen and was calculated by the following equation: S/P = (Sample mean − Negative control mean)/(Positive control mean − Negative control mean). As presented in Figure 6, under the concentration of 2 mg/mL, the S/P value of all of the tested groups were lower than the virus control, which means that the ALV p27 antigen of polysaccharides treated groups were dramatically lower than virus control. The effect of LGFP-2, LGFP-3, and LGFP-4 was better than that of GFP and LGFP-1, which possess higher Mw. Among them, LGFP-3 showed the best anti-ALV-J activity with a S/P value of 0.13, thus, LGFP-3 was cautiously chosen for further study.

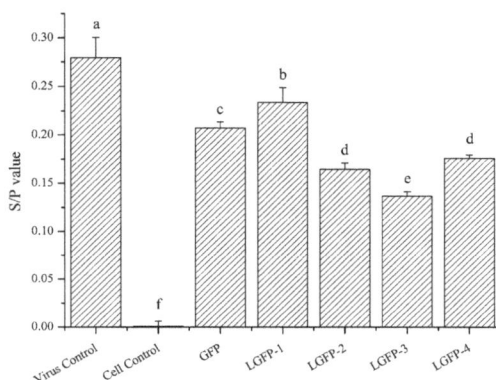

Figure 6. Expression of ALV-J p27 antigen. Antiviral activity of GFP and LGFPs were determined by ALV p27 antigen test kit. Results are recognized as positive when S/P value is greater than 0.2. Data are shown as the Mean + SD. Values with different letters in the same column (a–f) are significantly different ($p < 0.05$) from each other.

2.4.2. Action Phase of Polysaccharides

Different plant polysaccharides exerted various antiviral characteristics [1,27,28]. In order to explore the action phase of the samples, LGFP-3 was assayed following different conditions, as indicated in Section 4.7.2. All of the treatments were applied to cells and virus in vitro, and the antiviral activity was detected by ELISA. The results showed in Figure 7 illustrated that the S/P value of Ad group was 0.15, significantly lower than the virus control that was 0.24, and also lower than that of BA and AA. This means that the GFP-3 might inhibit virus adsorption onto the DF-1 cells. The p27 expression of BA and AA (both 0.23) groups was lower than the virus control but the difference was not significantly ($p > 0.05$), which indicated that the treatment with polysaccharides before or after virus inoculation did not significantly decrease the virus infection.

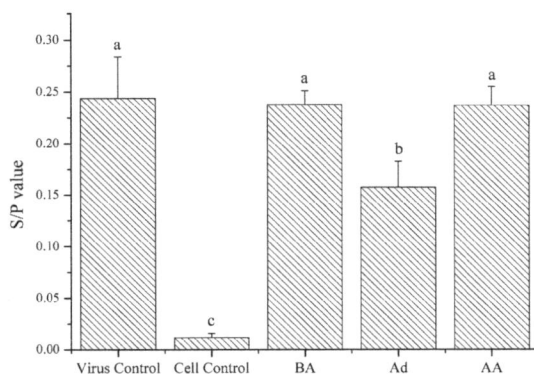

Figure 7. Expression of p27 after different modes of administration: LGFP-3 treated DF-1 cells before adsorption (BA); LGFP-3 treated virus at the adsorption phase (Ad); and, LGFP-3 treated DF-1 cells after adsorption (AA). The final concentrations of LGFP-3 were all 1 mg/mL in these three administration. DF-1 cells with and without inoculation were used as the virus and cell control, respectively. Data are shown as the Mean + SD. Values with different letters in the same column (a–c) are significantly different ($p < 0.05$) from each other.

2.4.3. Gene Relative Expression of ALV-J

Real time PCR was performed to evaluate the ALV-J gene expression with or without LGFP-3 treatment. The results were presented in Figure 8. The expression of ALV-J gene decreased significantly after treatment with LGFP-3 (1000 and 200 μg/mL), and the suppression was dose-dependent. Treatment with 1000 μg/mL LGFP-3 caused the strongest inhibition against ALV-J adsorption; the relative gene expression of 200 μg/mL in the experiment group was 42.52, weaker than 1000 μg/mL group but significantly lower than virus control. While handling with 40 μg/mL LGFP-3, the gene expression was higher than the other two concentrations and even higher than the virus control. This suggested that under a low concentration, the LGFP might promote ALV-J adsorption.

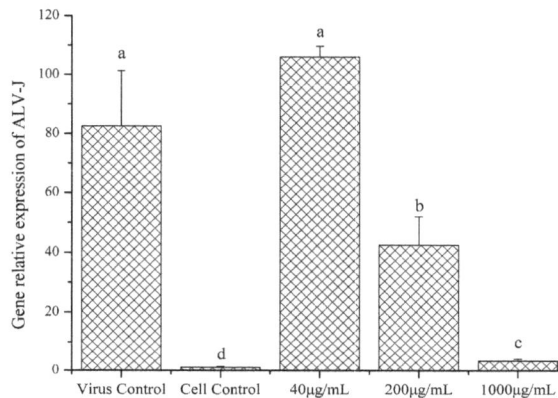

Figure 8. Gene relative expression of ALV-J measured with real-time PCR. DF-1 cells with or without inoculation were used as the virus and cell control, respectively. Data are shown as the Mean + SD. Values with different letters in the same column (a–d) are significantly different ($p < 0.05$) from each other.

2.4.4. Western-Blot and Indirect Immunofluorescence Assay (IFA) Analysis

To further investigate the variation of ALV-J when treated with LGFP-3 as compared to cell and virus control, western-blot and IFA were also used; the results were shown in Figures 9 and 10, respectively. As it happened with the results of real time PCR, in western-blot analysis, expression of ALV-J gp85 protein was almost the same as virus control when treated with 40 μg/mL LGFP-3. However, when the concentration was 1000 μg/mL, the gp85 protein expression significantly decreased and gained a much greater effect when compared with 200 μg/mL and 40 μg/mL LGFP-3. These results also showed that the antiviral effects were dose dependent.

A more direct observation of DF-1 cells infected by ALV-J could be obtained from IFA. Results showed that five days after inoculation, the green fluorescence signal intensity of 40 μg/mL group was the strongest and similar to that of the virus control. The majority of the DF-1 cells were also infected with ALV-J in the 200 μg/mL group but less in number than the virus control group. Cells that were treated with 1000 μg/mL LGFP-3 had the lowest virus expression. The antiviral effects increased following by the increase of LGFP-3 concentration, a reaction to that real time PCR and Western blot.

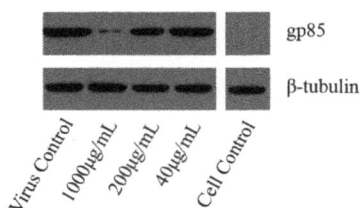

Figure 9. Expression of ALV-J gp85 protein evaluated by western-blot. DF-1 cells with or without inoculation were used as the virus and cell control, respectively.

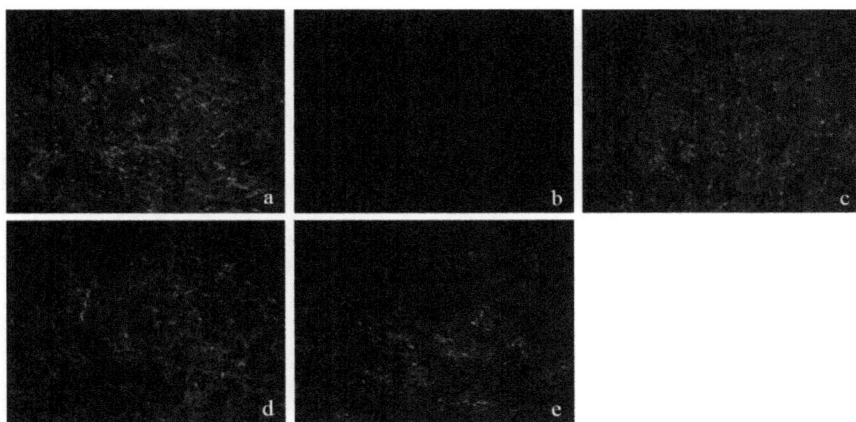

Figure 10. Expression of ALV-J gp85 protein evaluated by IFA. (a): virus control; (b): cell control; (c): ALV-J treated with 40 μg/mL LGFP-3; (d): ALV-J treated with 200 μg/mL LGFP-3; and, (e): ALV-J treated with 1000 μg/mL LGFP-3.

3. Discussion

The GFP was recognized for its various biological activities [14,29,30]. However, researches focusing on the degradation of GFP and antiviral activity of LGFP are still scarce. Therefore, it is necessary and significant to explore the influence of degradation conditions on the Mw of GFP and the antiviral activity of LGFP. For all of the degradation methods, H_2O_2 oxidative degradation was characterized for its mild reaction and low cost; besides, the unreacted excess H_2O_2 is easy to remove. Because of the advantages mentioned above, H_2O_2 oxidative degradation has been widely used in degrading polysaccharides such as chitosan, starch, and cellulose [31–33]. Further, the pH, H_2O_2 concentration and temperature were considered to have a great impact on the degradation rate [17,34].

In this study, the GFP degradation rate was shown to be positively related to temperature, H_2O_2 concentration, and negatively to pH in general. Probably, during the degradation process, H_2O_2 could generate free radicals such as HOO^-, $^\bullet O_2$, and $HO^\bullet$, which could remove a H atom bonded to contributing to the fracture of the glycosidic linkages [35]. Our results showed that the degradation for 0.03% and 0.15% H_2O_2 were slower than that of 0.3% in the first 90 min. This might suggest that a higher H_2O_2 concentration meant that more free radicals are produced in solution, therefore accelerating the degradation. When considering that the degradation results for 0.6% and 1.5% H_2O_2 were same as the 0.3% from the beginning to the end, so, concentration at 0.3% was enough for GFP degradation. However, after 90 min, the 0.03% and 0.15% groups were gradually approached to the 0.3% group, and the final results were basically the same for the three groups. This suggests that a high H_2O_2 concentration might speed up the degradation rate, but it dose not influence the final results. Another parameter that affects the degradation of GFP might be pH. For example, Iqbal et al. degraded dextran at 80 °C for 2 h, and verified that the Mw of the product was 270 kDa with pH of 1.8, while the Mw decreased to only 77 kDa for pH 1.4 [36]. In our study, compared with pH 3, 4, and 8, pH 1 and 2 could significantly accelerate the degradation probably because pH could also affect the generation rate of free radicals in the solution. Therefore, in an acidic environment provided by a low pH, polysaccharides could be degraded more easily. Another important factor in GFP degradation is temperature. As shown in Figure 3, 10 kDa products could be obtained within 60 min at 90 °C and 100 °C, but at 60 °C and 70 °C, the Mw of the products were 63 kDa and 16 kDa only after 240 min. Hence, the Mw of the degradation products decreased significantly with the increase of temperature. There, we adopted 90 °C to prepare LGFPs (Table 1).

Based on the degradation results, several LGFPs were prepared and the chemical composition of GFP and LGFPs were also characterized. According to the results, we found that the protein content were reduced with the decrease of Mw in general, which indicated that the degradation was a benefit for protein removal. Along with the decrease of Mw, and the total sugar content also decreased, it might because of the comprehensive effect, which is caused by the temperature, H^+, and free radicals in the degradation process. The monosaccharide composition results showed that all of the samples were mainly composed of galactose, the other monosaccharide content was extremely low. The pattern mentioned in FT-IR indicated that GFP and LGFPs were all sulfated polysaccharides and the degradation process did not change the chemical structure of GFP.

ALV-J has gradually spread in China and caused great losses to poultry breeding industry [37–39]. When compared with other ALVs, high recombination of ALV-J and selection pressure lead to the faster variation rate [40,41], which caused great difficulties in preventing and treating ALV-J outbreaks.

Polysaccharides from the algae were found to exert inhibition to many viruses [42–44]. To verify the suppression of GFP and LGFPs to ALV-J, the ALV p27 antigen was detected in the presence or absence of the polysaccharides. P27 antigen is a common protein in ALVs, but it is often used to determine the ALV-J infection levels in the laboratory when the pathogen is ensured. Results in the study showed that all of the samples could significantly inhibit the expression of p27 antigen, and LGFP-2, 3, and 4 were better than GFP and LGFP-1. Among them, LGFP-3 with a Mw of 8.7 kDa exerted the best inhibition effect.

In order to determine the function time-point of polysaccharides on ALV-J, LGFP-3 was chosen for further experiments. Figure 5 demonstrated that p27 antigen levels reduced only when the virus and LGFP-3 were administered simultaneously. Further, the BA and AA treatments were similar to the virus control. Therefore, the polysaccharides might mainly inhibit ALV-J adsorption on host cells. This was also observed in some other reports. Bouhlal et al., for example, found that polysaccharides extracted from *Sphaerococcus coronopifolius* and *Boergeseniella thuyoides* inhibited HSV-1 adsorption onto host cells [28], and Bourgougnon et al. suggested that the polysaccharides from *Schizymenia dubyi* could suppress HIV and the activity was mainly attributed to the inhibition of virus to cell attachment [45]. While Wang et al. found that fucoidan from *Kjellmaniella crassifolia* could inhibit the influenza A virus (IAV) treated with fucoidan before or after inoculation [27]. It seems that the action phase of fucoidan was not the same as LGFP. This is probably because the monosaccharide composition and the chemical structure of fucoidan, and also the viruses were different from this study. In addition, many researches showed that the sulfate content played a very important role in polysaccharides activities [46–49]. However, in this research, although the inhibition effect of LGFP-3 was the best, the sulfate content of LGFP-3 was not the highest. Perhaps, the inhibition effect of GFP and LGFPs to ALV-J was not only related to the sulfate content but also to the Mw of the polysaccharides or the sulfate pattern of distribution. According to the results, we may deduce that LGFP with a Mw of about 8.7 kDa might bind to some specific receptors on the cells with a more suitable structure and ionic interactions between polysaccharides and cells, or it could fully bind to ALV-J particles and block the specific recognition between virus and cells, thereby preventing the virus from adsorption. In order to further prove the LGFP-3 anti-ALV-J effect, virus genes, and specific protein expression were tested by real-time PCR, Western-blot and IFA. A significant suppression of ALV-J gene and gp85 protein expression were detected. Based on the real-time PCR results, gene expression of virus control were 24-times higher than that of 1000 μg/mL LGFP-3 group, thus the virus replication was significantly decreased. From the western-blot and IFA results, it can be observed that the virus protein expression and the number of ALV-J decreased significantly in the groups under 1000 and 200 μg/mL treatment. Thus the quantity of viral protein synthesis was also reduced. All of the above results showed that LGFP-3 could inhibit the ALV-J adsorption and reduce the infection probability of virus. Similar results were obtained by Yu et al., who discovered that the polysaccharides from Taishan *Pinus massoniana* pollen could directly coat to the virus envelope protein, which would bind to the cell surface receptor, thereby affecting the virus adsorption [1]. Because LGFP-3 had no impact on ALV-J when applied before adsorption (BA), we speculated that our samples could not bind to the cell membrane, and the possible way was to interact with the virus envelope protein just like the polysaccharides from Taishan *Pinus massoniana* pollen. But, considering the complex configuration of our samples, the action targets might be more and the specific mechanism needs to be elucidated in the future.

4. Materials and Methods

4.1. Seaweeds Samples and Reagents

G. filicina was collected from the number two bathing beach of Qingdao, China in January 2016. The seaweeds were washed with distilled water to remove salt, sediment, and other impurities, then dried at 50 °C in an oven to constant weight and stored at room temperature. All of the reagents used were of analytical grade.

4.2. Extraction of GFP

Traditional hot water extraction and alcohol precipitation method [26] was used with some modifications. Briefly, the dried algae were mixed with 60-fold volume of distilled water and maintained at 100 °C for 4 h. After cooling down, the polysaccharide solution was filtered and condensed to 1/2 volume by a rotavapor. Then, the supernatant was dialyzed at tap water for 24 h and then changed to distilled water for 24 h, respectively, using dialysis tube with a 3.5 kDa Mw cut off. The liquid was concentrated to 1/4 volume and after that 3-fold volume anhydrous ethanol was added to precipitate polysaccharides. The mixture was placed over-night at 4 °C and then centrifuged. The precipitate was freeze-dried and the yield (%) of the polysaccharide was calculated.

4.3. Influence of H_2O_2 Oxidation on GFP Degradation and Preparation of LGFP

A 0.8 g mass of GFP was dissolved in 40 mL distilled water, added with H_2O_2 solution, and adjusted pH by hydrochloric acid or sodium hydroxide, and then heated in a water bath with stirring. Every 30 min, 0.5 mL samples were removed from the reaction solution. After being filtrated with a 0.22 μm polyethersulfone filter (Tianjin Jinteng Experiment Equipment Co., Ltd, Tianjin, China), the Mw of the samples was determined by high performance gel permeation chromatography (HPGPC) to investigate the effect of pH (1, 2, 3, 4, and 8), final concentration of H_2O_2 (0.03%, 0.15%, 0.3%, 0.6%, and 1.5%) in reaction solution and temperature (60 °C, 70 °C, 80 °C, 90 °C, and 100 °C) on the GFP degradation. Afterwards, according to the above results, LGFPs were prepared under different conditions.

4.4. Chemical Characterization

The Mw of GFP and LGFPs was measured by HPGPC method with a TSK gel G3000PWxl column, and using 0.1 mol/L Na_2SO_4 as the mobile phase on Agilent 1260 HPLC systerm equipped with a refractive index detector. The column temperature was 35 °C and flow rate was 0.5 mL/min. Dextran standards (Mw 1000, 5000, 12,000, 50,000, 80,000, 270,000, and 670,000, Sigma, Mendota Heights, MN, USA) were used to calibrate the column.

Total sugars were analyzed by phenol-sulphuric acid method [50] using galactose as the standard. Bradford's method [51] was used to detect the protein content and using bovine serum albumin (BSA) as the standard. Sulfated content was determined by barium chloride gelatin method [52]. The Fourier transform infrared (FT-IR) spectra of the samples were recorded by Nicolet-360 FT-IR spectrometer (Thermo Fisher, Waltham, MA, USA) in KBr disks using a scan range from 500 to 4000 cm^{-1}.

The monosaccharide composition (molar ratio) was analyzed using 1-phenyl-3-methyl-5-pyrazolone (PMP) pre-column derivation HPLC [53]. Briefly, 10 mg samples were put into the ampoule and 1 mL distilled water was added then the mixture was hydrolyzed in 4 mol/L trifluoroacetic acid for 4 h at 110 °C in an oven followed by neutralization with sodium hydroxide to pH 5–6. Later, the pre-column derivatives were carried out with PMP and separated by HPLC using a YMC Pack ODS AQ column. The mannose, rhamnose, fucose, galactose, xylose, glucose, and glucuronic acid standards were obtained from Sigma Aldrich (St. Louis, MO, USA).

4.5. Cell Lines, Cell Culture, Virus and Antibodies

A DF-1 cell line, a NX0101 strain of ALV-J, and the ALV-J gp85-specific monoclonal antibody were kindly gifted by Prof. Cheng, Shandong Agricultural University. The DF-1 cells were cultured in Dulbecco's modified Eagle's medium (DMEM) provided with 100 units/mL of penicillin and 100 µg/mL of streptomycin, and supplemented with 10% (v/v) or 1% (v/v) fetal bovine serum (FBS) as the growth medium (GM) or maintenance medium (MM), respectively. The ALV-J titer was tested using DF-1 monolayers and expressed by tissue culture infectious dose 50 (TCID$_{50}$) using the Reed-Muench formula. 100 TCID$_{50}$ of ALV-J was used in subsequent experiments.

4.6. Determination of Cytotoxic Activity

MTT assay was used to evaluate the cytotoxic activity of GFP and LGFP on DF-1 cells. DF-1 cell monolayers grown in 96-well plate were supplied with the 100 µL MM as the cell control or 100 µL polysaccharides solution that were dissolved in MM (the maximum concentration was 2 mg/mL and diluted in 2-fold steps to 0.03125 mg/mL). All of the treatments were performed in triplicate. After 24 h, the liquid was removed; 20 µL MTT was added and the cells were incubated for another 4 h. Then, the MTT was discarded and 100 µL DMSO was added. Finally, the enzyme-linked immunosorbent assay reader (iMark^{TM} BIO-RAD, Hercules, CA, USA) was used to determine the absorbance of each well at 490 nm. The cytotoxic activity was represented by the relative survival rate of DF-1 cells and calculated by the formula: survival rate (%) = (As/Ac) × 100%, where Ac and As were the absorbance of cell control (Ac) and experimental group (As), respectively. When the survival rates were higher than 85%, it was considered that the samples had no cytotoxicity on DF-1 cells.

4.7. Anti-ALV-J Activity In Vitro

4.7.1. ALV Specific Antigen Detection

DF-1 cell monolayers grown in 96-well plate were infected with 100 TCID$_{50}$ of ALV-J, meanwhile, different polysaccharides solutions at a final concentration of 2 mg/mL, which were dissolved in MM were added. After 2 h incubation at 37 °C, the cells were washed and covered with MM containing corresponding polysaccharides at a concentration of 2 mg/mL, subsequently, the cell control (without infection of ALV-J) and virus control (without exposure to polysaccharides) were also set. After 24 h of incubation, the viral titers were measured using the ALV p27 antigen test kit (Beijing IDEXX-Yuanheng Laboratories, Co. Ltd., Beijing, China) following by the instructions of the manufacturer. The relative expressions of p27 antigen were calculated by the following equation: S/P = (Sample mean − Negative control mean)/(Positive control mean − Negative control mean).

4.7.2. Detection of the Polysaccharides Action Phase

DF-1 cells grown in 96-well plate were infected with 100 TCID$_{50}$ of ALV-J. Then, they were treated with MM containing polysaccharides at a final concentration of 1 mg/mL in different ways. According to the artificially divided viral infection phase, the specific operations were as follows:

Before Adsorption (BA): The DF-1 cells were covered with MM containing polysaccharides for 2 h before the inoculation. After that, the cells were washed and then incubation with ALV-J for 2 h. After inoculation, the cells were washed again and maintained with MM for 24 h.

Adsorption (Ad): The cells were treated with MM containing polysaccharides at the time they were infected with ALV-J. After incubation at 4 °C for 2 h, the supernatant was removed and replaced by simple MM.

After Adsorption (AA): The DF-1 cells were infected with ALV-J and incubated for 2 h, and then the cells were washed and recovered with polysaccharides dissolved in MM.

The cell control and virus control were also served. After 24 h incubation, all of the supernatants were collected and the viral titers were measured by ALV p27 antigen test kit.

4.7.3. Viral Gene Relative Expression and Protein Expression

DF-1 cells grown in 12-well plate to monolayers in simple MM medium were infected with 100 TCID$_{50}$ of ALV-J and treated with MM containing polysaccharides at a final concentration of 1000 μg/mL, 200 μg/mL, and 40 μg/mL. After 2 h adsorption at 37 °C, the supernatant was discarded and replaced by simple MM, then the cells were incubated at 37 °C with 5% CO$_2$ for about 24 h.

Real-time PCR: The ALV-J gene replication levels in DF-1 cells were determined by real-time PCR. In brief, RNAprep Pure Cell/Bacteria Kit (TIANGEN BIOTECH Co. Ltd., Beijing, China) was used to extract the total RNA from the treated DF-1 cells following the manufacturer's instructions. Then, PrimeScript^{TM} RT reagent Kit with gDNA Eraser (Takara BIO INC, Dalian, China) was applied for reverse transcription immediately. The cDNA produced by reverse transcription was used for real-time PCR as performed by SYBR$^{®}$ *Premix ExTaq*^{TM} Kit (Takara BIO INC, Dalian, China). The forward primer (5′-GCGTGCGTGGTTATTATTTC-3′) and reverse primer (5′-AATGGTGAGGTCGCTGACTGT-3′) were used as the ALV-J primers, the forward primer for internal control GAPDH was (5′-GAACATCATCCCAAGCGTCCA-3′) and reverse primer (5′-CGGCAGGTCAGGTCAACAAC-3′). The amplification cycles were carried out as follows: 95 °C for 30 s; 95 °C for 5 s and 60 °C 34 s (34 cycles); 95 °C for 15 s and 60 °C for 60 s. The results were calculated by the formula $2^{-\Delta\Delta Ct}$ [54,55].

The expression of ALV-J gp85 protein was detected by western blot and IFA.

Western-blot: After 24 h incubation, the cells were lysed and separated by SDS-PAGE (12% acrylamide gels). Proteins were blocked with Tris-buffered saline (TBS) buffer containing 5% (*w/v*) dry skimmed milk, and 0.1% (*v/v*) Tween 20 for 1 h at room temperature. Membranes were then incubated overnight at 4 °C with mouse ALV-J gp85-specific monoclonal antibody or β-tubulin antibody as control. The membranes were washed with TBS-Tween and then incubated with HRP-linked secondary antibody (1:10,000) for 1 h at room temperature. Membranes were washed and protein densities were detected using chemiluminesence detection reagents.

IFA: After inoculation, the DF-1 cells were incubated at 37 °C with 5% CO$_2$ for five days and then IFA was executed as follows. The cells were fixed with cold acetone and ethanol (3:2 for *v/v*) for about 10 min washed with PBS and the gp85-specific monoclonal antibody was added. The mixture was incubated overnight at 4 °C and washed thoroughly with PBS to eliminate the uncombined antibodies. Afterwards, the cells were treated with goat anti-mouse IgG-FITC at 37 °C for 1 h, washed with PBS for five times and then photographed on an inverted fluorescence microscope.

4.8. Statistical Analysis

Statistical analysis was performed using SPSS and the difference among groups was analyzed by one-way ANOVA.

5. Conclusions

In this study, the polysaccharides (Mw 216 kDa) extracted from *G. filicina* were degraded by H_2O_2 oxidative method. The degradation rate was positively related to temperature, H_2O_2 concentration, and negatively to high pH. Several LGFPs (Mw 40, 14, 8.7, and 2.7 kDa) were prepared under different conditions and their anti-ALV-J activity were also investigated. The results showed that initial GFP and LGFPs all possessed antiviral activity, and low Mw is better for the inhibition of ALV-J, LGFP-3 (Mw 8.7 kDa) had the best effect. The action phase of our samples was also determined. The results showed that LGFP-3 were mainly take effect when treated by Ad administration the time that ALV-J adsorbing on host cells. Further experiments indicated that after being treated with LGFP-3, the ALV-J gene and protein expression decreased significantly: it revealed that LGFP-3 inhibited virus adsorption onto DF-1 cells and contributed to the decrease of ALV-J expression. The results demonstrated that polysaccharides from *G. filicina* have great potential in developing as an anti-ALV-J drug. However, the antiviral activity of LGFPs in vivo and complicated mechanism in detail require further study.

Acknowledgments: This work was supported by the Commonweal Item of the State Oceanic Administration of the People's Republic of China (201505033), NSFC-Shandong joint Fund (U1606403), Shandong Province Key Research and Development Project (2016YYSP010) and Qingdao People's Livelihood Science and Technology Projects (16-6-2-41-nsh).

Author Contributions: Conceived and designed the experiment: Yuhao Sun, Xiaolin Chen, Pengcheng Li. Performed the experiments: Yuhao Sun. Analyzed the data: Yuhao Sun, Xiaolin Chen. Contributed reagents/materials/analysis tools: Xiaolin Chen, Ziqiang Cheng, Song Liu, Huahua Yu, Xueqin Wang, Pengcheng Li. Wrote the paper: Yuhao Sun, Xiaolin Chen, Pengcheng Li.

Conflicts of Interest: The author declares on conflict of interest.

References

1. Yu, C.L.; Wei, K.; Liu, L.P.; Yang, S.F.; Hu, L.P.; Zhao, P.; Meng, X.Y.; Shao, M.X.; Wang, C.W.; Zhu, L.J.; et al. Taishan *Pinus massoniana pollen* polysaccharide inhibits subgroup J avian leucosis virus infection by directly blocking virus infection and improving immunity. *Sci. Rep.* **2017**, *7*, 44353. [CrossRef] [PubMed]
2. Cheng, J.; Wen, S.; Wang, S.; Hao, P.; Cheng, Z.; Liu, Y.; Zhao, P.; Liu, J. gp85 protein vaccine adjuvanted with silica nanoparticles against ALV-J in chickens. *Vaccine* **2017**, *35*, 293–298. [CrossRef] [PubMed]
3. Silva, R.F.; Fadly, A.M.; Hunt, H.D. Hypervariability in the envelope genes of subgroup J avian leukosis viruses obtained from different farms in the United States. *Virology* **2000**, *272*, 106–111. [CrossRef] [PubMed]
4. Dou, W.W.; Li, H.M.; Cheng, Z.Q.; Zhao, P.; Liu, J.Z.; Cui, Z.Z.; Liu, H.G.; Jing, W.F.; Guo, H.J. Maternal antibody induced by recombinant gp85 protein vaccine adjuvanted with CpG-ODN protects against ALV-J early infection in chickens. *Vaccine* **2013**, *31*, 6144–6149. [CrossRef] [PubMed]
5. Liu, C.; Dong, J.; Waterhouse, G.I. N.; Cheng, Z.; Ai, S. Electrochemical immunosensor with nanocellulose-Au composite assisted multiple signal amplification for detection of avian leukosis virus subgroup J. *Biosens. Bioelectron.* **2018**, *101* (Suppl. C), 110–115. [CrossRef] [PubMed]
6. Wang, Z.F.; Cui, Z.Z. Evolution of gp85 gene of subgroup J avian leukosis virus under the selective pressure of antibodies. *Sci. China Ser. C-Life Sci.* **2006**, *49*, 227–234. [CrossRef]
7. De Clerck, O.; Gavio, B.; Fredericq, S.; Barbara, I.; Coppejans, E. Systematics of *Grateloupia filicina* (Halymeniaceae, Rhodophyta), based on rbcL sequence analyses and morphological evidence, including the reinstatement of *G. minima* and the description of *G. capensis* sp. nov. *J. Phycol.* **2005**, *41*, 391–410.
8. Denis, C.; Morancais, M.; Gaudin, P.; Fleurence, J. Effect of enzymatic digestion on thallus degradation and extraction of hydrosoluble compounds from *Grateloupia turuturu*. *Bot. Mar.* **2009**, *52*, 262–267. [CrossRef]
9. Nikapitiya, C.; De Zoysa, M.; Jeon, Y.J.; Lee, J. Isolation of sulfated anticoagulant compound from fermented red seaweed *Grateloupia filicina*. *J. World Aquac. Soc.* **2007**, *38*, 407–417. [CrossRef]

10. Yu, Q.; Yan, J.; Wang, S.; Ji, L.; Ding, K.; Vella, C.; Wang, Z.; Hu, Z. Antiangiogenic effects of GFP08, an agaran-type polysaccharide isolated from *Grateloupia filicina*. *Glycobiology* **2012**, *22*, 1343–1352. [CrossRef] [PubMed]

11. Ye, D.Y.; Jiang, Z.B.; Zheng, F.C.; Wang, H.M.; Zhang, Y.M.; Gao, F.F.; Chen, P.H.; Chen, Y.C.; Shi, G.G. Optimized extraction of polysaccharides from *Grateloupia livida* (Harv.) Yamada and biological activities. *Molecules* **2015**, *20*, 16817–16832. [CrossRef] [PubMed]

12. Jiang, Z.B.; Chen, Y.C.; Yao, F.; Chen, W.Z.; Zhong, S.P.; Zheng, F.C.; Shi, G.G. Antioxidant, antibacterial and antischistosomal activities of extracts from *Grateloupia livida* (Harv). Yamada. *PLoS ONE* **2013**, *8*, e80413. [CrossRef] [PubMed]

13. Chattopadhyay, K.; Mateu, C.G.; Mandal, P.; Pujol, C.A.; Damonte, E.B.; Ray, B. Galactan sulfate of *Grateloupia indica*: Isolation, structural features and antiviral activity. *Phytochemistry* **2007**, *68*, 1428–1435. [CrossRef] [PubMed]

14. Wang, S.C.; Bligh, S.W.; Shi, S.S.; Wang, Z.T.; Hu, Z.B.; Crowder, J.; Branford-White, C.; Vella, C. Structural features and anti-HIV-1 activity of novel polysaccharides from red algae *Grateloupia longifolia* and *Grateloupia filicina*. *Int. J. Biol. Macromol.* **2007**, *41*, 369–375. [CrossRef] [PubMed]

15. Ramnani, P.; Chitarrari, R.; Tuohy, K.; Grant, J.; Hotchkiss, S.; Philp, K.; Campbell, R.; Gill, C.; Rowland, I. In vitro fermentation and prebiotic potential of novel low molecular weight polysaccharides derived from agar and alginate seaweeds. *Anaerobe* **2012**, *18*, 1–6. [CrossRef] [PubMed]

16. Sun, L.Q.; Wang, L.; Zhou, Y. Immunomodulation and antitumor activities of different-molecular-weight polysaccharides from *Porphyridium cruentum*. *Carbohydr. Polym.* **2012**, *87*, 1206–1210. [CrossRef]

17. Li, B.; Liu, S.; Xing, R.; Li, K.; Li, R.; Qin, Y.; Wang, X.; Wei, Z.; Li, P. Degradation of sulfated polysaccharides from *Enteromorpha prolifera* and their antioxidant activities. *Carbohydr. Polym.* **2013**, *92*, 1991–1996. [CrossRef] [PubMed]

18. Zhao, X.; Guo, F.; Hu, J.; Zhang, L.; Xue, C.; Zhang, Z.; Li, B. Antithrombotic activity of oral administered low molecular weight fucoidan from *Laminaria Japonica*. *Thromb. Res.* **2016**, *144*, 46–52. [CrossRef] [PubMed]

19. Tang, F.X.; Chen, F.; Li, F. Preparation and potential in vivo anti-influenza virus activity of low molecular-weight kappa-carrageenans and their derivatives. *J. Appl. Polym. Sci.* **2013**, *127*, 2110–2115. [CrossRef]

20. Yamada, T.; Ogamo, A.; Saito, T.; Uchiyama, H.; Nakagawa, Y. Preparation of O-acylated low-molecular-weight carrageenans with potent anti-HIV activity and low anticoagulant effect. *Carbohydr. Polym.* **2000**, *41*, 115–120. [CrossRef]

21. Lee, J.B.; Takeshita, A.; Hayashi, K.; Hayashi, T. Structures and antiviral activities of polysaccharides from *Sargassum trichophyllum*. *Carbohydr. Polym.* **2011**, *86*, 995–999. [CrossRef]

22. Kim, W.J.; Choi, J.W.; Jang, W.J.; Kang, Y.S.; Lee, C.W.; Synytsya, A.; Park, Y.I. Low-molecular weight mannogalactofucans prevent herpes simplex virus type 1 infection via activation of Toll-like receptor 2. *Int. J. Biol. Macromol.* **2017**, *103*, 286–293. [CrossRef] [PubMed]

23. Wang, L.; Liu, H.M.; Qin, G.Y. Structure characterization and antioxidant activity of polysaccharides from Chinese quince seed meal. *Food Chem.* **2017**, *234*, 314–322. [CrossRef] [PubMed]

24. Kolsi, R.B.; Ben Salah, H.; Jardak, N.; Chaaben, R.; Jribi, I.; El Feki, A.; Rebai, T.; Jamoussi, K.; Allouche, N.; Blecker, C.; et al. Sulphated polysaccharide isolated from *Sargassum vulgare*: Characterization and hypolipidemic effects. *Carbohydr. Polym.* **2017**, *170*, 148–159. [CrossRef] [PubMed]

25. Seo, Y.; Lee, B.-J.; Kim, Y.A.; Lee, H.-J.; Joo, Y.H. Effects of several salt marsh plants on mouse spleen and thymus cell proliferation using MTT assay. *Ocean Sci. J.* **2005**, *40*, 209–212. [CrossRef]

26. Song, L.; Chen, X.L.; Liu, X.D.; Zhang, F.B.; Hu, L.F.; Yue, Y.; Li, K.C.; Li, P.C. Characterization and comparison of the structural features, immune-modulatory and anti-avian influenza virus activities conferred by three algal sulfated polysaccharides. *Mar. Drugs* **2016**, *14*, 4. [CrossRef] [PubMed]

27. Wang, W.; Wu, J.D.; Zhang, X.S.; Hao, C.; Zhao, X.L.; Jiao, G.L.; Shan, X.D.; Tai, W.J.; Yu, G.L. Inhibition of influenza a virus infection by fucoidan targeting viral neuraminidase and cellular EGFR pathway. *Sci. Rep.* **2017**, *7*, 10760. [CrossRef] [PubMed]

28. Bouhlal, R.; Haslin, C.; Chermann, J.-C.; Colliec-Jouault, S.; Sinquin, C.; Simon, G.; Cerantola, S.; Riadi, H.; Bourgougnon, N. Antiviral activities of sulfated polysaccharides isolated from *Sphaerococcus coronopifolius* (Rhodophyta, Gigartinales) and *Boergeseniella thuyoides* (Rhodophyta, Ceramiales). *Mar. Drugs* **2011**, *9*, 1187–1209. [CrossRef] [PubMed]

29. Chen, X.; Yang, S.; Wang, J.; Song, L.; Xing, R.; Liu, S.; Yu, H.; Li, P. Sulfated polysaccharides isolated from cloned *Grateloupia filicina* and their anticoagulant activity. *BioMed Res. Int.* **2015**, *2015*, 612352. [PubMed]
30. Athukorala, Y.; Jung, W.-K.; Park, P.-J.; Lee, Y.-J.; Kim, S.-K.; Vasanthan, T.; No, H.-K.; Jeon, Y.-J. Evaluation of biomolecular interactions of sulfated polysaccharide isolated from *Grateloupia filicina* on blood coagulation factors. *J. Microbiol. Biotechnol.* **2008**, *18*, 503–511. [PubMed]
31. Qin, C.Q.; Du, Y.M.; Xiao, L. Effect of hydrogen peroxide treatment on the molecular weight and structure of chitosan. *Polym. Degrad. Stabil.* **2002**, *76*, 211–218. [CrossRef]
32. Parovuori, P.; Hamunen, A.; Forssell, P.; Autio, K.; Poutanen, K. Oxidation of potato starch by hydrogen-peroxide. *Starch-Starke* **1995**, *47*, 19–23. [CrossRef]
33. Zeronian, S.H.; Inglesby, M.K. Bleaching of cellulose by hydrogen peroxide. *Cellulose* **1995**, *2*, 265–272. [CrossRef]
34. Hien, N.Q.; Phu, D.V.; Duy, N.N.; Nguyen, T.K.L. Degradation of chitosan in solution by gamma irradiation in the presence of hydrogen peroxide. *Carbohydr. Polym.* **2012**, *87*, 935–938. [CrossRef]
35. Hou, Y.; Wang, J.; Jin, W.H.; Zhang, H.; Zhang, Q.B. Degradation of Laminaria japonica fucoidan by hydrogen peroxide and antioxidant activities of the degradation products of different molecular weights. *Carbohydr. Polym.* **2012**, *87*, 153–159. [CrossRef]
36. Iqbal, S.; Marchetti, R.; Aman, A.; Silipo, A.; Ul Qader, S.A.; Molinaro, A. Enzymatic and acidic degradation of high molecular weight dextran into low molecular weight and its characterizations using novel Diffusion-ordered NMR spectroscopy. *Int. J. Biol. Macromol.* **2017**, *103*, 744–750. [CrossRef] [PubMed]
37. Xu, B.R.; Dong, W.X.; Yu, C.M.; He, Z.Q.; Lv, Y.L.; Sun, Y.H.; Feng, X.Y.; Li, N.; Lee, L.F.; Li, M. Occurrence of avian leukosis virus subgroup J in commercial layer flocks in China. *Avian Pathol.* **2004**, *33*, 13–17. [CrossRef] [PubMed]
38. Cheng, Z.Q.; Zhang, L.; Liu, S.D.; Zhang, L.J.; Cui, Z.Z. Emerging of fibro subgroup J in a flock of Chinese local breed. *Acta Microbiol. Sin.* **2005**, *45*, 584–587.
39. Sun, S.H.; Cui, Z.Z. Epidemiological and pathological studies of subgroup J avian leukosis virus infections in Chinese local "yellow" chickens. *Avian Pathol.* **2007**, *36*, 221–226. [CrossRef] [PubMed]
40. Venugopal, K.; Smith, L.M.; Howes, K.; Payne, L.N. Antigenic variants of J subgroup avian leukosis virus: sequence analysis reveals multiple changes in the env gene. *J. Gen. Virol.* **1998**, *79*, 757–766. [CrossRef] [PubMed]
41. Zavala, G.; Cheng, S.; Jackwood, M.W. Molecular epidemiology of avian leukosis virus subgroup j and evolutionary history of its 3′ untranslated region. *Avian Dis. Dig.* **2007**, *51*, 942–953. [CrossRef]
42. Cao, Y.G.; Hao, Y.; Li, Z.H.; Liu, S.T.; Wang, L.X. Antiviral activity of polysaccharide extract from *Laminaria japonica* against respiratory syncytial virus. *Biomed. Pharmacother.* **2016**, *84*, 1705–1710. [CrossRef] [PubMed]
43. Trejo-Avila, L.M.; Elizondo-Gonzalez, R.; Rodriguez-Santillan, P.; Aguilar-Briseno, J.A.; Ricque-Marie, D.; Rodriguez-Padilla, C.; Cruz-Suarez, L.E. Innocuity and anti-Newcastle-virus-activity of *Cladosiphon okamuranus* fucoidan in chicken embryos. *Poult. Sci.* **2016**, *95*, 2795–2802. [CrossRef] [PubMed]
44. Gheda, S.F.; El-Adawi, H.I.; El-Deeb, N.M. Antiviral profile of brown and red seaweed polysaccharides against hepatitis C virus. *Iran. J. Pharm. Res.* **2016**, *15*, 483–491. [PubMed]
45. Bourgougnon, N.; Lahaye, M.; Quemener, B.; Chermann, J.C.; Rimbert, M.; Cormaci, M.; Furnari, G.; Kornprobst, J.M. Annual variation in composition and in vitro anti-HIV-1 activity of the sulfated glucuronogalactan from *Schizymenia dubyi* (Rhodophyta, Gigartinales). *J. Appl. Phycol.* **1996**, *8*, 155–161. [CrossRef]
46. Chen, Y.; Song, M.Y.; Wang, Y.X.; Xiong, W.; Zeng, L.; Zhang, S.B.; Xu, M.Y.; Du, H.X.; Liu, J.G.; Wang, D.Y.; et al. The anti-DHAV activities of *Astragalus* polysaccharide and its sulfate compared with those of BSRPS and its sulfate. *Carbohydr. Polym.* **2015**, *117*, 339–345. [CrossRef] [PubMed]
47. Yu, M.; Ji, Y.B.; Qi, Z.; Cui, D.; Xin, G.S.; Wang, B.; Cao, Y.; Wang, D.D. Anti-tumor activity of sulfated polysaccharides from *Sargassum fusiforme*. *Saudi Pharm. J.* **2017**, *25*, 464–468. [CrossRef] [PubMed]
48. Xie, J.H.; Wang, Z.J.; Shen, M.Y.; Nie, S.P.; Gong, B.; Li, H.S.; Zhao, Q.; Li, W.J.; Xie, M.Y. Sulfated modification, characterization and antioxidant activities of polysaccharide from *Cyclocarya paliurus*. *Food Hydrocoll.* **2016**, *53*, 7–15. [CrossRef]
49. Li, J.; Chi, Z.; Yu, L.; Jiang, F.; Liu, C. Sulfated modification, characterization, and antioxidant and moisture absorption/retention activities of a soluble neutral polysaccharide from *Enteromorpha prolifera*. *Int. J. Biol. Macromol.* **2017**, *105*, 1544–1553. [CrossRef] [PubMed]

50. Dubois, M.; Gilles, K.A.; Hamilton, J.K.; Rebers, P.A.; Smith, F. Colorimetric method for determination of sugars and related substances. *Anal. Chem.* **1956**, *28*, 350–356. [CrossRef]

51. Bradford, M.M. Rapid and sensitive method for quantitation of microgram quantities of protein utilizing principle of protein-dye binding. *Anal. Biochem.* **1976**, *72*, 248–254. [CrossRef]

52. Kawai, Y.; Seno, N.; Anno, K. A modified method for chondrosulfatase assay. *Anal. Biochem.* **1969**, *32*, 314–321. [CrossRef]

53. Zhang, J.; Zhang, Q.; Wang, J.; Shi, X.; Zhang, Z. Analysis of the monosaccharide composition of fucoidan by precolumn derivation HPLC. *Chin. J. Oceanol. Limn.* **2009**, *27*, 578–582. [CrossRef]

54. Livak, K.J.; Schmittgen, T.D. Analysis of relative gene expression data using real-time quantitative PCR and the $2^{-\Delta\Delta CT}$ method. *Methods* **2001**, *25*, 402–408. [CrossRef] [PubMed]

55. Shandiz, S.A.S.; Khosravani, M.; Mohammadi, S.; Noorbazargan, H.; Mirzaie, A.; Inanlou, D.N.; Jalali, M.D.; Jouzaghkar, H.; Baghbani-Arani, F.; Keshavarz-Pakseresht, B. Evaluation of imatinib mesylate (Gleevec) on KAI1/CD82 gene expression in breast cancer MCF-7 cells using quantitative real-time PCR. *Asian Pac. J. Trop. Biomed.* **2016**, *6*, 159–163. [CrossRef]

marine drugs

MDPI

Article

Immunomodulatory and Anti-IBDV Activities of the Polysaccharide AEX from *Coccomyxa gloeobotrydiformis*

Qiang Guo [1,†], Qiang Shao [1,†], Wenping Xu [1], Lei Rui [1], Ryo Sumi [2], Fumio Eguchi [3] and Zandong Li [1,*]

[1] State Key Laboratory for Agrobiotechnology, College of Biological Sciences, China Agricultural University, Beijing 100193, China; fzgq249@163.com (Q.G.); shaoqiang19880316@126.com (Q.S.); xwp120@126.com (W.X.); ruilei@cau.edu.cn (L.R.)
[2] Nikken Sohonsha Corporation, Gifu 501-6255, Japan; ryosumi_nikken@yahoo.co.jp
[3] Faculty of Regional Environment Science, Tokyo University of Agriculture, Tokyo 156-8502, Japan; f1eguchi@nodai.ac.jp
* Correspondence: lzdws@cau.edu.cn; Tel.: +86-10-6273-2144
† Both authors contributed equally to this work.

Academic Editor: Orazio Taglialatela-Scafati
Received: 24 September 2016; Accepted: 3 February 2017; Published: 10 February 2017

Abstract: A number of polysaccharides have been reported to show immunomodulatory and antiviral activities against various animal viruses. AEX is a polysaccharide extracted from the green algae, *Coccomyxa gloeobotrydiformis*. The aim of this study was to examine the function of AEX in regulating the immune response in chickens and its capacity to inhibit the infectious bursal disease virus (IBDV), to gain an understanding of its immunomodulatory and antiviral ability. Here, preliminary immunological tests in vitro showed that the polysaccharide AEX can activate the chicken peripheral blood molecular cells' (PBMCs) response by inducing the production of cytokines and NO, promote extracellular antigen presentation but negatively regulate intracellular antigen presentation in chicken splenic lymphocytes, and promote the proliferation of splenic lymphocytes and DT40 cells. An antiviral analysis showed that AEX repressed IBDV replication by the deactivation of viral particles or by interfering with adsorption in vitro and reduced the IBDV viral titer in the chicken bursa of Fabricius. Finally, in this study, when AEX was used as an adjuvant for the IBDV vaccine, specific anti-IBDV antibody (IgY, IgM, and IgA) titers were significantly decreased. These results indicate that the polysaccharide AEX may be a potential alternative approach for anti-IBDV therapy and an immunomodulator for the poultry industry. However, more experimentation is needed to find suitable conditions for it to be used as an adjuvant for the IBDV vaccine.

Keywords: *Coccomyxa*; polysaccharide; IBDV; AEX; cytokine

1. Introduction

Polysaccharides present an enormous variety of structures in organisms and are an under-exploited novel source of natural compounds for drug discovery. Plant and marine microalgae-derived polysaccharides have been shown to have a variety of bioactivities, such as immune-modulatory, anti-viral, anti-tumor, anticoagulant, and antioxidant properties [1]. The polysaccharide from Astragalus radix can increase the production of TNF, GM-CSF, and NO in mice [2]. The sulfated polysaccharide, p-KG103, which is purified from the marine microalgae, *Gyrodinium impudium*, can activate NO production in a JNK-dependent manner to stimulate the production of cytokines, such as interleukin-1 (IL-1), IL-6, and TNF-α, in macrophages, thereby preventing tumor cell growth both in vitro and in vivo [3,4]. Polysaccharides are also crucial regulatory factors for the adaptive immune response, especially humoral immunity. Jingjing Yang et al. reported that a water-soluble polysaccharide (WSPA), which is isolated from the stem of *Physalis alkelengi* L., significantly enhanced specific antibody IgG titers in mice [5]. The polysaccharides derived from Taishan *Pinus massoniana* pollen (TPPPS) also improved the effects of different vaccines when used as an immunoadjuvant [6–8]. In addition, several types of polysaccharides (such as carrageenan, alginate, fucan, laminarin, ulvan, dextran sulfate, heparin, and fucoidan (fuc)) possess a broad-spectrum of antiviral activity in vitro against dengue virus, herpes simplex virus (HSV), human immunodeficiency virus (HIV), and influenza virus by targeting several steps of the viral cycle [9–16].

Infectious bursal disease virus (IBDV) is the etiological agent of infectious bursal disease (IBD), which is an acute, highly contagious disease in young chickens [17] that contributes to a huge economic loss in the public poultry industry. IBDV is a non-enveloped, double-stranded RNA (dsRNA) virus belonging to the *Birnaviridae* family [18–20]. IBDV infection causes a massive destruction of B cells in lymphoid organs, resulting in lymphopenia (immunosuppression) [21]. This leads to an increased susceptibility to secondary infection [22].

Coccomyxa gloeobotrydiformis is a species of green algae that was first isolated in 1969 by Reisigl [23]. It belongs to the family of Coccomyxaceae and distributes in Iceland, Himalayas, Japan, New Zealand and Antarctica [24]. Previous studies have shown that *Coccomyxa gloeobotrydiformis* has a variety of bioactivities in a rat model, such as neuroprotective effects in ischemic stroke [25], learning and memory improvement effects in intrinsic aging rats [26], inhibitory effects in benign prostate hyperplasia [27], and a protective effect in cerebral ischemia-reperfusion injury [28]. AEX is an acidic polysaccharide isolated from *Coccomyxa gloeobotrydiformi*. Monosaccharide composition analysis revealed that AEX contained galactose (Gal), mannose (Man), glucose (Glc), arabinose (Ara), xylose (Xyl) and rhamnose (Rha) [29]. Takayuki Komatsu et al. have demonstrated that AEX possesses antiviral activity against human influenza A virus infection in vitro [29]. However, its bioactivities, such as the immunomodulatory and antiviral activities of AEX in chicken, are still unknown.

This study analyzed the in vitro effects of AEX on cytokine production in chicken PBMCs and spleen lymphocytes, spleen lymphocyte proliferation, NO production and cytotoxicity. We also evaluated the in vitro and in vivo effects of AEX on IBDV replication in the chicken bursa of Fabricius and specific antibody levels in peripheral blood.

2. Results

2.1. Effects of AEX on Cytokine and NO Production in Chicken PBMCs

Cytokines play a critical role in innate immunity and adaptive activation during infection. Inflammatory stimulation or cytokines can induce the expression of inducible nitric oxide synthase, which uses NADPH, O_2, and Arg to synthesize NO. To assess the effect of AEX on cytokine and NO production in chicken PBMCs, the PBMCs were isolated from the peripheral blood of 4-week-old SPF chickens and then treated with AEX, LPS, or PBS at the indicated concentrations. Inflammatory cytokines (IFN-β, IL-1β, IL-6, TNF-α), T-helper cell differentiation cytokines (IL-10, IL-12p40), and iNOS mRNA levels were analyzed by qRT-PCR. Figure 1A–D shows that AEX significantly increased IL-1β, IL-6, and TNF-α mRNA levels in a dose-dependent manner but did not affect IFN-β, whereas LPS increased IFN-β mRNA expression by 1.8-fold. As shown in Figure 1E,F, IL-10 and IL-12p40 mRNA levels were also remarkably increased by AEX in a dose-dependent manner. In addition, the mRNA expression of iNOS was significantly increased by AEX (Figure 1G). Moreover, the NO concentration analysis showed that AEX also induced NO production in a dose-dependent manner (Figure 1H). These observations suggest that AEX can up-regulate various pro-inflammatory cytokines, T-helper cell differentiation cytokines, and NO production by activating chicken PBMCs.

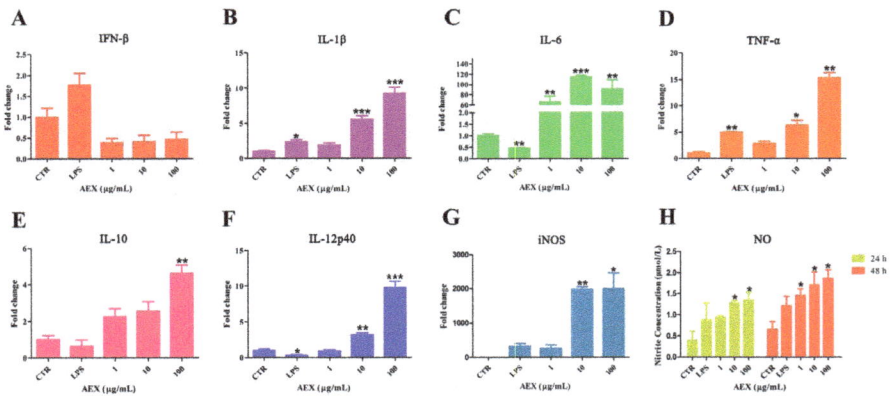

Figure 1. Multiple cytokines and iNOS expression were upregulated and NO production were increased by AEX in PBMCs. (**A–G**) PBMCs were isolated and cultured with AEX (0–100 μg/mL) or LPS (100 ng/mL) for 24 h. Then total RNA was extracted and analyzed by qRT-PCR for IFN-β, IL-1β, IL-6, TNF-α, IL-10, IL-12p40, and iNOS; (**H**) PBMCs were isolated and cultured with AEX (0–100 μg/mL) or LPS (100 ng/mL) for 24 h and 48 h. Then, the culture supernatants were collected and nitrite contents were determined by Griess reaction. Data represent means ± SEM from three wells per group. * $p \leq 0.05$; ** $p \leq 0.01$; *** $p \leq 0.001$. Results are representative of two independent experiments.

2.2. Effect of AEX on Gene Expression in Splenic Lymphocytes

As the major lymphoid organ for antigen presentation, the spleen contains a mass of DCs and T and B lymphoid cells. To investigate the effect of AEX on inflammatory cytokines (IL-1β and IL-6), major histocompatibility complexes (MHC I and MHC II), and T lymphocyte surface markers (CD3, CD4, CD8) mRNA expression in chicken splenic lymphocytes, the lymphoid cells were isolated from the chicken spleen and treated with AEX, LPS, or PBS at the indicated concentrations. As shown in Figure 2A,B, AEX significantly increased IL-1β and IL-6 mRNA expression in a dose-dependent fashion. MHC I mRNA levels were reduced by AEX treatment, whereas the expression of MHC II was not (Figure 2C,D), implying that AEX may negatively regulate the antigen presentation of viral infection between the antigen presenting cells (APCs) and the CD8[+] T lymphoid cells. However, the expression of CD3, CD4, and CD8 mRNA was not affected by AEX treatment (Figure 2E–G). These data suggest that AEX can up-regulate inflammatory cytokines (IL-1β and IL-6) mRNA levels but down-regulate MHC I expression.

2.3. Effect of AEX on Lymphocyte Proliferation

To assess the effect of AEX on lymphocyte proliferation, we isolated splenic lymphocytes from 4-week-old SPF chickens and co-cultured them with AEX, Con A or PBS. The MTT assays indicate that AEX only slightly promoted splenic lymphocyte proliferation at 24 h and 72 h, but a significant promotion of proliferation was observed at 48 h in a dose-dependent manner (Figure 3B). The lymphocyte proliferation assay was also performed on the DT40 cells (chicken bursa of Fabricius derived cell line) and the peripheral blood lymphocytes. AEX significantly promoted the proliferation of the DT40 cells (Figure 3A). However, the proliferation of peripheral blood lymphocytes was reduced by AEX treatment in a dose-dependent manner (Figure 3C). Consistent with the lymphocyte proliferation effect, AEX up-regulated IL-2 mRNA levels in chicken splenic lymphocytes but down-regulated it in chicken peripheral blood lymphocytes (Figure 3D,E), suggesting that AEX may promote splenic lymphocyte proliferation by up-regulating IL-2 expression and inhibit peripheral blood lymphocyte proliferation by down-regulating IL-2 expression.

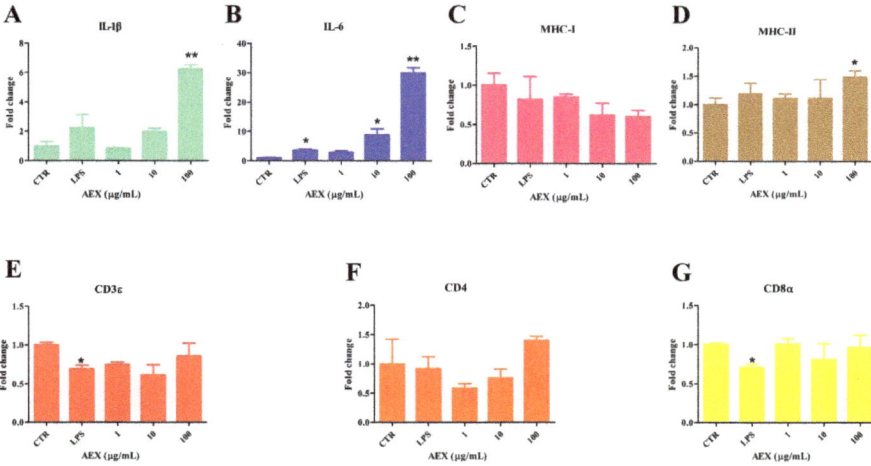

Figure 2. Inflammatory cytokines and surface molecules on splenic lymphocytes were regulated by AEX. Splenic lymphocytes were isolated and cultured with AEX (0–100 μg/mL) or LPS (100 ng/mL) for 24 h. Then, total RNA was extracted and analyzed by qRT-PCR for IL-1β (**A**), IL-6 (**B**), MHC I (**C**), MHC II (**D**), CD3ε (**E**), CD4 (**F**), and CD8α (**G**). Data represent means ± SEM from three wells per group. * $p \leq 0.05$; ** $p \leq 0.01$. Results are representative of two independent experiments.

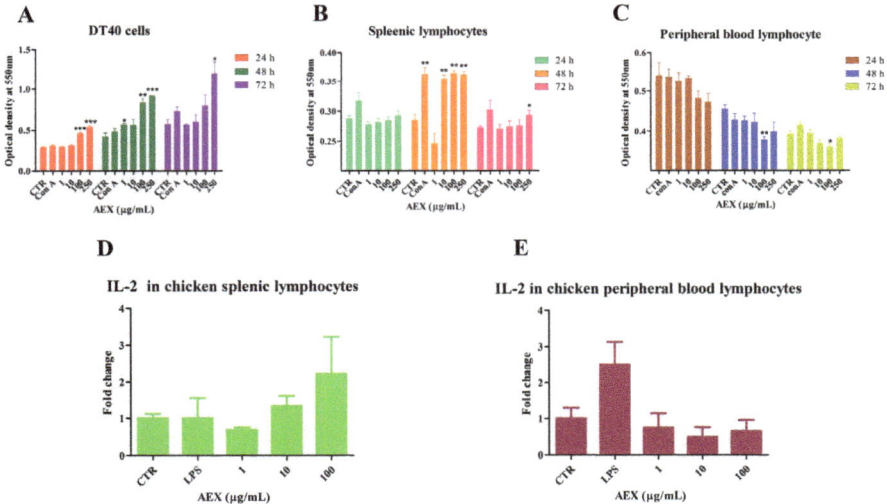

Figure 3. AEX promoted DT40 and splenic lymphocytes proliferation, but reduced peripheral blood lymphocyte proliferation in vitro. (**A–C**) DT40 cells, splenic lymphocytes and peripheral blood lymphocytes were cultured in 96-well plates. After being stimulated by AEX (0–250 μg/mL) or Con A (40 μg/mL) for 24 h, 48 h, and 72 h, respectively, the proliferation was examined by MTT method as described in the Materials and Methods section; (**D,E**) Splenic lymphocytes and peripheral blood lymphocytes were isolated and cultured with AEX (0–100 μg/mL) or LPS (100 ng/mL) for 24 h, respectively. Then, total RNA was extracted and analyzed by qRT-PCR for IL-2. Data represent means ± SEM from three wells per group. * $p \leq 0.05$; ** $p \leq 0.01$; *** $p \leq 0.001$. Results are representative of two independent experiments.

2.4. In Vitro Anti-Viral Effect of AEX on the IBDV TS Strain Virus

First, we investigated the cytotoxicity of AEX on the metabolism of the Vero cells. The MTT assay showed that there was no observable difference in the metabolic activity of the Vero cells treated with 62.5–1000 µg/mL of AEX and the untreated cells. However, at a concentration of 5000 µg/mL, the AEX treatment reduced the relative metabolic activity by 62% (Figure 4A).

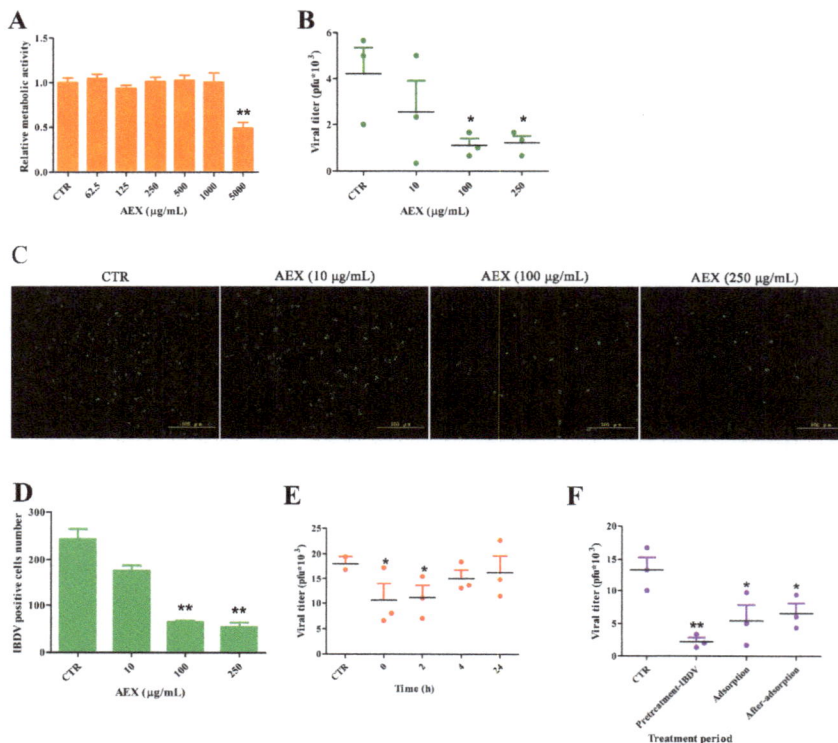

Figure 4. AEX inhibited IBDV replication by deactivating viral particle and interfering with adsorption. (**A**) The Vero cells were treated with different concentrations of AEX for 24 h and cell viability was measured by MTT assay; (**B**) The Vero cells infected with IBDV (MOI = 1.0) were treated with different concentrations of AEX, and viral titers in the supernatant were identified by TCID$_{50}$ assays on the Vero cells. PBS treatment wells served as the control; (**C,D**) The Vero cells infected with IBDV (MOI = 1.0) were treated with different concentrations of AEX, and immunofluorescence staining was performed using anti-IBDV antibodies. (**C**) Scale bar represents 100 µm; (**D**) IBDV positive cells number was calculated; (**E**) The Vero cells were incubated with IBDV (MOI = 1.0) for 1 h and treated with AEX at the indicated time points (0, 2, 4, 24 h p.i.). Viral titers in the supernatant were determined by TCID$_{50}$ assays on the Vero cells at 48 h p.i. (**F**) The Vero cells were infected with IBDV (MOI = 1.0) via different polysaccharide treatment method (IBDV pretreatment, adsorption, and after adsorption) and viral titers in the supernatant were determined by TCID$_{50}$ assays on the Vero cells at 48 h p.i. Data represent means ± SEM from three wells per group. * $p \leq 0.05$; ** $p \leq 0.01$. Results are representative of two independent experiments.

Then, we tested the in vitro anti-IBDV activity of AEX on the Vero cells. The Vero cells were first infected with the IBDV TS strain (MOI = 1.0) and then treated with compounds at the indicated concentrations after removal of the viral inoculum. After 48 h, the virus titers and the IBDV positive cells were identified by PFU and IFA assays. As shown in Figure 4B–D, AEX significantly reduced viral titers when used at a concentration greater than 100 μg/mL ($p < 0.05$). AEX reduced the IBDV positive cell number from 250 (CTR) to 50 at a concentration of 250 μg/mL. Furthermore, the influence of the treatment duration with AEX on IBDV multiplication was examined.

AEX was added to the Vero cells at different times after IBDV adsorption. As shown in Figure 4E, the viral titers were decreased by AEX only when it was added to the Vero cells at 0 and 2 h, whereas AEX showed no anti-IBDV effect when it was added at 4 and 24 h. Moreover, when IBDV was pretreated with AEX, the inhibitory effect was more than 70%. However, there was less of an inhibitory effect when AEX was added during adsorption (50%) or after-adsorption (50%) (Figure 4F). These data indicate that the antiviral activity of AEX is largely related to its deactivation of viral particles and inhibition of virus life-cycle early events that occur 0–2 h after adsorption.

2.5. In Vivo Anti-Viral Effect of AEX on the IBDV TS Strain Virus

To further explore the therapeutic potential of AEX, SPF chickens were infected with $10^{5.5}$TCID$_{50}$ of the IBDV TS strain and treated with AEX at the concentrations indicated in Figure 5. Oral administration of AEX was started 24 h post infection and repeated once daily for 3 days. As shown in Figure 5A,B, IBDV replicated efficiently in chicken bursa of Fabricius in the CTR group. AEX slightly decreased *VP2* mRNA expression at concentrations of 62.5 and 125 mg/kg body weight. When 250 mg/kg body weight of AEX was used, *VP2* mRNA was significantly reduced (Figure 5A). The viral titer in the bursa of Fabricius detection also showed that AEX decreased the IBDV viral titer at concentrations of 62.5 to 500 mg/kg body weight. (Figure 5B). 125 mg/kg body weight of AEX decreased the viral titer from $10^{5.5}$TCID$_{50}$ (CTR) to $10^{3.5}$TCID$_{50}$. These data suggest that AEX can suppress the replication of the IBDV TS strain in chicken bursa of Fabricius.

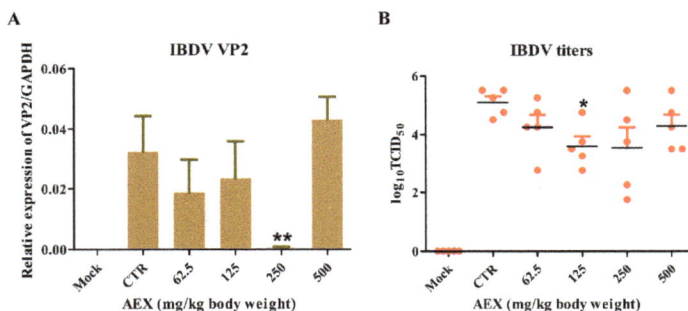

Figure 5. AEX inhibited IBDV replication in chicken bursa of Fabricius. SPF chickens (6 chickens per group) were oral inoculated with different concentrations of AEX (0, 62.5, 125, 250 and 500 mg/kg body weight) per day. After three days, chickens were eye-dropped with IBDV ($10^{5.5}$TCID$_{50}$). Bursa of Fabricius were collected at 5 d.p.i. and divided into two parts. (**A**) Half of the bursa of Fabricius were grinded and virus titers were determined by TCID$_{50}$; (**B**) The other half bursa of Fabricius were subjected to isolate total RNA and the expression lever of IBDV *VP2* was evaluated by qRT-PCR. Data represent means ± SEM from five wells per group. * $p \leq 0.05$; ** $p \leq 0.01$. Results are representative of two independent experiments.

2.6. Effect of AEX on Anti-IBDV Antibodies in Immunized Chickens

The B cells in chicken bursa of Fabricius are the major target of IBDV, which causes B cell apoptosis and depletion in the bursa of Fabricius to repress humoral immunity. To investigate the regulation of AEX on specific antibody titers against IBDV, we immunized SPF chickens with an intermediate IBDV vaccine every two weeks for a total of three times and detected the IBDV antibody titers (IgY, IgM, IgA) in the sera with an indirect ELISA. As shown in Figure 6, both the IgY and IgM antibody titers significantly increased after the first and second immunization and peaked 14 days after the secondary immunization. However, IgA antibody titers were only slightly increased after 7 days after the secondary immunization. Surprisingly, the IgY and IgM antibody titers in the IBDV immunized chickens were slightly decreased by AEX at concentrations from 12.5 to 50 mg/mL, and the IgY and IgM antibody titers of the FA + 100AEX groups were significantly lower than those of the FA + IBDV group (Figure 6A,B). In addition, AEX showed an inhibitory effect on the IgA antibody titers in a dose-dependent manner (Figure 6C). These results suggest that in some case AEX may negatively regulate specific antibody production when we immunized the chickens with live attenuated vaccines.

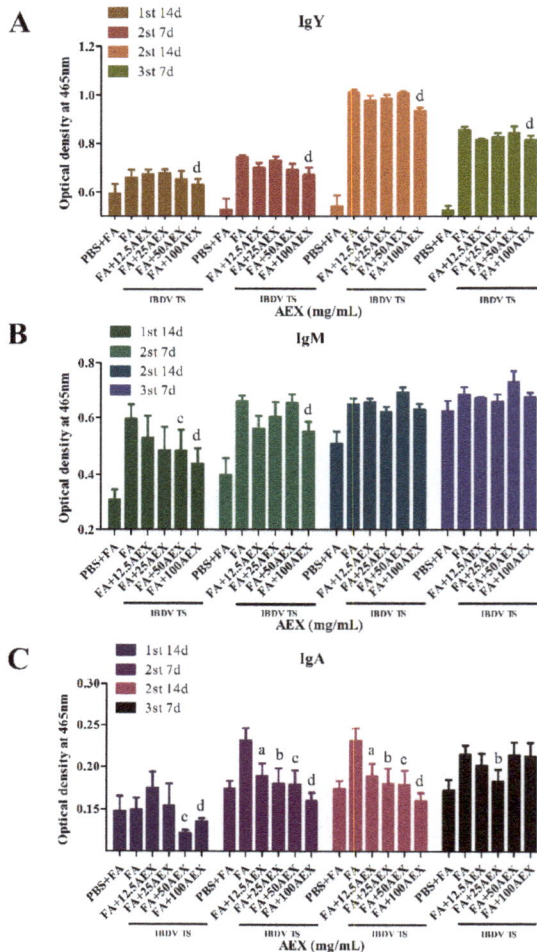

Figure 6. AEX reduced specific anti-IBDV antibody (IgY, IgM, and IgA) titers in the sera of chickens immunized with IBDV. SPF chickens (6 chickens per group) were immunized with IBDV vaccine by subcutaneous injection in the absence or presence of AEX (PBS used as control) thrice at a 14-day interval. The sera were collected (14 days after the first immunization, 7 days and 14 days after secondary immunization, and 7 days after tertiary immunization) and specific antibody (**A**) IgY; (**B**) IgM; (**C**) IgA were detected by ELISA. Data represent means ± SEM from five wells per group. [a,b,c,d] $p \leq 0.05$ vs. *FA*. Results are representative of two independent experiments.

3. Discussion

AEX was previously reported to exert an antiviral effect on the influenza virus [29]. In this study, we investigated the immunoregulatory effect of AEX on chicken immune cells and the antiviral effect of AEX against IBDV, as well as the adjuvant effect of AEX.

Immunomodulators are natural or synthetic substances that regulate or modify the function of the immune system [30]. According to their effects, Immunomodulators can be divided into immunostimulants and immunosuppressants. It has been reported that many polysaccharides have immunoregulatory activities [31]. Here, we observed that AEX significantly increased IL-1β, IL-6, TNF-α, and iNOS mRNA levels, and promoted the secretion of NO in chicken PBMCs. The activation of chicken PBMCs was characterized by the release of inflammatory cytokines and NO production [32–34], which implied that AEX could effectively activate chicken PBMCs. So AEX is an immunostimulant.

Interestingly, though, the mRNA expression levels of IFN-β were not regulated by AEX. As IRF3 is a transcription factor essential for IFN-β [35], we theorize that it may not be involved in the AEX receptor signaling pathway.

The activation of T cells generally requires a signal delivered by the interaction of the TCR with a specific antigen on the MHC molecules (MHC I and MHC II). By interacting with CD8 molecules on surfaces of cytotoxic T cells, MHC I mediates the destruction of intracellular antigens, especially virus infections, and establishes cellular immunity. MHC II mediates humoral immunity by interacting with CD4 molecules on the surface of T-helper cells. In this study, the expression of MHC I mRNA was down-regulated by AEX treatment whereas MHC II mRNA levels were increased, which implies that AEX promotes extracellular antigen presentation but negatively regulates intracellular antigen presentation.

Lymphocytes proliferate extensively before they differentiate into functional effector cells of a particular specificity. Tang et al. demonstrated that Sophy β-glucans significantly improved the lymphocyte proliferative response of duck PBMCs and potentially enhanced the cellular immune response [36]. In the current study, the proliferation of both chicken primary splenic lymphocytes and DT40 cells, which are a chicken bursa of Fabricius lymphoma cell line, was significantly improved by AEX treatment. However, peripheral blood lymphocyte proliferation was decreased by AEX treatment. Considering that DT40 is a B lymphocyte line and splenic lymphocytes consists of both B and T lymphocytes [37], whereas most peripheral blood lymphocytes are T lymphocytes [38], AEX could improve the B lymphocyte proliferation but suppress the proliferation of T lymphocytes.

Astragalus polysaccharides, fucoidan, rape pollen polysaccharide, epimedium polysaccharide, and echinacea polysaccharide have been reported to show anti-IBDV activity and improve chicken immune response [39–41]. Here, we observed that AEX showed an in vitro anti-IBDV effect, especially when added at 0–2 h post infection or when pretreated with IBDV particles, which suggests that AEX can inhibit replication by inactivating IBDV particles. Our in vivo anti-IBDV assay also indicated that AEX can down-regulate IBDV *VP2* mRNA expression and viral titers when AEX was delivered by oral administration. Therefore, AEX could be a suitable alternative approach to therapy meant to address anti-IBDV.

Immunomodulators are potential agents for improving poultry immunity and growth performance [42]. Several glucans have been used as an adjuvant for vaccines [43–45]. Sulfated polysaccharides isolated from both *Grateloupia filicina*, *Ulva pertusa* and *Sargassum qingdaoense*, could increase H9N2-specific antibody titers in the sera of mice immunized with inactivated avian influenza virus [46]. Despite the fact that chicken PBMCs activation and lymphocyte proliferation were improved by AEX treatment, specific antibodies against IBDV in the sera of immunized chicken were decreased instead of showing improvement. That may due to the direct virucidal action of AEX. The vaccine we used in this experiment is an intermediate vaccine which has the similar action as normal IBDV. The AEX, as an effective virucidal substance, may affect the immunogenicity of the vaccine particles or block its surface antigen, which results in lower specific antibodies titers. To avoid that negative effect, we could try to find a suitable dosage between the promotion of producing antibody and the inhibition of vaccine particles, or we could inject AEX adjuvant before vaccines, lowering the negative impact on vaccines activity. In addition, we suppose that AEX may be a good choice as adjuvant for an inactivated IBDV vaccines.

4. Materials and Methods

4.1. Animals, Cells, Compounds and Reagents

Ten-day- and 4-week-old specific pathogen-free (SPF) White Leghorn chickens were purchased from Meria (Meria, Beijing, China) and raised in a laboratory animal house. Peripheral blood lymphocytes, spleen lymphocytes and DT 40 cells were maintained in RPMI-1640 medium (HyClone, Logan, UT, USA) supplemented with 10% fetal bovine serum (FBS; HyClone, Logan, UT, USA) and 1% antibiotics (Sigma, St. Louis, MO, USA). The Vero cells were purchased from ATCC and maintained in DMEM medium (HyClone, Logan, UT, USA) supplemented with 10% FBS and 1% antibiotics. The acidic polysaccharide of the *Coccomyxa gloeobotrydiformis* Nikken strain (AEX) was supplied by the Nikken Sohonsha Corporation (Hashima, Gifu, Japan). LPS and Concanavalin A (Con A) were purchased from Sigma-Aldrich (St. Louis, MO, USA).

4.2. Cell Isolation

The peripheral blood collected from 4-week-old SPF chickens was diluted with Ca^{2+}- and Mg^{2+}-free Hank's balanced salt solution (1:1, vol/vol), carefully layered onto the Histopaque-1077 (Sigma) in 15 mL conical centrifuge tubes, and centrifuged at $400\times g$ for 30 min at room temperature. The lymphocyte layer was collected, washed 3 times in Hank's solution, and spun at $250\times g$ for 10 min. Then, the cells were resuspended in RPMI-1640 medium containing 10% FBS and cultured for 2 h. The suspension cells (peripheral blood lymphocytes) were collected and used for the lymphocyte proliferation assay, whereas the anchorage-dependent cells (peripheral blood molecular cells) were used for the cytokine and NO analyses.

The spleens from the 4-week-old SPF chickens, sacrificed under aseptic conditions, washed with Hank's solution and crushed to isolate the spleen cells. The spleen cell mass was passed through a 200 mesh copper sieve to obtain a homogeneous cell suspension. Spleen lymphocytes were isolated from the suspension using the same method as that used for the isolation of the peripheral blood lymphocytes.

4.3. Lymphocyte Proliferation Assay

Isolated peripheral blood lymphocytes, spleen lymphocytes and DT 40 cells were resuspended in RPMI-1640 medium containing 10% FBS and adjusted to 1×10^7 cells/mL. The lymphocytes were distributed (100 µL per well) onto 96-well plates (Costar, Corning, Grand Island, NY, USA) and treated with different concentrations of AEX (1, 10, 100, and 250 µg/mL) for 24 h, 48 h and 72 h, respectively. Con A served as the positive control. Lymphocyte proliferation activity was tested by a 3-(4,5-dimethylthiazol-2-yl)-2,5-diphenyl tetrazolium bromide (MTT; Sigma, St. Louis, MO, USA) assay [47]. 10 µL of PBS containing MTT (final concentration: 0.5 mg/mL) was added to each well. After a 4 h incubation at 37°C, the absorbance values were measured in a microplate reader (Bio-Rad, Hercules, CA, USA) at 550 nm.

4.4. Nitric Oxide (NO) Production Assay

After a 24 h treatment with various concentrations of AEX, the culture supernatants of the PBMCs were collected, and the nitrite contents were determined by the Griess reaction using the Griess Reagent System (Promega, Madison, WI, USA) according to the manufacturer's protocol.

4.5. qRT-PCR Analysis

Total RNA was first extracted from the AEX-treated PBMCs, spleen lymphocytes and bursa of Fabricius using a TRIgene (GeneStar, Beijing, China) and treated with Dnase I (Promega, Madison, WI, USA) to remove the DNA. Then, 1 µg of RNA was reverse-transcribed into cDNA using a GoScript reverse transcription system (Promega, Madison, WI, USA) in a 20 µL reaction mixture. Quantitative real-time PCR amplification was performed in 15 µL of the LightCycler® 480 SYBR Green I Master Mix (Roche, Rotkreuz, Switzerland) with the LightCycler® 480 Real-time PCR System (Roche, Rotkreuz, Switzerland). The individual primers used were designed by Primer Express 3.0 and are shown in Table 1. The qRT-PCR was performed under the following cycling conditions: 95 °C for 10 min, 40 cycles of 95 °C for 15 s and 60 °C for 1 min, and one cycle of 95 °C for 15 s, 60 °C for 15 s, 95 °C for 15 s and 40 °C for 30 s. The relative mRNA abundances were calculated using the $2^{-\Delta\Delta Ct}$ method with GAPDH as a reference and plotted as the fold changed relative to the control samples.

Table 1. Sequence of the primers used in qRT-PCR.

Genes	Direction	Sequence	Accession NO. in GenBank
GAPDH [a]	Forward	TGCCATCACAGCCACACAGAAG	AF047874.1
	Reverse	ACTTTCCCCACAGCCTTAGCAG	
IFN-β	Forward	ACAACTTCCTACAGCACAACAACTA	X92479.1
	Reverse	GCCTGGAGGCGGACATG	
IL-1β	Forward	TTGCTGGTTTCCATCTCGTATGTA	NM_204524
	Reverse	CCCAGAGCGGCTATTCCA	
IL-2 [b]	Forward	TTCTGGGACCACTGTATGCTCTT	AF000631.1
	Reverse	TACCGACAAAGTGAGAATCAATCAG	
IL-4 [b]	Forward	AATGACATCCAGGGAGAGGTTTC	AJ621249.1
	Reverse	AGGCTTTGCATAAGAGCTCAGTTT	
IL-6 [a]	Forward	GACGAGGAGAAATGCCTGACG	AJ309540.1
	Reverse	CCGAGTCTGGGATGACCACTTC	
IL-10 [b]	Forward	GCTGAGGGTGAAGTTTGAGGAA	AF000631.1
	Reverse	GAAGCGCAGCATCTCTGACA	
IL-12p40 [b]	Forward	CGAAGTGAAGGAGTTCCCAGAT	AY262752.1
	Reverse	GACCGTATCATTTGCCCATTG	
TNF-α	Forward	GGAATGAACCCTCCGCAGTA	AY765397.1
	Reverse	CACTACGGGTTGCTGCACAT	
iNOS	Forward	GCCCCTCCAGCTGATCAGA	D85422.1
	Reverse	AGGCCTGTGAGAGTGTGCAA	
CD3ε	Forward	TGCCAAAGTGTGTGCAAACTG	NM_206904.1
	Reverse	AGTACCCCCAAGGTGATGAGAA	
CD4	Forward	GCTGTGTGTTTGCGGTCATC	Y12012.1
	Reverse	CCTTTCCTGCAATCCCAATC	
CD8α	Forward	CGACAATGGTGTCTCCTGGAT	NM_205235.1
	Reverse	GGGAAAGTGGTCCGGGATAA	
MHC-I	Forward	TGGTTGGTGTTGGATTCATCAT	KF294514.1
	Reverse	GCTGGATCCACCTTCCTTGTC	
MHC-II	Forward	GTGGGCTCAGTTCGGTTTTC	DQ207939.1
	Reverse	AATTCGGGCAGCCTCCATA	
TS vp2 [b]	Forward	ACCGGCACCGACAACCTTA	AF076230.1
	Reverse	CCCTGCCTGACCACCACTT	

[a] Primers from Reference [48]; [b] Primers from Reference [49].

4.6. Viruses and Infection

The IBDV Ts strain was amplified in the Vero cells, and the viral titers were determined as $TCID_{50}$ by IFA on the Vero cells as previously described [50]. For the virus infection, the virus propagation solution was diluted in DMEM and was added to the cells at the indicated multiplicity of infection (MOI). The virus was allowed to adsorb for 1 h at 37 °C. After removing the virus inoculum, the cells were maintained in infective media (DMEM) at 37 °C in 5% CO_2. All of the experiments with the IBDV Ts strain were performed under biosafety level 2 (BSL-2) conditions with investigators wearing the appropriate protective equipment and complying with the general biosafety standards for microbiological and biomedical laboratories of the Ministry of Health of the People's Republic of China (WS 233-2002).

4.7. Cytotoxicity Assay

Cell viability was measured by an MTT assay. The Vero cells were seeded into 96-well plates (1×10^5 cells/well) overnight. Subsequently, the medium was supplied with different concentrations of AEX or LPS. After a 24 h incubation at 37 °C and 5% CO_2, 10 µL of PBS containing MTT (at a final concentration of 0.5 mg/mL) was added to each well for 4 h. After 4 h of incubation at 37 °C, the supernatant was removed, and 200 µL of DMSO was added to each well to solubilize the formazan crystals. After a vigorous shaking, the absorbance values were measured in a microplate reader at 570 nm.

4.8. Antiviral Assay (TCID$_{50}$)

Antiviral activity was evaluated by the $TCID_{50}$ assay in Vero cells. The Vero cells were incubated with the IBDV at an MOI of 1 for 1 h at 37 °C and washed to remove the unbound virus; then, the infecting media containing different concentrations of AEX was added to the cells, which were incubated at 37 °C for 48 h. Subsequently, the supernatant was removed and collected, 10-fold serially diluted, and incubated with the Vero cells for 36 h. The $TCID_{50}$ was determined according to Reed and Muench (1938) [51].

4.9. Immunization of SPF Chicken

Five-day-old SPF chickens were randomly divided into 6 groups ($n = 6$/group) and immunized with different concentrations of AEX and IBDV live vaccine (strain B87, Lufang Biology, Yangling, Shaanxi, China). Immunization of chickens with IBDV and AEX was performed as described previously [52]. For the first injection, vaccine (following the dosing recommendations in the labeling) and different doses of AEX (0, 12.5, 25, 50, 100 mg/mL) were solved in PBS and emulsified with an equal volume of complete Freund's adjuvant (FA, Sigma). Each chicken received 0.1 mL emulsion by subcutaneous injection in the neck. Booster injections were given subcutaneously in the neck region 2 week and 4 week after the first injection respectively with the same dose emulsified with incomplete FA (Sigma).

4.10. Measurement of Specific Antibody in the Sera of Immunized Chickens

The sera samples were collected from immunized chickens 14 days after the first immunization, 7 days and 14 days after the secondary immunization, and 7 days after the tertiary immunization. Then, specific antibodies were detected by ELISA. Briefly, flat-bottom 96-well ELISA plates (Costar, Coring, NY, USA) were coated with IBDV ($TCID_{50}$ of $10^{5.5}$/0.1 mL) in 0.05 mol/L carbonate-bicarbonate buffer pH 9.6 for 24 h at 4 °C. The wells were washed three times with phosphate-buffered saline (PBS) containing 0.05% (v/v) Tween 20 (PBST), and blocked with PBS containing 1% gelatin at 37 °C for 1 h. After washing the wells with PBST three more times, 100 µL of a series of diluted sera from immunized chickens or PBS containing 0.1% gelatin were added to the triplicate wells as control. The plates were then incubated for 1 h at 37 °C, followed by being washed with PBST, 100 µL of goat anti-chicken

IgG-HRP, IgM-HRP, and IgA-HRP (Cwbio, Beijing, China) (diluted 1:5000 with PBS containing 0.1% gelatin, respectively) were added to each plate. The plates were further incubated for 1 h at 37 °C. Substrate 3,3′,5,5′-tetramethylbenzidine (TMB) was added to each well after being washed with PBST, and the plate was incubated for 10 min at room temperature. Reaction was terminated by adding 50 μL of 2 mol/L H_2SO_4 to each well, and optical density (OD) was detected at 465 nm with a microplate reader (Bio-Rad, Hercules, CA, USA).

4.11. Statistics

All data analyses were performed using SPSS 16.0 (SPSS Inc., Chicago, IL, USA). One-way ANOVA was used to detect significant differences between compound-treated groups and control groups. p values < 0.05 were considered significant. Data are presented as mean ± SEM from at least three wells per group. Results are representative of two independent experiments.

4.12. Ethics Statements

This study was approved by the Laboratory Animal Care and Use Committee of the China Agricultural University (Permit Number: SKLAB-2014-06-06). All efforts were made to minimize animal suffering and the number of animals used.

5. Conclusions

In summary, our study first reported that AEX, derived from *Coccomyxa gloeobotrydiformis*, activates chicken PBMCs by up-regulating cytokine and NO production, promotes chicken splenic lymphocyte proliferation, and inhibits IBDV replication by interfering with the early events of IBDV replication or inactivating IBDV particles in vitro and in vivo. AEX is a promising antiviral candidate for the prevention or treatment of IBDV in the poultry industry. However, it has some limitations and still needs more testing to find suitable conditions for use as an adjuvant for the IBDV vaccine.

Acknowledgments: This work was supported by the National Basic Research Program (973 Program, number 2013CB945000), Chinese Universities Scientific Fund (Project No. 2015jk009). We also thank Weili Kong and Huijie Gao for tissue homogenate and valuable advice.

Author Contributions: Q.G., Q.S. and Z.L. conceived and designed the experiments; Q.G., Q.S. and W.X. performed the experiments; Q.G. and Q.S. analyzed the data; Q.G., Q.S., R.S., F.E. and Z.L. contributed reagents/materials/analysis tools; Q.G., Q.S. and L.R. wrote the paper.

References

1. Wang, W.; Wang, S.X.; Guan, H.S. The antiviral activities and mechanisms of marine polysaccharides: An overview. *Mar. Drugs* **2012**, *10*, 2795–2816. [CrossRef] [PubMed]
2. Li, R.-J.; Qiu, S.-D.; Chen, H.-X.; Tian, H.; Wang, H.-X. The immunotherapeutic effects of astragalus polysaccharide in type 1 diabetic mice. *Biol. Pharm. Bull.* **2007**, *30*, 470–476. [CrossRef] [PubMed]
3. Bae, S.-Y.; Yim, J.H.; Lee, H.K.; Pyo, S. Activation of murine peritoneal macrophages by sulfated exopolysaccharide from marine microalga *Gyrodinium impudicum* (strain *KG03*): Involvement of the NF-κB and JNK pathway. *Int. Immunopharmacol.* **2006**, *6*, 473–484. [CrossRef] [PubMed]
4. Yim, J.H.; Son, E.; Pyo, S.; Lee, H.K. Novel sulfated polysaccharide derived from red-tide microalga *Gyrodinium impudicum* strain KG03 with immunostimulating activity in vivo. *Mar. Biotechnol.* **2005**, *7*, 331–338. [CrossRef] [PubMed]
5. Yang, J.; Yang, F.; Yang, H.; Wang, G. Water-soluble polysaccharide isolated with alkali from the stem of physalis alkekengi l.: Structural characterization and immunologic enhancement in DNA vaccine. *Carbohydr. Polym.* **2015**, *121*, 248–253. [CrossRef] [PubMed]

6.   Wei, K.; Sun, Z.; Yan, Z.; Tan, Y.; Zhu, X.; Wang, X.; Sheng, P.; Zhu, R. Effects of Taishan *Pinus massoniana* pollen polysaccharide on immune response of rabbit haemorrhagic disease tissue inactivated vaccine and on production performance of Rex rabbits. *Vaccine* **2011**, *29*, 2530–2536. [CrossRef] [PubMed]

7.   Guo, F.; Xue, C.; Wu, C.; Zhao, X.; Qu, T.; He, X.; Guo, Z.; Zhu, R. Immunoregulatory effects of Taishan *Pinus massoniana* pollen polysaccharide on chicks co-infected with *avian leukosis* virus and *Bordetella avium* early in ovo. *Res. Vet. Sci.* **2014**, *96*, 260–266. [CrossRef] [PubMed]

8.   Zhao, X.; Liang, M.; Yang, P.; Guo, F.; Pan, D.; Huang, X.; Li, Y.; Wu, C.; Qu, T.; Zhu, R. Taishan *Pinus massoniana* pollen polysaccharides promote immune responses of recombinant *Bordetella avium* ompa in BALB/c mice. *Int. Immunopharmacol.* **2013**, *17*, 793–798. [CrossRef] [PubMed]

9.   Hidari, K.I.; Takahashi, N.; Arihara, M.; Nagaoka, M.; Morita, K.; Suzuki, T. Structure and anti-dengue virus activity of sulfated polysaccharide from a marine alga. *Biochem. Biophys. Res. Commun.* **2008**, *376*, 91–95. [CrossRef] [PubMed]

10.   Hosoya, M.; Balzarini, J.; Shigeta, S.; De Clercq, E. Differential inhibitory effects of sulfated polysaccharides and polymers on the replication of various myxoviruses and retroviruses, depending on the composition of the target amino acid sequences of the viral envelope glycoproteins. *Antimicrob. Agents Chemother.* **1991**, *35*, 2515–2520. [CrossRef] [PubMed]

11.   Lee, J.B.; Hayashi, K.; Hirata, M.; Kuroda, E.; Suzuki, E.; Kubo, Y.; Hayashi, T. Antiviral sulfated polysaccharide from *Navicula directa*, a diatom collected from deep-sea water in Toyama Bay. *Biol. Pharm. Bull.* **2006**, *29*, 2135–2139. [CrossRef] [PubMed]

12.   Talarico, L.B.; Pujol, C.A.; Zibetti, R.G.; Faria, P.C.; Noseda, M.D.; Duarte, M.E.; Damonte, E.B. The antiviral activity of sulfated polysaccharides against dengue virus is dependent on virus serotype and host cell. *Antivir. Res.* **2005**, *66*, 103–110. [CrossRef] [PubMed]

13.   Queiroz, K.; Medeiros, V.; Queiroz, L.; Abreu, L.; Rocha, H.; Ferreira, C.; Juca, M.; Aoyama, H.; Leite, E. Inhibition of reverse transcriptase activity of HIV by polysaccharides of brown algae. *Biomed. Pharmacother.* **2008**, *62*, 303–307. [CrossRef] [PubMed]

14.   Xin, X.; Geng, M.; Guan, H.; Li, Z. Study on the mechanism of inhibitory action of 911 on replication of HIV-1 in vitro. *Chin. J. Mar. Drugs* **1999**, *19*, 15–18.

15.   Muto, S.; Niimura, K.; Oohara, M.; Oguchi, Y.; Matsunaga, K.; Hirose, K.; Kakuchi, J.; Sugita, N.; Furusho, T. Polysaccharides from Marine Algae and Antiviral Drugs Containing the Same as Active Ingredients. Eur. Patent EP295956, 21 Decmber 1988.

16.   Carlucci, M.; Scolaro, L.; Noseda, M.; Cerezo, A.; Damonte, E. Protective effect of a natural carrageenan on genital herpes simplex virus infection in mice. *Antivir. Res.* **2004**, *64*, 137–141. [CrossRef]

17.   Mundt, E.; Köllner, B.; Kretzschmar, D. VP5 of infectious bursal disease virus is not essential for viral replication in cell culture. *J. Virol.* **1997**, *71*, 5647–5651. [PubMed]

18.   Lasher, H.; Shane, S. Infectious bursal disease. *Worlds Poult. Sci. J.* **1994**, *50*, 133–166. [CrossRef]

19.   Abdu, P.; Abdullahi, S.; Adesiyun, A.; Ezeokoli, C. Infectious bursal disease. *Worlds Poult. Sci. J.* **1986**, *42*, 219–231. [CrossRef]

20.   Berg, T.P.V.D. Acute infectious bursal disease in poultry: A review. *Avian Pathol.* **2000**, *29*, 175–194. [CrossRef] [PubMed]

21.   Käufer, I.; Weiss, E. Significance of bursa of fabricius as target organ in infectious bursal disease of chickens. *Infect. Immun.* **1980**, *27*, 364–367. [PubMed]

22.   Kibenge, F.; Dhillon, A.; Russell, R. Biochemistry and immunology of infectious bursal disease virus. *J. Gen. Virol.* **1988**, *69*, 1757–1775. [CrossRef] [PubMed]

23.   Reisigl, H. Bodenalgen-studien II. *Plant Syst. Evol.* **1969**, *116*, 492–506. [CrossRef]

24.   Guiry, M.D.; Guiry, G.M. *Coccomyxa gloeobotrydiformis* Reisigl. Available online: http://www.algaebase.org/search/species/detail/?species_id=H72dc4e4819414c40 (accessed on 1 January 2017).

25.   Sun, L.; Jin, Y.; Dong, L.; Sumi, R.; Jahan, R.; Li, Z. The neuroprotective effects of *Coccomyxa gloeobotrydiformis* on the ischemic stroke in a rat model. *Int. J. Biol. Sci.* **2013**, *9*, 811–817. [CrossRef] [PubMed]

26.   Sun, L.; Jin, Y.; Dong, L.; Sui, H.J.; Sumi, R.; Jahan, R.; Hu, D.; Li, Z. *Coccomyxa gloeobotrydiformis* improves learning and memory in intrinsic aging rats. *Int. J. Biol. Sci.* **2015**, *11*, 825–832. [CrossRef] [PubMed]

27.   Dong, L.M.; Jin, Y.; Liu, Y.L. Inhibitory effect of *Cocoomyxa gloeobotrydifomis* on benign prostate hyperplasia in aged rats and its action mechanism. *Zhonghua Nan Ke Xue* **2013**, *19*, 506–510. [PubMed]

28. Liu, D.; Sui, H.; Zhang, L.; Jin, Y.; Pharmacology, D.O.; University, L.M. Protective mechanism of *Cocoomyxa gloeobotrydifomis* in cerebral ischemia-reperfusion injury model rats. *Pharmacol. Clin. Chin. Mater. Med.* **2014**, *6*, 036.

29. Komatsu, T.; Kido, N.; Sugiyama, T.; Yokochi, T. Antiviral activity of acidic polysaccharides from *Coccomyxa gloeobotrydiformi*, a green alga, against an in vitro human influenza a virus infection. *Immunopharmacol. Immunotoxicol.* **2013**, *35*, 1–7. [CrossRef] [PubMed]

30. Patil, U.S.; Jaydeokar, A.V.; Bandawane, D.D. Immunomodulators: A pharmacological review. *Int. J. Pharm. Pharm. Sci.* **2012**, *4*, 30–36.

31. Jiang, M.H.; Zhu, L.; Jiang, J.G. Immunoregulatory actions of polysaccharides from Chinese herbal medicine. *Expert Opin. Ther. Targets* **2010**, *14*, 1367–1402. [CrossRef] [PubMed]

32. Duerksen-Hughes, P.J. Both tumor necrosis factor and nitric oxide participate in lysis of simian virus 40-transformed cells by activated macrophages. *J. Immunol.* **1992**, *149*, 2114–2122. [PubMed]

33. Farias-Eisner, R.; Chaudhuri, G. Nitric oxide is an important mediator for tumoricidal activity in vivo. *Proc. Natl. Acad. Sci. USA* **1994**, *91*, 9407–9411. [CrossRef] [PubMed]

34. Lorsbach, R.B.; Murphy, W.J.; Lowenstein, C.J.; Snyder, S.H.; Russell, S.W. Expression of the nitric oxide synthase gene in mouse macrophages activated for tumor cell killing. Molecular basis for the synergy between interferon-gamma and lipopolysaccharide. *J. Biol. Chem.* **1993**, *268*, 1908–1913. [PubMed]

35. Baigent, S.J.; Zhang, G.; Fray, M.D.; Flick-Smith, H.; Goodbourn, S.; McCauley, J.W. Inhibition of beta interferon transcription by noncytopathogenic bovine viral diarrhea virus is through an interferon regulatory factor 3-dependent mechanism. *J. Virol.* **2002**, *76*, 8979–8988. [CrossRef] [PubMed]

36. Tang, X.Y.; Gao, J.S.; Yuan, F.; Zhang, W.X.; Shao, Y.J.; Sakurai, F.; Li, Z. Effects of Sophy β-glucan on growth performance, carcass traits, meat composition, and immunological responses of Peking ducks. *Poult. Sci.* **2011**, *90*, 737–745. [CrossRef] [PubMed]

37. Hoffmannfezer, G.; Rodt, H.; Götze, D.; Thierfelder, S. Anatomical distribution of T and B lymphocytes identified by immunohistochemistry in the chicken spleen. *Int. Arch. Allergy Appl. Immunol.* **1977**, *55*, 86–95. [CrossRef]

38. Fair, J.M.; Taylor-McCabe, K.J.; Shou, Y.; Marrone, B.L. Immunophenotyping of chicken peripheral blood lymphocyte subpopulations: Individual variability and repeatability. *Vet. Immunol. Immunopathol.* **2008**, *125*, 268–273. [CrossRef] [PubMed]

39. Sun, H.L.; Miao, D.Y.; Gong, Y.M.; Zhang, P.J. Effect of three kinds of plant amylose on the immunity with infectious bursal disease live vccine in chickens. *Hua Bei Nong Xue Bao* **2006**, *21*, 106–109.

40. Kuang, L.; Zhan-Qiang, S.U.; Lei, C.H.; Long, L.I.; Shi-Wei, L.I.; Jian, L.I.; Ning-Ning, S.I.; Zhang, Y.H. Study on the immune-promoting effects of Astragalus polysaccharide and epimedium polysaccharide on living vaccine against IBD in chicks. *J. Tradit. Chin. Vet. Med.* **2013**, *2*, 009.

41. Gai-Ru, L.U.; Wei, S.P.; Yao-Di, N.I.; Liu, R.X. Study on active components in Echinacea-Astragalus on relieving immune suppression of IBD chickens. *Dongwu Yixue Jinzhan* **2010**, *5*, 002.

42. Cobb, B.A.; Wang, Q.; Tzianabos, A.O.; Kasper, D.L. Polysaccharide processing and presentation by the mhcii pathway. *Cell* **2004**, *117*, 677–687. [CrossRef] [PubMed]

43. Bohn, J.A.; Bemiller, J.N. (1→3)-β-D-glucans as biological response modifiers: A review of structure-functional activity relationships. *Carbohydr. Polym.* **1995**, *28*, 3–14. [CrossRef]

44. Hashimoto, T.; Ohno, N.; Adachi, Y.; Yadomae, T. Enhanced production of inducible nitric oxide synthase by beta-glucans in mice. *FEMS Immunol. Med. Microbiol.* **1997**, *19*, 131–135. [CrossRef]

45. Chen, J.; Seviour, R. Medicinal importance of fungal beta-(1→3), (1→6)-glucans. *Mycol. Res.* **2007**, *111*, 635–652. [CrossRef] [PubMed]

46. Song, L.; Chen, X.; Liu, X.; Zhang, F.; Hu, L.; Yue, Y.; Li, K.; Li, P. Characterization and comparison of the structural features, immune-modulatory and anti-avian influenza virus activities conferred by three algal sulfated polysaccharides. *Mar. Drugs* **2016**, *14*, 4. [CrossRef] [PubMed]

47. Mosmann, T. Rapid colorimetric assay for cellular growth and survival: Application to proliferation and cytotoxicity assays. *J. Immunol. Methods* **1983**, *65*, 55–63. [CrossRef]

48. Withers, D.R.; Young, J.R.; Davison, T.F. Infectious bursal disease virus-induced immunosuppression in the chick is associated with the presence of undifferentiated follicles in the recovering bursa. *Viral Immunol.* **2005**, *18*, 127–137. [CrossRef] [PubMed]

49. Liu, H.; Zhang, M.; Han, H.; Yuan, J.; Li, Z. Comparison of the expression of cytokine genes in the bursal tissues of the chickens following challenge with infectious bursal disease viruses of varying virulence. *Virol. J.* **2010**, *7*, 364. [CrossRef] [PubMed]
50. Shao, Q.; Xu, W.; Yan, L.; Liu, J.; Rui, L.; Xiao, X.; Yu, X.; Lu, Y.; Li, Z. Function of duck RIG-I in induction of antiviral response against IBDV and avian influenza virus on chicken cells. *Virus Res.* **2014**, *191*, 184–191. [CrossRef] [PubMed]
51. Reed, L.J.; Muench, H. A simple method of estimating fifty per cent endpoints. *Am. J. Epidemiol.* **1938**, *27*, 493–497.
52. Hida, S.; Miura, N.N.; Adachi, Y.; Ohno, N. Effect of candida albicans cell wall glucan as adjuvant for induction of autoimmune arthritis in mice. *J. Autoimmun.* **2005**, *25*, 93–101. [CrossRef] [PubMed]

marine drugs

MDPI

Article

Structural Analysis of a Heteropolysaccharide from *Saccharina japonica* by Electrospray Mass Spectrometry in Tandem with Collision-Induced Dissociation Tandem Mass Spectrometry (ESI-CID-MS/MS)

Weihua Jin [1,2], Jing Wang [1,3], Sumei Ren [4], Ni Song [4] and Quanbin Zhang [1,*]

[1]   Institute of Oceanology, Chinese Academy of Science, Qingdao 266071, China;
      jwh.054130305@yahoo.com.cn (W.J.); jingwang@qdio.ac.cn (J.W.)
[2]   University of Chinese Academy of Sciences, Beijing 100049, China
[3]   Nantong Marine Science and Technology R & D Center, IOCAS, Jiangsu 226006, China
[4]   College of Medicine and Pharmaceutics, Ocean University of China, Qingdao 266003, China;
      rensumei@ouc.edu.cn (S.R.); nisong1975@ouc.edu.cn (N.S.)
*   Author to whom correspondence should be addressed; qbzhang@ms.qdio.ac.cn;
      Tel.: +86-532-82898703; Fax: +86-532-82898703.

Received: 17 August 2012; in revised form: 10 September 2012; Accepted: 13 September 2012; Published: 25 September 2012

Abstract: A fucoidan extracted from *Saccharina japonica* was fractionated by anion exchange chromatography. The most complex fraction F0.5 was degraded by dilute sulphuric acid and then separated by use of an activated carbon column. Fraction Y1 was fractionated by anion exchange and gel filtration chromatography while Fraction Y2 was fractionated by gel filtration chromatography. The fractions were determined by ESI-MS and analyzed by ESI-CID-MS/MS. It was concluded that F0.5 had a backbone of alternating 4-linked GlcA and 2-linked Man with the first Man residue from the nonreducing end accidentally sulfated at C6. In addition, F0.5 had a 3-linked glucuronan, in accordance with a previous report by NMR. Some other structural characteristics included GlcA 1→3 Man 1→4 GlcA, Man 1→3 GlcA 1→4 GlcA, Fuc 1→4 GlcA and Fuc 1→3 Fuc. Finally, it was shown that fucose was sulfated at C2 or C4 while galactose was sulfated at C2, C4 or C6.

Keywords: ESI-CID-MS/MS; fucoidan; oligosaccharide; *Saccharina japonica*

1. Introduction

Fucoidan, which is a family of sulfated heteropolysaccharides extracted from brown algae and invertebrates, has been increasingly studied because of its biological activity, including antitumor activity [1,2], protective effects against γ-radiation-induced blood cell damage [3], antiangiogenic activity [4], immunodulating activity [5], inhibition of colony formation in human melanoma and colon cancer cells [6] and the most studied anticoagulant activity [7–11]. However, the relationships between fucoidan structure and biological activity are poorly understood due to the complexity of fucoidan structure. Research on structural features has traditionally been conducted using chemical methods, such as methylation analysis and NMR [7,9,12]; however, structural characteristics are still largely unknown. It has been reported that fucoidan contains two different main backbones, one made up of (1→3)-linked fucopyranose residues with fucose branches [12] and another with repeating (1→3) and (1→4) glycosidic bonds [13]. It was also demonstrated that the positions of the sulfate groups and of the branching units, including fucose residues, sulfated fucose residues, galactose residues and glucuronic acid residues, were different. These results are similar to previous studies [14,15].

Sulfated and fucosylated glucuronomannan was also found in *Kjellmaniella crassifolia* and investigated by Sakai *et al.* [16]. Later, it was reported [17] that fucoidan extracted from *Hizikia fusiforme* was composed of →2) α-D-Man (1→ and →4) β-D-GlcA (1→ alternatively while slight →4) β-D-Gal (1→ was mixed in the main chain. The branched points were at C-3 of →2) α-D-Man (1→, C-2 of →4) β-D-Gal (1→ and C-2 of →6) β-D-Gal (1→. About 2/3 of fucose were at the nonreducing ends, and left of them were 1→4, 1→3 and 1→2 glycosidic linkages. About 2/3 of xylose were at the nonreducing ends, and left of them were 1→4 glycosidic linkage. It was sulfated at C-6 of →2,3) α-D-Man (1→, C-4 and C-6 of →2) α-D-Man (1→, C-3 of →6) β-D-Gal (1→, C-2, C-3 or C-4 of fucose, while some fucose had two sulfate groups. In addition, Bilan *et al.* [12] also found the existence of glucuronomannan in *Saccharina latissima*.

However, mass spectrometry (MS) applied to the structural analysis of saccharides has allowed the determination of minor structural components. Along with NMR spectroscopy, electrospray mass spectrometry (ESI) and Matrix-Assisted Laser Desorption/Ionization Time of Flight mass spectrometry (MALDI-TOF) have become primary analytical tools, providing high sensitivity and selectivity. In addition, electrospray mass spectrometry in tandem with collision-induced dissociation tandem mass spectrometry (ESI-CID-MS/MS) has yielded information on the type of linkages, sulfation, and the backbone sequences of saccharides. Through electrospray ion trap mass spectrometry and capillary electrophoresis, it has been possible to differentiate the three isomers 2-O-, 3-O- and 4-O-sulfated fucose [18]. Finally, oligosaccharide mixtures derived from sulfated carrageenan-derived oligosaccharides [19] and fucoidan of *Ascophyllum nodosum* [20], *Fucus evanescens* [13] and *Laminaria cichorioides* [21] have been analyzed by ESI-CID-MS/MS to elucidate the structural features of oligosaccharides. Recently, it was reported [22] that a fucoidan isolated from Hizikia fusiforme was analyzed by partial acid hydrolysis followed by characterization of the oligosaccharide fragments using ESI-CID-MS/MS.

This study is also dedicated to structural characteristics of fucoidan extracted from *Saccharina japonica* using NMR, ESI-MS and ESI-MS/MS.

2. Results and Discussion

2.1. Preparation of Fucoidan and Oligosaccharides

In this paper, we separate three fractions. With respect to monosaccharide and sulfate content (Table 1), the first fraction (F0.5) contained a large amount of uronic acid and a small amount of sulfate while the subsequent two fractions (F1 and F2) contained trace amounts of uronic acid and large amounts of sulfate. In other words, F0.5 consisted mainly of sulfated fucomannoglucuronan whereas F1 and F2 consisted mainly of sulfated galactofucan and sulfated fucan. To further study its structure, F0.5 was degraded with 4% sulphuric acid, and the resulting mixture separated by 95% ethanol precipitation and ran through the activated carbon column to yield two main fractions, Y1 and Y2. Y1 was fractioned by anion exchange chromatography to produce three fractions: YF (water), YD (0.05 M NaCl) and YT (0.1 M NaCl). Fraction YD and YT were purified by gel filtration chromatography with an elution of water (Supplementary data). YD had two peaks named YD-1 and YD-2 while YT showed a symmetric peak. In addition, Y2 was presumed to be a mixture of neutral oligosaccharides, it was immediately fractionated by gel filtration chromatography to yield six fractions: G1–G6 (Figure 1). The molecular weight and degree of polymerization (DP) of all fractions were determined by negative-ion ESI-MS. The apparent structural composition of each fraction is summarized in Table 2.

72

Table 1. Chemical composition (%, dry weight) of the degraded polysaccharides.

| Sample | Fuc (%) | U A (%) | SO4 (%) | Monosaccharides (molar ratio) | | | | | | | Mw |
				Fuc	Gal	Man	Glc A	Rha	Xyl	Glc	
F0.5	13.77	20.34	29.07	1	0.98	0.80	0.95	0.12	0.30	0.40	5954
F1	54.84	7.3	32.26	1	0.36	0.13	0.10	0.02	0.04	0.10	8436
F2	35.04	0.71	53.40	1	0.07	0.03	0.01	0	0	0.02	12586
YF	-	-	-	1	0.43	0.32	0.28	0.41	0	0.26	-
YD-1	-	-	-	0	1.10	11.14	12.59	0.17	0	1.08	-
YD-2	-	-	-	1	0.28	3.53	1.96	0	0	0.29	-
YT	-	-	-	0	0	8.21	9.30	0	0	1.48	-
G5	-	-	-	1	0.11	0.02	0	0	0	0.07	-
G6	-	-	-	1	0.04	0	0	0	0	0.05	-

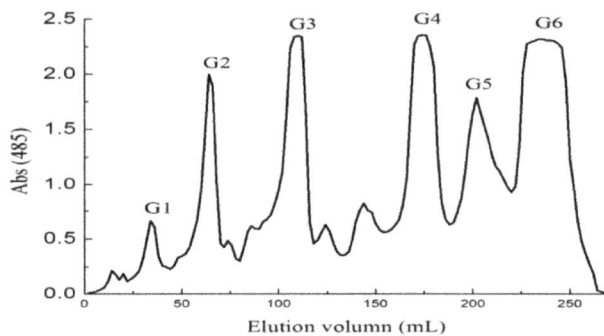

Figure 1. Gel filtration chromatography of Y2 oligosaccharide on a Bio-Gel P-4 Gel column.

Table 2. Structural compositions of all fractions separated from degraded F0.5.

Samples	DP	Ions (charges)	*m/z*	Predicted structural compositions
YF	1	193.035(−1)	193.035	GlcA
	2	355.086(−1)	355.086	GlcAMan
	2	369.065(−1)	369.065	GlcA$_2$
	2	339.091(−1)	339.091	GlcAFuc
	3	531.117(−1)	531.117	GlcA$_2$Man
	4	346.081(−2) 693.170(−1)	693.170	GlcA$_2$Man$_2$
	6	515.123(−2)	1031.249	GlcA$_3$Man$_3$

Table 2. *Cont.*

	Samples	DP	Ions (charges)	*m/z*	Predicted structural compositions
YD	YD-1	2	184.032(−2) 369.065(−1)	369.065	$GlcA_2$
		2	217.018(−2)	435.030	$GlcAManSO_3H$
		3	265.055(−2) 531.117(−1)	531.117	$GlcA_2Man$
		4	346.081(−2) 693.170(−1)	693.170	$GlcA_2Man_2$
		4	386.059(−2)	773.117	$GlcA_2Man_2SO_3H$
		6	515.123(−2)	1031.249	$GlcA_3Man_3$
	YD-2	1	193.035(−1)	193.035	GlcA
		1	243.017(−1)	243.017	$FucSO_3H$
		1	259.012(−1)	259.012	$GalSO_3H$
		2	339.231(−1)	339.231	GlcAFuc
		2	355.086(−1)	355.086	GlcAMan
		2	217.072(−2)	435.033	$GlcAManSO_3H$
YT		2	217.072(−2)	435.033	$GlcAManSO_3H$
		3	257.037(−3)	545.096	$GlcA_3$
		3	305.033(−2)	611.068	$GlcA_2ManSO_3H$
		4	386.059(−2)	773.118	$GlcA_2Man_2SO_3H$
		5	474.073(−2)	949.147	$GlcA_3Man_2SO_3H$
		6	369.731(−3) 555.101(−2)	1111.201	$GlcA_3Man_3SO_3H$
G1		8	1369.333(−1)	1369.333	$GlcA_4Man_4$
G2		6	343.081(−3) 515.125(−2) 1031.249(−1)	1031.249	$GlcA_3Man_3$
G3		4	346.082(−2) 693.173(−1)	693.173	$GlcA_2Man_2$
G4		2	355.086(−1)	355.086	GlcAMan

2.2. Analysis of the Oligosaccharides of All Fractions by ESI-MS

YF was obtained by anion exchange chromatography on a DEAE-Bio Gel Agarose FF gel with elution by water. Thus YF was a mixture of neutral oligosaccharides, which was approved by the results in Table 2 and Figure 2a.

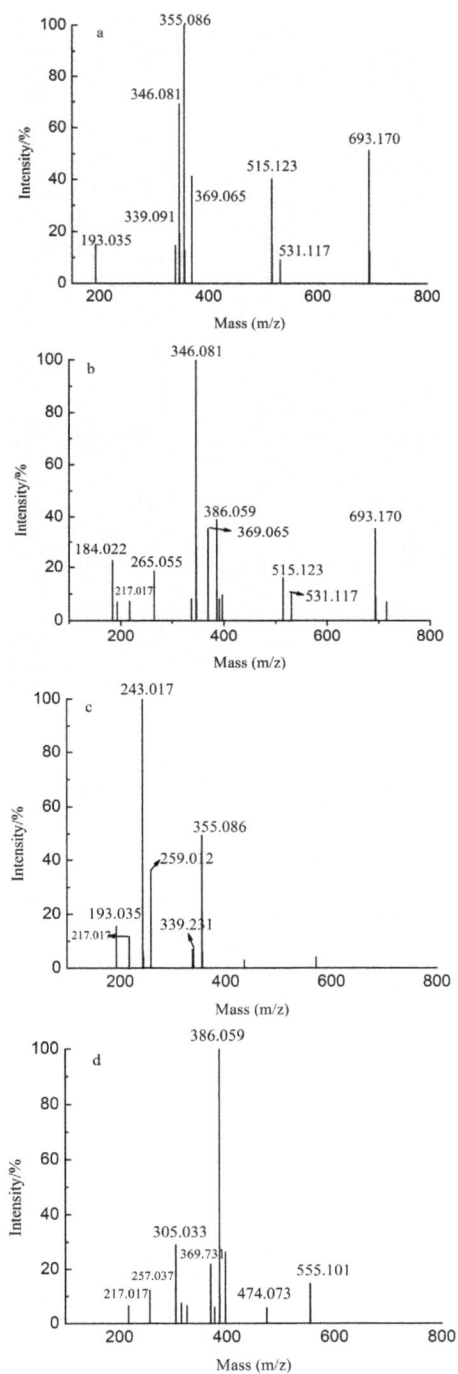

Figure 2. Negative-ion mode electrospray mass spectrometry (ESI-MS) spectra of YF (**a**), YD-1 (**b**), YD-2 (**c**) and YT (**d**).

In the ESI-MS spectrum of YD-1 (Figure 2b), there were six apparent structural compositions (GlcA$_2$, GlcAManSO$_3$H, GlcA$_2$Man, GlcA$_2$Man$_2$, GlcA$_2$Man$_2$SO$_3$H and GlcA$_3$Man$_3$) corresponding to the ions at m/z 184.032 (-2)/369.065 (-1), 217.017 (-2), 265.055 (-2)/531.117 (-1), 346.081 (-2)/693.170 (-1), 386.059 (-2) and 515.123 (-2), respectively. Six ions at m/z 193.035, 243.017, 259.012, 339.231, 355.086 and 435.033 were detected in the ESI-MS spectrum of YD-2 (Figure 2c), suggesting the presence of GlcA, FucSO$_3$H, GalSO$_3$H, GlcAFuc, GlcAMan and GlcAManSO$_3$H, respectively. Inference of structural composition was based on the analysis of the molar ratio of monosaccharide (Table 1). Interestingly, the main fraction of YD-2 was FucSO$_3$H while the main fraction of YD-1 was GlcA$_2$Man$_2$; these data might suggest that the eluting power of 0.05 M NaCl (Fraction YD) was equivalent to the substances containing one sulfate group (such as FucSO$_3$H) or two uronic acid residues (such as GlcA$_2$Man$_2$).

Our data further suggested that the eluting power of 0.1 M NaCl (Fraction YT) was equivalent to the substances consisting of two sulfate groups (such as Fuc(SO$_3$H)$_2$), one sulfate group and two uronic acid residues (GlcA$_2$ManSO$_3$H) or four uronic acid residues (GlcA$_4$). In the ESI-MS spectrum of fraction YT, the main ion at m/z 386.059 (-2) (Figure 2d) was identified as GlcA$_2$Man$_2$SO$_3$H while the less intensive ions detected at m/z 217.017 (-2), 257.037 (-3), 305.033 (-2), 386.059 (-2), 474.073 (-2), 369.731 (-3)/555.101 (-2) corresponded to GlcAManSO$_3$H, GlcA$_3$, GlcA$_2$ManSO$_3$H, GlcA$_2$Man$_2$SO$_3$H, GlcA$_3$Man$_2$SO$_3$H, and GlcA$_3$Man$_3$SO$_3$H, respectively. These results helped us to confirm the above speculation. The absorptive capacity of the DEAE-Bio Gel Agarose FF gel permitted detection of ions at m/z 386.059 (-2) (Figure 2b,d).

Y2 was fractionated by gel filtration chromatography to obtain six fractions (G1–G6). The results (Figure 3) of HPLC analysis confirmed that G1, G2, G3 and G4 were relatively pure. Fractions G1, G2, G3 and G4 were identified by NMR (supplementary data) as a series of oligosaccharides consisted of a repeating disaccharide unit of GlcA and Man. The degrees of polymerization (DP) of each fraction were 8, 6, 4 and 2, respectively (Figure 3). The results suggest that the bonds between mannose and glucuronic acid were stable against the acid condition and that F0.5 had a backbone of –[Man-GlcA]$_n$–. Owing to the separation of gel filtration chromatography, it was indicated that Fraction G5 consisted of disaccharides and monosaccharides while G6 consisted mainly of fucose and a smaller quantity of galactose and glucose (Table 1). To further study the exact structural features of F0.5, some samples were subjected to ESI-CID-MS/MS.

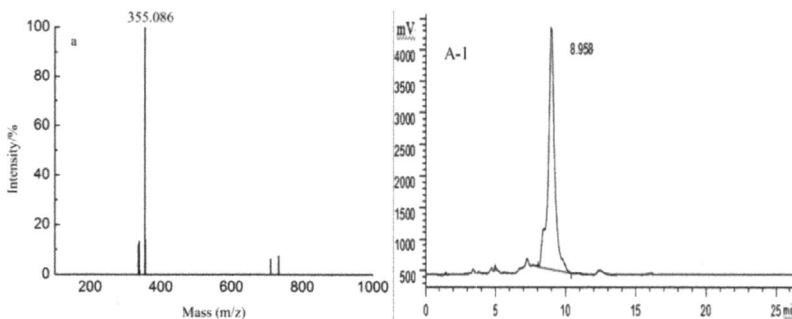

Figure 3. *Cont.*

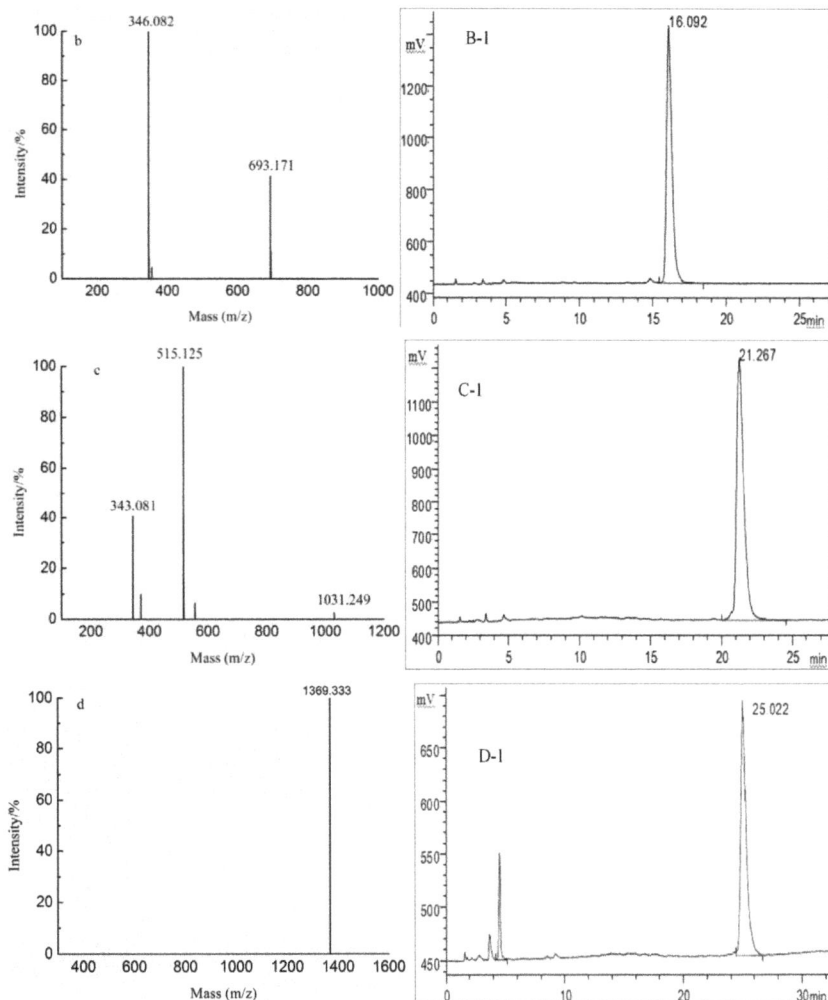

Figure 3. Negative-ion mode ESI-MS spectra and HPLC spectra of G1 (**d,D-1**), G2 (**c,C-1**), G3 (**b,B-1**) and G4 (**a,A-1**).

2.3. Analysis of the Structural Features of the Fractions by ESI-CID-MS/MS

The ^{0,2}X (m/z 138.970) and ^{0,2}A (m/z 182.996) ions indicated sulfation of Fuc at C2 or C4 in the ESI-CID-MS/MS spectrum of the ion at m/z 243.017 (not shown), whereas the ^{0,2}A (m/z 198.991), ^{2,5}A (m/z 180.980) and ^{0,3}A (m/z 168.981) ions suggested sulfation of Gal at C4 or at C6 in the ESI-CID-MS/MS spectrum of the ion at m/z 259.011 (not shown).

In fraction G5, the ion at m/z 309.112 (Figure 4), corresponding to [Fuc$_2$ − H]$^-$ (supplementary data), was detected. The ions at m/z 145.048 and 163.058 were assigned as B$_1$/Z$_1$ and C$_1$/Y$_1$, respectively. The low intensity fragmentation ion at m/z 249.093 was ^{0,2}A$_2$, suggesting the prevalence of 1→3 linkage between the two Fuc residues and minor presence of 1→4 linkage. This has been confirmed by previous study [15]. Therefore, the ion (m/z 309.112), corresponded to disaccharide, was major Fuc 1→3 Fuc and minor Fuc 1→4 Fuc.

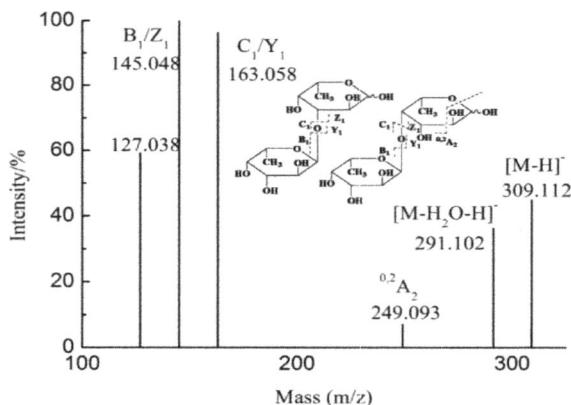

Figure 4. Negative-ion mode electrospray mass spectrometry in tandem with collision-induced dissociation tandem mass spectrometry (ESI-CID-MS/MS) spectrum of the ion at m/z 309.112.

The ion detected at m/z 339.230 (Figure 5) confirmed that the oligosaccharides with the GlcA residue were more stable in the acid condition. The C_1 (m/z 163.112) and Y_1 (193.034) ions corresponded to Fuc and GlcA residues, respectively. The characteristic ion at m/z 235.044, assigned to the $^{0,2}X_1$ ion, suggested that the reducing terminal was a GlcA residue. Therefore, the ion at m/z 339.230 was Fuc-GlcA. In addition, the $^{2,5}A_2$ (m/z 261.060) and $^{0,2}A_2$ (m/z 279.070) ions indicated that the linkage between the Fuc and GlcA residues was a 1→4 linkage. Moreover, we could also speculate that the ions at m/z 163.112 (Y_1), 175.024 (B_1), 193.034 (C_1) and 235.044 ($^{0,2}A_2$) suggested that the structural feature of the ion at m/z 339.230 was GlcA 1→2 Fuc.

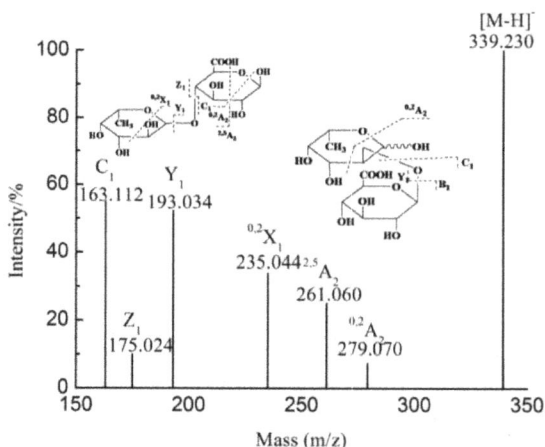

Figure 5. Negative-ion mode ESI-CID-MS/MS spectrum of the ion at m/z 339.230.

The ions at m/z 193.035, 184.032 (−2)/369.065 (−1) and 257.037 (−3) corresponded to GlcA, GlcA$_2$ and GlcA$_3$, suggesting the presence of glucuronan. To elucidate the fine structure of glucuronan, the ion at m/z 545.096 was analyzed by ESI-CID-MS/MS (Figure 6a). Four series of B/C-type and

Y/Z-type ions suggested that glucuronan was a trisaccharide with GlcA-GlcA-GlcA. The low intensity characteristic ions at m/z 309.043 and 485.075, corresponding to $^{0,2}A_2$ and $^{0,2}A_3$ ions, respectively, indicated the prevalence of a 3-linked GlcA. These data suggest a backbone of 1→3 linked glucuronan.

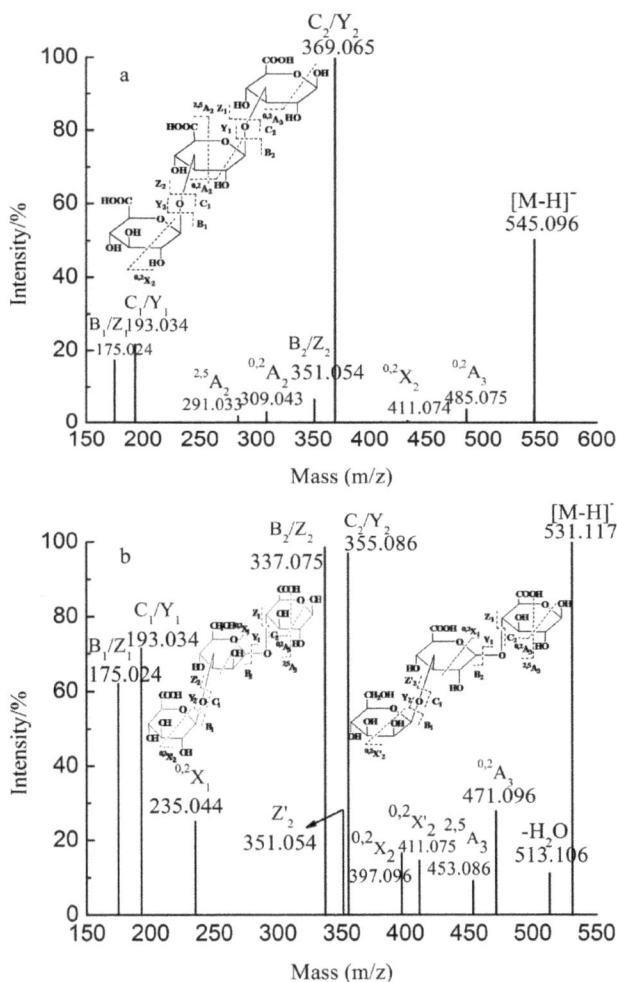

Figure 6. Negative-ion mode ESI-CID-MS/MS spectra of the ions at m/z 545.096 (**a**) and 531.117 (**b**).

The analyses of glucuronomannan and sulfated glucuronomannan suggest that the linkage of 1→4 was more sensitive in the acid condition than that of linkage of 1→2. However, we also detected the less intensive ion at m/z 531.117 (Figure 6b), assigned as [GlcA$_2$Man − H]$^−$. The ions at m/z 175.024 and 193.034 are indicative of dehydrated GlcA and GlcA residues from the nonreducing or reducing end, corresponding to the B_1/C_1-type or Z_1/Y_1-type ions, respectively. The $^{0,2}X_1$-type ion suggests that the reducing end consisted of a GlcA residue. However, we found two $^{0,2}X_2$-type ions at m/z 397.096 and 411.075, named $^{0,2}X_2$ and $^{0,2}X_2$, respectively, which suggests two structural sequences of trisaccharide. As the $^{0,2}X_1$-type ion suggests a reducing end of GlcA residue, we conclude that one sequence was Man-GlcA-GlcA and the other was GlcA-Man-GlcA. The identity of the former was confirmed by the presence of the ion at *m/z* 351, corresponding to a Z_2-type ion. The ions at *m/z* 337.075 and 355.086 were identified as B_2/Z_2-type and C_2/Y_2-type ions, respectively. The characteristic ions at *m/z* 453.086 and 471.096 corresponded to $^{2,5}A_3$-type and $^{0,2}A_3$-type ions, respectively, suggesting that the two residues from reducing terminal had a 1→4 glycosidic bond. $^{0,2}A_2$-type ions were not found, indicating that the two residues from the nonreducing terminal had a 1→3 glycosidic bond. In summary, the two structural sequences of trisaccharide were Man 1→3 GlcA 1→4 GlcA and GlcA 1→3 Man 1→4 GlcA.

The fragmentation pattern of ions at *m/z* 355.086, 346.082 (−2) /693.173 (−1) and 1369.333 (not shown) were similar to that of the ion at *m/z* 343.081 (−3)/ 515.125 (−2)/ 1031.249 (−1) (Figure 7a). The fragment ion at *m/z* 1031 corresponded to [GlcA$_4$Man$_4$ − H]$^-$. The ions at *m/z* 175.024 and 193.249 were the characteristic fragments, corresponding to fragment B_1 ion and C_1 ion, respectively. The results suggest that the nonreducing terminal was a GlcA residue. Two series of B-type and C-type ions from the nonreducing end were found with Y-type and Z-type ions from the reducing end; this pattern confirmed a linear backbone of alternating GlcA and Man. In addition, the characteristic ions of $^{0,2}A$-type were detected. However, the ions at *m/z* 235 and 573 do not appear in (Figure 7a) because of their lower intensity. The $^{0,2}A$-type [23] ions were the cross-ring cleavage by releasing the C1-C2 portion of the reducing-ring, leading to the loss of $C_4H_8O_4$ at *m/z* 120. However, no characteristic ion of $^{0,2}A$-trpe was detected. And there was no information obtained regarding the linkage of GlcA residue in the spectra of ESI-CID-MS/MS. The linkage of GlcA residue was therefore determined by NMR (supplementary data) and confirmed the results of ESI-CID-MS/MS. Finally, it was concluded that F0.5 had a backbone of repeating 4-linked GlcA and 2-linked Man, in accordance with Bilan *et al.* [12] and Sakai *et al.* [16].

Bilan *et al.* [12] also reported the presence of sulfated glucuronomannan. YT was primarily composed of a sulfated tetrasaccharide. The ion at *m/z* 555.100 (−2) corresponded to [GlcA$_3$Man$_3$SO$_3$H − H]$^-$ (Figure 7b). The ions at *m/z* 175.024 and 193.034 were identified as dehydrated GlcA and GlcA residues, suggesting that the nonreducing end was GlcA residue. The reducing terminal was identified as Man based on the presence of ions at *m/z* 931.131/465.069 (−2) and 474.072 (−2), assigned as B_5 and C_5, respectively. A series of B-type ions at *m/z* 175.024 (B_1), 417.022 (B_2), 377.045 (−2) (B_4), 931.131 (B_5) along with C-type ions at *m/z* 193.034 (C_1), 773.111 (C_4) indicated that the linear sequence was GlcA-ManSO$_3$H-GlcA-Man-GlcA-Man. The ion at *m/z* 259.011 (ManSO$_3$H) was also confirmed in the above analysis. The ions at *m/z* 513.102 (B'_3), 693.161 (C'_4) and 851.173 (B'_5) suggested that the sulfate group was lost whereas the ions at *m/z* 407.055 [($^{0,2}A_6$-GlcA)$^{2-}$] and 597.077 (C_4-GlcA) suggested that the GlcA residue from the nonreducing end was sensitive in the negative mode. The characteristic ion at *m/z* 495.079 (−2) was detected, suggesting that the glycosidic bond between the GlcA and Man residues was 1→2 linkage. However, the ion at *m/z* 326 (−2) did not appear due to its low intensity. Although the linkage of the Man residue was identified as a 2-linked Man, the linkage of the GlcA residue was not determined by ESI-CID-MS/MS. With the assistances of NMR (supplementary data) and the above results, the linkage of the GlcA residue was determined to be a 4-linked GlcA and the substitution of sulfation of Man was at C6.

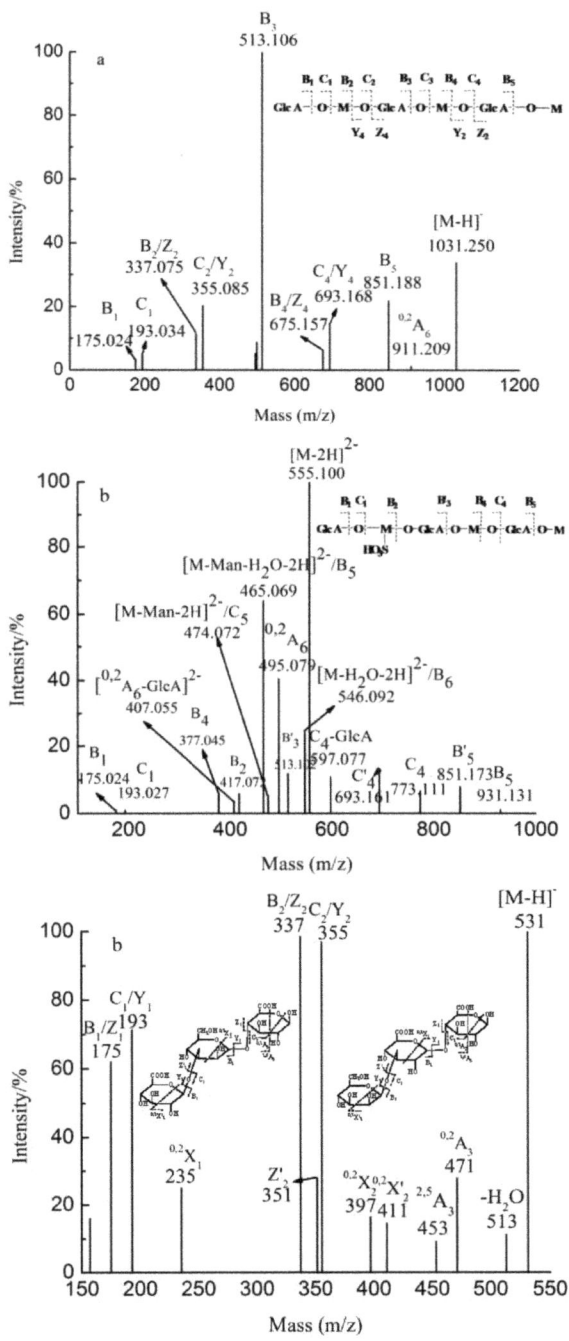

Figure 7. Negative-ion mode ESI-CID-MS/MS spectra of the ions at *m/z* 1031.250 (**a**) and 555.100 (−2) (**b**).

3. Experimental Section

3.1. Materials

The brown algae *S. japonica* was obtained from Shazikou, Qingdao, China in May 2011. Seven standard monosaccharides (L-fucose (Fuc), D-galactose (Gal), D-mannose (Man/M), D-glucuronic acid (GlcA), L-rhamnose monohydrate (Rha), D-xylose (Xyl) and D-glucose (Glc)) were purchased from Sigma-Aldrich. 3-Methyl-1-phenyl-2-pyrazolin-5-one (99%) was purchased from Aldrich chemistry.

3.2. Preparation and Purification of Fucoidans

Crude fucoidan was extracted following Wang *et al.* [14] and degraded using hydrogen dioxide and ascorbic acid to obtain a fucoidan of low molecular weight. Briefly speaking, crude fucoidan (1 g) was dissolved in water (100 mL). Then ascorbic acid (0.5 g) and hydrogen dioxide (0.3 mL) were added and the solution was stirred for 1 h at room temperature. Then, it was ultrafiltrated, concentrated and precipitated by 72% ethanol before further precipitation with 95% ethanol. Finally, the degraded polysaccharides were dried by using an infrared lamp. The degraded polysaccharides (60 g) underwent anion exchange chromatography on a DEAE-Bio Gel Agarose FF gel (12 cm × 70 cm) with elution by 0.5 M (35L) (F0.5), 1 M (35L) (F1) and 2 M NaCl (35L) (F2). The polysaccharides were then ultrafiltrated, concentrated and precipitated by 72% ethanol before further precipitation with 95% ethanol. Finally, the F0.5, F1 and F2 factions were dried by using an infrared lamp.

3.3. Preparation and Purification of Oligosaccharides from F0.5

Polysaccharide F0.5 was dissolved with reflux in 4% sulphuric acid (60 mg mL^{-1}) for 5 h and then neutralized with barium hydroxide after cooling to room temperature. The solution was centrifugated and the supernate was concentrated. The concentrated solution was fractionated using the activated carbon column (2.6 cm × 30 cm) with a gradient elution from 50% ethanol to 95% ethanol. Then, eluent (Y1) was concentrated and precipitated in 95% ethanol. The activated carbon column was washed with a gradient elution from 50% ethanol to 95% ethanol. The elution (Y2) was combined, concentrated and freeze-dried. Y1 (1.0 g) was separated by anion exchange chromatography on a DEAE-Bio Gel Agarose FF gel (2.6 cm × 30 cm) with elution by water (YF), 0.05 M NaCl (YD) and 0.1 M NaCl (YT). The concentrated solutions of fractions YD and YT were desalted and separated on a Sephadex G-10 column (2.6 cm × 100 cm) with an elution of water. Y2 (0.5 g) was separated on a Bio-Gel P-4 Gel (Extra Fine, <45 μM) column (2.6 cm × 100 cm) (Figure 1) eluted with 0.5 M NH$_4$HCO$_3$ at a flow rate 0.14 mL min^{-1}. It was collected every 14 min per pipe after 27 h. Six fractions were collected and lyophilized.

3.4. Composition Analysis

The sulfated content was determined by ion chromatography on Shodex IC SI-52 4E column (4.0 × 250 mm) eluted with 3.6 mM Na$_2$CO$_3$ at a flow rate of 0.8 mL min^{-1} at 45 °C. The molar ratio of monosaccharide composition and the content of fucose were determined following Zhang *et al.* [24]. Briefly speaking, a solution of sample (10 mg mL^{-1}) was hydrolyzed in 2 M trifluoroacetic acid in a 10 mL ampoule. The ampoule was sealed in a nitrogen atmosphere and hydrolyzed for 4 h at 110 °C. Then the hydrolyzed mixture was neutralized to pH 7 with sodium hydroxide. Later the mixture was converted into its 1-phenyl-3-methyl-5-pyrazolone derivatives and separated by HPLC chromatography. Uronic acid was analyzed by a modified carbazole method [25]. Molecular weight was determined by GPC-HPLC on TSK gel PWxl 3000 column (7 μm 7.8 × 300 mm) eluted with 0.2 M Na$_2$SO$_4$ at a flow rate of 0.5 mL min^{-1} at 30 °C.

3.5. MS Analysis of Oligosaccharides

ESI-MS was performed on a Micromass Q-Tof Ultima instrument (Waters, Manchester, UK).

Samples were dissolved in CH_3CN-H_2O (1:1, v/v). Mass spectra were registered in the negative ion mode at a flow rate of 5 μL min^{-1}. The capillary voltage was set to -3000 V, and the cone voltage was set at -50 V. The source temperature was 80 °C, and the desolvation temperature was 150 °C. The collision energy was optimized between 10 and 50 eV. All spectra were analyzed by MassLynx software.

3.6. Condition of HPLC

The fractions were analyzed by HPLC with an ELSD detector, performed on a "click" maltose column (10 μm, 10 × 150 mm) at 1 mL min^{-1} in a gradient solution. Gradient 1: 0–30 min, water-acetonitrile-ammonium formate buffer (100 mM, pH 3.0): 10:80:10 (v/v/v)→40:50:10(v/v/v). Gradient 2: 30–35 min, water-acetonitrile-ammonium formate buffer (100 mM, pH 3.0): 40:50:10 (v/v/v)→75:15:10(v/v/v).

4. Conclusions

To further determine the precise structure of F0.5, it was degraded with partial acid hydrolysis (due to the absence of available enzymes). The depolymerized mixture was then separated by use of an activated carbon column. Y1 was purified using anion exchange and gel filtration chromatography while Y2 was purified using gel filtration chromatography. The results suggest that F0.5 consisted of two types of polysaccharides: (1) a glucuronomannan and a sulfated glucuronomannan with the same backbone of repeating 4-linked GlcA and 2-linked Man, and the latter with the first mannopyranose residue from the nonreducing terminus sulfated at C-6 and (2) a glucuronan with a backbone of 3-linked GlcA. There were also some other structural fragments, including GlcA 1→3 Man 1→4 GlcA, Man 1→3 GlcA 1→4 GlcA, Fuc 1→4 GlcA and Fuc 1→3 Fuc. Combined with the results obtained in this study and a previous study [15], it was concluded that it might provide the whole structural model of fucoidan.

Acknowledgments: This study was supported by the Ocean Public Welfare Scientific Research Project, the State Oceanic Administration of the People's Republic of China (No. 201005024), the Doctorial Fund of Shandong Province (No. BS2011YY066), the Science and Technology Project of Shandong Province (No. 2011GHY11529) and the Prospective Joint Research Projects of Jiangsu Province (No. BY2011189).

References

1. Sokolova, R.V.; Ermakova, S.P.; Awada, S.M.; Zvyagintseva, T.N.; Kanaan, H.M. Composition, structureal characteristics, and antitumor properties of polysaccharides from the brown algal *Dictyopteris polypodioides* and *Sargassum* sp. *Chem. Nat. Compd.* **2011**, *47*, 329–334. [CrossRef]
2. Synytsya, A.; Kim, W.-J.; Kim, S.-M.; Pohl, R.; Synytsya, A.; Kvasnička, F.; Čopíková, J.; Park, Y.I. Structure and antitumour activity of fucoidan isolated from sporophyll of Korean brown seaweed *Undaria pinnatifida*. *Carbohydr. Polym.* **2010**, *81*, 41–48. [CrossRef]
3. Rhee, K.; Lee, K. Protective effects of fucoidan against γ-radiation-induced damage of blood cells. *Arch.Pharm. Res.* **2011**, *34*, 645–651. [CrossRef]
4. Koyanagi, S.; Tanigawa, N.; Nakagawa, H.; Soeda, S.; Shimeno, H. Oversulfation of fucoidan enhances its anti-angiogenic and antitumor activity. *Biochem. Pharmacol.* **2003**, *65*, 173–179.
5. Caipang, C.M.A.; Lazado, C.C.; Berg, I.; Brinchmann, M.F.; Kiron, V. Influence of alginic acid and fucoidan on the immune responses of head kidney leukocytes in cod. *Fish Physiol. Biochem.* **2010**, *37*, 603–612.
6. Ermakova, S.; Sokolova, R.; Kim, S.-M.; Um, B.-H.; Isakov, V.; Zvyagintseva, T. Fucoidans from Brown Seaweeds *Sargassum hornery, Eclonia cava, Costaria costata*: Structural characteristics and anticancer activity. *Appl. Biochem. Biotechnol.* **2011**, *164*, 841–850. [CrossRef]
7. Bilan, M.I.; Grachev, A.A.; Ustuzhanina, N.E.; Shashkov, A.S.; Nifantiev, N.E.; Usov, A.I. Structure of a fucoidan from the brown seaweed *Fucus evanescens* C.Ag. *Carbohydr. Res.* **2002**, *337*, 719–730. [CrossRef]
8. Croci, D.O.; Cumashi, A.; Ushakova, N.A.; Preobrazhenskaya, M.E.; Piccoli, A.; Totani, L.; Ustyuzhanina, N.E.; Bilan, M.I.; Usov, A.I.; Grachev, A.A.; *et al.* Fucans, but not Fucomannoglucuronans, determine the biological activities of sulfated polysaccharides from *Laminaria saccharina* brown seaweed. *PLoS ONE* **2011**, *6*, e17283.

9. Duarte, M.E.R.; Cardoso, M.A.; Noseda, M.D.; Cerezo, A.S. Structural studies on fucoidans from the brown seaweed *Sargassum stenophyllum*. *Carbohydr. Res.* **2001**, *333*, 281–293. [CrossRef]

10. Mestechkina, N.M.; Shcherbukhin, V.D. Sulfated polysaccharides and their anticoagulant activity: A review. *Appl. Biochem. Microbiol.* **2010**, *46*, 267–273. [CrossRef]

11. Jiao, G.; Yu, G.; Zhang, J.; Ewart, H.S. Chemical structures and bioactivities of sulfated polysaccharides from marine algae. *Mar. Drugs* **2011**, *9*, 196–223. [CrossRef]

12. Bilan, M.I.; Grachev, A.A.; Shashkov, A.S.; Kelly, M.; Sanderson, C.J.; Nifantiev, N.E.; Usov, A.I. Further studies on the composition and structure of a fucoidan preparation from the brown alga *Saccharina latissima*. *Carbohydr. Res.* **2010**, *345*, 2038–2047. [CrossRef]

13. Anastyuk, S.D.; Shevchenko, N.M.; Nazarenko, E.L.; Dmitrenok, P.S.; Zvyagintseva, T.N. Structural analysis of a fucoidan from the brown alga *Fucus evanescens* by MALDI-TOF and tandem ESI mass spectrometry. *Carbohydr. Res.* **2009**, *344*, 779–787. [CrossRef]

14. Wang, J.; Zhang, Q.; Zhang, Z.; Li, Z. Antioxidant activity of sulfated polysaccharide fractions extracted from *Laminaria japonica*. *Int. J. Biologic. Macromol.* **2008**, *42*, 127–132. [CrossRef]

15. Wang, J.; Zhang, Q.; Zhang, Z.; Zhang, H.; Niu, X. Structural studies on a novel fucogalactan sulfate extracted from the brown seaweed *Laminaria japonica*. *Int. J. Biologic. Macromolec.* **2010**, *47*, 126–131. [CrossRef]

16. Sakai, T.; Kimura, H.; Kojima, K.; Shimanaka, K.; Ikai, K.; Kato, I. Marine bacterial sulfated fucoglucuronomannan (SFGM) lyase digests brown algal SFGM into trisaccharides. *Mar. Biotechnol.* **2003**, *5*, 70–78. [CrossRef]

17. Li, B.; Wei, X.-J.; Sun, J.-L.; Xu, S.-Y. Structural investigation of a fucoidan containing a fucose-free core from the brown seaweed, *Hizikia fusiforme*. *Carbohydr. Res.* **2006**, *341*, 1135–1146. [CrossRef]

18. Tissot, B.; Salpin, J.; Martinez, M.; Gaigeot, M.; Daniel, R. Differentiation of the fucoidan sulfated l-fucose isomers constituents by CE-ESIMS and molecular modeling. *Carbohydr. Res.* **2006**, *341*, 598–609. [CrossRef]

19. Yu, G.; Zhao, X.; Yang, B.; Ren, S.; Guan, H.; Zhang, Y.; Lawson, A.M.; Chai, W. Sequence determination of sulfated carrageenan-derived oligosaccharides by high-sensitivity negative-ion electrospray tandem mass spectrometry. *Anal. Chem.* **2006**, *78*, 8499–8505.

20. Daniel, R.; Chevolot, L.; Carrascal, M.; Tissot, B.; Mourão, P.A.S.; Abian, J. Electrospray ionization mass spectrometry of oligosaccharides derived from fucoidan of *Ascophyllum nodosum*. *Carbohydr. Res.* **2007**, *342*, 826–834. [CrossRef]

21. Anastyuk, S.D.; Shevchenko, N.M.; Nazarenko, E.L.; Imbs, T.I.; Gorbach, V.I.; Dmitrenok, P.S.; Zvyagintseva, T. Structural analysis of a highly sulfated fucan from the brown alga *Laminaria cichorioides* by tandem MALDI and ESI mass spectrometry. *Carbohydr. Res.* **2010**, *345*, 2206–2212. [CrossRef]

22. Wang, P.; Zhao, X.; Lv, Y.; Liu, Y.; Lang, Y.; Wu, J.; Liu, X.; Li, M.; Yu, G. Analysis of structural heterogeneity of fucoidan from *Hizikia fusiforme* by ES-CID-MS/MS. *Carbohydr. Polym.* **2012**, *90*, 602–607. [CrossRef]

23. Saad, O.M.; Leary, J.A. Delineating mechanisms of dissociation for isomeric heparin disaccharides using isotope labeling and ion trap tandem mass spectrometry. *J. Am. Soc. Mass Spectrom.* **2004**, *15*, 1274–1286. [CrossRef]

24. Zhang, J.J.; Zhang, Q.B.; Wang, J.; Shi, X.L.; Zhang, Z.S. Analysis of the monosaccharide composition of fucoidan by precolumn derivation HPLC. *Chin. J. Oceanol. Limnol.* **2009**, *27*, 578–582. [CrossRef]

25. Bitter, T.; Muir, H.M. A modified uronic acid carbazole reaction. *Anal. Biochem.* **1962**, *4*, 330–334. [CrossRef]

Samples Availability: Available from the authors.

marine drugs

MDPI

Article

Structural Analysis and Anti-Complement Activity of Polysaccharides from *Kjellmaniella crsaaifolia*

Wenjing Zhang [1,2], Weihua Jin [1], Delin Sun [3], Luyu Zhao [3], Jing Wang [1,4], Delin Duan [1] and Quanbin Zhang [1,*]

[1] Institute of Oceanology, Chinese Academy of Sciences, Qingdao 266071, China; wenjingwing@126.com (W.Z.); jinweihua@qdio.ac.cn (W.J.); jingwang@qdio.ac.cn (J.W.); dlduan@qdio.ac.cn (D.D.)
[2] College of Earth Science, University of Chinese Academy of Sciences, Beijing 100049, China
[3] Heze Juxinyuan Food Co. Ltd., Heze 274400, China; zhou56018@163.com (D.S.); luyuzhao519@163.com (L.Z.)
[4] Nantong Branch, Institute of Oceanology, Chinese Academy of Sciences, Nantong 226006, China
* Author to whom correspondence should be addressed; qbzhang@qdio.ac.cn; Tel./Fax: +86-532-8289-8703.

Academic Editor: Paola Laurienzo
Received: 15 January 2015; Accepted: 5 March 2015; Published: 16 March 2015

Abstract: Two polysaccharides, named KCA and KCW, were extracted from *Kjellmaniella crassifolia* using dilute hydrochloric acid and water, respectively. Composition analysis showed that these polysaccharides predominantly consisted of fucose, with galactose, mannose and glucuronic acid as minor components. After degradation and partial desulfation, electrospray ionization mass spectrometry (ESI-MS) was performed, which showed that the polysaccharides consisted of sulfated fucooligosaccharides, sulfated galactofucooligosaccharides and methyl glycosides of mono-sulfated/multi-sulfated fucooligosaccharides. The structures of the oligomeric fragments were further characterized by electrospray ionization collision-induced dissociation tandem mass spectrometry (ESI-CID-MS^2 and ESI-CID-MS^3). Moreover, the activity of KCA and KCW against the hemolytic activity of both the classical and alternative complement pathways was determined. The activity of KCA was found to be similar to KCW, suggesting that the method of extraction did not influence the activity. In addition, the degraded polysaccharides (DKCA and DKCW) displayed lower activity levels than the crude polysaccharides (KCA and KCW), indicating that molecular weight had an effect on activity. Moreover, the desulfated fractions (ds-DKCA and ds-DKCW) showed less or no activity, which confirmed that sulfate was important for activity. In conclusion, polysaccharides from *K. crassifolia* may be good candidates for the treatment of diseases involving the complement pathway.

Keywords: polysaccharide; *Kjellmaniella crassifolia*; anti-complement

1. Introduction

Kjellmaniella crassifolia is a brown alga that is widely distributed in the waters around the southern area of Hokkaido, Japan. Compared to other seaweeds, *K. crassifolia* contains high levels of polysaccharides (fucoidan) and the water extract shows very high viscoelasticity. The polysaccharide extract from *K. crassifolia* is an effective immunomodulator and a potent immune adjuvant [1,2]. The structural features of the polysaccharide (fucoidan, not alginate) from *K. crassifolia* have been previously analyzed by extracellular enzyme assays [3] and NMR [4]. The former reported that the novel polysaccharide was fucoglucuronomannan, with a backbone consisting of alternating 4-linked GlcA and 2-linked Man, and was branched at the C-3 position of mannose by a fucopyranose residue. The latter study suggested that the polysaccharide was sulfated fucan, with a backbone of 3-linked fucopyranose sulfated at C-2 and C-4. Because of its accuracy, sensitivity and selectivity, mass spectroscopy (MS) with electrospray ionization (ESI-MS) is an important tool for the analysis of polysaccharides; this technique has been used to successfully elucidate the structures of other heteropolysaccharides [5–7].

The complement pathway is an important part of the immune system, playing an essential role in host defense against pathogens. However, inappropriate activation of the complement pathway has been implicated in certain diseases, such as rheumatoid arthritis, Alzheimer's disease, ischemia-reperfusion injury and systemic lupus erythematosus [8]. Polysaccharides from alga were non-toxic. Numerous natural or semi-synthetic polyanions, such as derivatized dextran, chondroitin sulfate, dextral sulfate, fucoidan and heparin, have been reported to inhibit complement activation [9–14].

Recently, *K. crassifolia* was introduced into the Rongcheng coastal area in Shandong Province, China. Thus, in the present study, the structural features of the polysaccharides from *K. crassifolia* cultured in Rongcheng were elucidated by ESI-MS, ESI-CID-MS^2 and ESI-CID-MS^3, and the potential activity of the polysaccharides against the classical and alternative complement pathways were investigated.

2. Results and Discussion

2.1. Preparation of Polysaccharides

The molar ratios of monosaccharides, contents of uronic acid (UA), fucose (Fuc) and sulfate, and average molecular weights of all samples are shown in Table 1. KCA and KCW were found to contain high levels of Fuc together with a small amount of mannose (Man), glucuronic acid (GlcA) and galactose (Gal). With respect to the contents of sulfate and Fuc, KCA contained a higher amount than KCW. In addition, the molecular weight of KCW was higher than KCA, suggesting that KCA degraded slightly during the extraction process. Only trace contents of UA and protein were found. Interestingly, the molecular weight of DKCA was higher than DKCW, indicating that KCW was more sensitive to degradation. Desulfation resulted in a substantial decrease in the molecular weights for both ds-DKCW and ds-DKCA, indicating that the parent compounds were likely to be highly sulfated. In addition, the ratios of other monosaccharides, such as rhamnose (Rha), glucose (Glc) and xylose (Xyl), were increased.

Table 1. Chemical composition (%, dry weight) of KCA, KCW and their derived fractions.

Sample	Total Sugar	Fuc (%)	UA (%)	SO$_4$ (%)	Protein (%)	Monosaccharides (Molar Ratio)							Mw (kDa)
						Man	Rha	GlcA	Glc	Gal	Xyl	Fuc	
KCW	58.06	37.07	–	24.21	0.45	0.12	0.05	0.09	0.06	0.08	0.02	1	168.7
DKCW	53.56	30.75	–	22.52	0.74	0.12	0.10	0.12	0.06	0.06	–	1	5.2
ds-DKCW	82.42	38.79	3.72	8.93	1.22	0.23	0.11	0.19	0.16	0.21	0.20	1	3.5
KCA	53.40	40.43	–	35.49	–	0.05	0.02	0.06	0.02	0.03	0.02	1	153.7
DKCA	45.76	35.87	–	32.37	–	0.04	0.04	0.05	0.02	0.02	–	1	9.7
ds-DKCA	65.58	48.97	3.51	12.72	0.77	0.13	0.05	0.12	0.27	0.14	0.03	1	2.3

2.2. IR Analysis

The IR spectra (Figure 1) showed that KCA, DKCA, KCW and DKCW had the same infrared absorption properties, suggesting that they contained the same functional groups. The band at 1260 cm^{-1} corresponded to the S=O stretching vibration, and the band at approximately 845 cm^{-1} was assigned to the C–O–S vibration, suggesting that the presence of the sulfate group was mainly at the C-4 axial position on fucose [15–17]. Thus, it was concluded that KCW and KCA were mainly sulfated at C-4 on fucose. After desulfation, the intense band at 1260 cm^{-1} vanished, indicating that ds-DKCW and ds-DKCA lost a high amount of sulfate during the process of desulfation, which was confirmed by the results in Table 1.

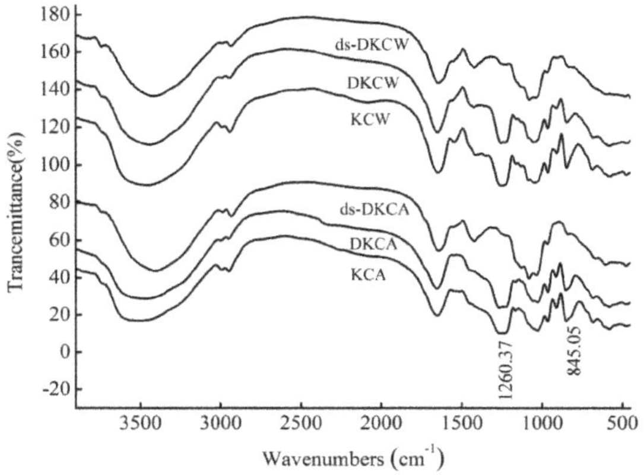

Figure 1. The IR spectra of polysaccharides.

2.3. MS Analysis of Structure

MS is an important tool for the analysis of heteropolysaccharides because of its speed and sensitivity. Though there have been many studies on the structural features of heteropolysaccharides [5, 6,18–23], it was previously not possible to analyze heteropolysaccharides that contained large, highly charged molecules. Thus, heteropolysaccharides needed to be degraded. The ESI-MS spectrum of ds-DKCW determined in the present study was shown in Figure 2a. There were five types of fragment ions. The most intense peaks at m/z 257.042, 403.103, 549.164, 695.226, 841.288, 987.350 and 1133.415 were determined to be $[MeFuc_nSO_3Na\text{-}Na]^-$ (n = 1–7). The less intense fragment ions at m/z 389.089, 535.148, 681.210, 827.272 and 973.333 revealed a distribution of singly charged ions, corresponding to sulfated fucooligosaccharides $[Fuc_nSO_3Na\text{-}Na]^-$ (n = 2–6). In addition, the set of fragment ions at m/z 314.089, 387.089, 460.120, 533.148, 606.184 and 679.194 was determined to be methyl glycosides of di-sulfated fucooligosaccharides $[MeFuc_n(SO_3Na)_2\text{-}2Na]^{2-}$ (n = 3–8), and the fragment ions at m/z 307.051, 380.089, 453.112, 526.141 and 599.174 corresponded to di-sulfated fucooligosaccharides $[Fuc_n(SO_3Na)_2\text{-}2Na]^{2-}$ (n = 3–7). Finally, the last set of doubly charged ions at m/z 275.069, 348.100, 421.129, 494.161 and 567.193 was determined to be $[Gal(Fuc)_nSO_3Na\text{-}Na]^{2-}$ (n = 2–7) (the hexose was determined to be D-Gal based on the analysis of monosaccharides).

The ESI-MS spectrum of ds-DKCA is shown in Figure 2b. The fragment ions were similar to those of ds-DKCW. However, ds-DKCA mainly consisted of methyl glycosides of mono-sulfated fucooligosaccharides and had few multi-sulfated fucooligosaccharides or sulfated galactofucooligosaccharides.

The fragmentation pattern for the singly-charged ion at m/z 1133.415 (−1), assigned as $[MeFuc_7SO_3Na\text{-}Na]^-$, was displayed in Figure 3. Singly-charged fragment ions at m/z 371.075, 517.136, 663.197, 809.258, 955.319 and 1101.381 corresponded to B-type ions, arising from the glycosidic bond cleavage from the reducing end, suggesting that the sulfate was located at the non-reducing end. Another set of fragment ions at m/z 403.075, 549.063, 841.285 and 987.347 (a loss of fucopyranose residue (146 Da)) arose from the glycosidic bond cleavage from the non-reducing end, indicating that the sulfate was located at the reducing end. Thus it was concluded that the ion $[MeFuc_7SO_3Na\text{-}Na]^-$ at m/z 1133.415 (−1) was a mixture of isomers Fuc(SO$_3$Na)→Fuc→Fuc→Fuc→Fuc→Fuc→Fuc-OMe and Fuc→Fuc→Fuc→Fuc→Fuc→Fuc→Fuc(SO$_3$Na)-OMe. This finding was in agreement with previous studies on the structural features of polysaccharides from *Saccharina japonica* and *Sargassum fusiforme* [22–24], suggesting that polysaccharides from *K. crassifolia* might have the same backbone as the fucoidan from *Saccharina japonica* and *Sargassum fusiforme*.

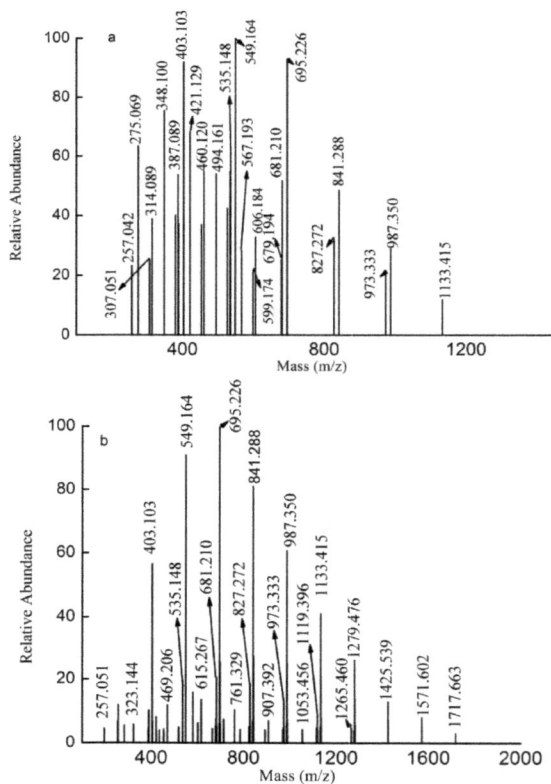

Figure 2. Negative ion mode ESI-MS spectra of ds-DKCW (**a**) and ds-DKCA (**b**).

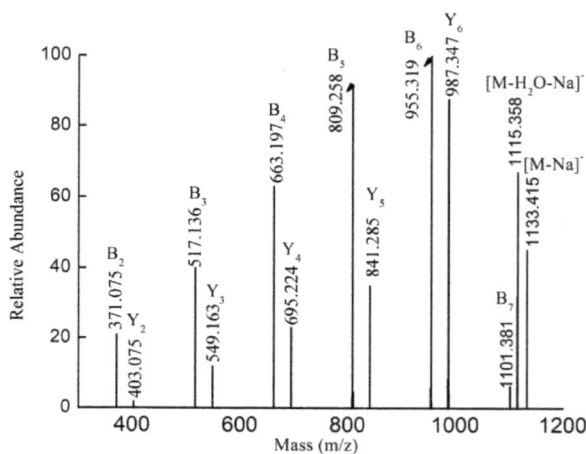

Figure 3. Negative ion mode ESI-CID-MS^2 spectrum of the ion [MeFuc$_7$SO$_3$Na-Na]$^-$ at *m/z* 1133.415 (−1).

The fragmentation pattern for the doubly charged ion at m/z 421.129 (-2) assigned to the ion $[Gal(Fuc)_4SO_3Na-Na-H]^{2-}$ was displayed in Figure 4a. The fragment ion at m/z 225.014 resulted from breakage of glycosidic bonds and cleavage of the dehydrated, sulfated fucose residue. No ions at m/z 241 or 259 were detected, suggesting that the sulfate group was substituted at the Fuc residue. One set of fragment ions at m/z 389.086, 535.147 and 681.208, assigned as C- or Y-type fragment ions, corresponded to $[(Fuc)_2SO_3Na-Na]^-$, $[(Fuc)_3SO_3Na-Na]^-$ and $[(Fuc)_4SO_3Na-Na]^-$, respectively. Another series of less intense fragment ions at m/z 551.143(C'_3/Y'_3) and 697.204(C'_4/Y_4) corresponded to $[Gal(Fuc)_2SO_3Na-Na]^-$ and $[Gal(Fuc)_3SO_3Na-Na]^-$, respectively. In addition, the characteristic fragment ion at m/z 383.103 (-2) corresponded to $^{2,5}A_5$, suggesting that the linkage between Fuc and Gal was a 1→4 linkage or a 1→3 linkage. Moreover, no characteristic ion of $^{0,2}A$ was detected, suggesting that the linkage between Fuc and Fuc was mainly 3-linked. Thus, it was hypothesized that $Gal(Fuc)_4SO_3Na$ consisted of Fuc(SO$_3$Na)→Fuc→Fuc→Fuc→Gal, Fuc(SO$_3$Na)→Fuc→Fuc→Gal→Fuc, Fuc(SO$_3$Na)→Fuc→Gal→Fuc→Fuc, Fuc(SO$_3$Na)→Gal→Fuc→Fuc→Fuc, Fuc→Gal→Fuc→Fuc→Fuc(SO$_3$Na), Gal→Fuc→Fuc→Fuc→Fuc(SO$_3$Na), Fuc→Fuc→Gal→Fuc→Fuc(SO$_3$Na) and Fuc→Fuc→Fuc→Gal→Fuc(SO$_3$Na).

To confirm the above hypothesis, the fragmentation pattern for the singly-charged ion at m/z 681.208 was characterized by ESI-CID-MS^3 and was found to correspond to the ion $[Fuc_4SO_3Na-Na]^-$, as shown in Figure 4b. One set of fragment ions at 225.014, 371.075 and 517.136 assigned as B-type ions arose from the glycosidic bond cleavage from the reducing end, suggesting that the sulfate was located at the non-reducing end. Anastyuk *et al.* and Saad and Leary [18,25] reported that, when an oligosaccharide's sulfate group was close to the glycosidic linkage spatially, it would undergo easier B-type fragmentation, which indicated that the sulfate group was substituted at C-2. Another set of less intense fragment ions, 389.086 and 535.147, corresponded to Y-type ions, suggesting the sulfate was substituted at the reducing end. In addition, the characteristic ion at m/z 607.170 was a $^{0,3}X_3$-type ion, suggesting that the linkage was a 1→3 linkage. No $^{0,2}A$-type ions were detected, which confirmed that the linkage between Fuc and Fuc was 3-linked. In sum, Fuc$_4$SO$_3$Na consisted primarily of Fuc(2SO$_3$Na)→Fuc→Fuc→Fuc, and, to a lesser degree, of Fuc→Fuc→Fuc→Fuc(2 or 4SO$_3$Na).

The fragmentation pattern for the doubly charged ion at m/z 383.103 (-2) was also elucidated by ESI-CID-MS^3 in Figure 4c. The characteristic ion at m/z 361.090 (-2) ($^{2,4}A_5$) arose from the loss of C_2H_4O (44 Da) from the ion at m/z 383.103 (-2) ($^{2,5}A_5$), suggesting that the reducing end was a Gal residue. In addition, it also indicated that the linkage between Fuc and Gal was a 1→4 linkage. Moreover, a set of fragment ions at m/z 243.014, 389.086 and 535.147 were determined to be C-type ions, suggesting that the sulfate was located at the non-reducing end. This glycosidic cleavage observation was not consistent with the previous study [18,25], indicating that the sulfate was at C-4. Therefore, it was concluded that $Gal(Fuc)_4SO_3Na$ was Fuc(4SO$_3$Na)→Fuc→Fuc→Fuc→Gal.

Thus, it was concluded that $Gal(Fuc)_4SO_3Na$ was mainly made of Fuc(2 or 4SO$_3$Na)→Fuc→Fuc→Fuc→Gal. In addition, it also might contain Gal→Fuc→Fuc→Fuc→Fuc(2 or 4SO$_3$Na), Fuc(SO$_3$Na)→Fuc→Fuc→Gal→Fuc, Fuc(SO$_3$Na)→Fuc→Gal→Fuc→Fuc, Fuc(SO$_3$Na)→Gal→Fuc→Fuc→Fuc, Fuc→Gal→Fuc→Fuc→Fuc(SO$_3$Na), Fuc→Fuc→Gal→Fuc→Fuc(SO$_3$Na) and Fuc→Fuc→Fuc→Gal→Fuc(SO$_3$Na).

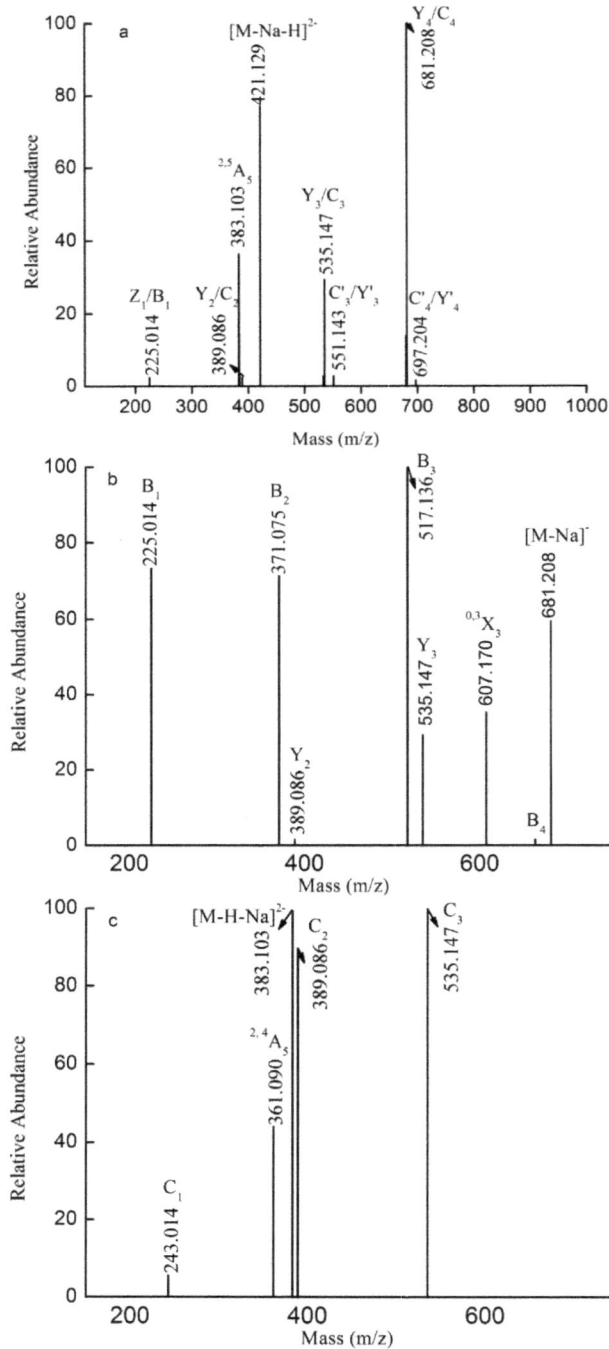

Figure 4. Negative ion mode ESI-CID-MS^2 spectrum of the ion at *m/z* 421.129 (−2) (**a**) and negative ion mode ESI-CID-MS^3 spectra of the ions at *m/z* 681.208(−1) (**b**) and 383.103 (−2) (**c**).

The fragmentation for the doubly charged ion at m/z 460.120 (-2), corresponding to the ion [MeFuc$_5$(SO$_3$Na)$_2$-2Na]$^{2-}$, was shown in Figure 5. Five types of fragments ions were found: (1) doubly charged fragment ions at m/z 225.014 (B$_2$), 298.075 (B$_3$), 371.063 (B$_4$) and 444.106 (B$_5$) corresponded to [Fuc$_2$(SO$_3$Na)$_2$-2Na]$^{2-}$, [Fuc$_3$(SO$_3$Na)$_2$-2Na]$^{2-}$, [Fuc$_4$(SO$_3$Na)$_2$-2Na]$^{2-}$ and [Fuc$_5$(SO$_3$Na)$_2$-2Na]$^{2-}$, respectively. No fragment ion at m/z 152 ([Fuc(SO$_3$Na)$_2$-2Na]$^{2-}$) was observed, suggesting that one isomer of MeFuc$_5$(SO$_3$Na)$_2$ was Fuc(SO$_3$Na)→Fuc(SO$_3$Na)→Fuc→Fuc→Fuc-OMe; (2) Less intense, doubly charged fragment ions at m/z 314.058 (Y$_3'$) and 387.089 (Y$_4'$) were determined to be [MeFuc$_3$(SO$_3$Na)$_2$-2Na]$^{2-}$ and [MeFuc$_4$(SO$_3$Na)$_2$-2Na]$^{2-}$, most likely having arisen from glycosidic bond cleavage at the non-reducing terminus. Thus, it was hypothesized that MeFuc$_5$(SO$_3$Na)$_2$ contained Fuc→Fuc→Fuc→Fuc(SO$_3$Na)→Fuc(SO$_3$Na)-OMe; (3) Singly charged fragment ions at m/z 225.014, 371.063, 517.136 and 663.198 (namely, B$_1$", B$_2$", B$_3$" and B$_4$", respectively) arose from the loss of a methyl glycoside of sulfated fucose (257 Da) from the reducing terminus, suggesting that MeFuc$_5$(SO$_3$Na)$_2$ might consist of Fuc(SO$_3$Na)→Fuc→Fuc→Fuc→Fuc(SO$_3$Na)-OMe or Fuc→Fuc→Fuc→Fuc(SO$_3$Na)→Fuc(SO$_3$Na)-OMe; (4) A singly-charged fragment ion at m/z 695.225 (namely, Y$_4$"), assigned as [MeFuc$_4$SO$_3$Na-Na]$^-$, arose from the loss of a sulfated fucopyranose residue (225 Da) from the side chain or from the non-reducing terminus. A singly-charged fragment ion at 549.163 (namely, Y$_3$"), corresponding to [MeFuc$_3$SO$_3$Na-Na]$^-$, arose from the loss of fucopyranose after the loss of the sulfated fucopyranose. Thus it was hypothesized that MeFuc$_5$(SO$_3$Na)$_2$ was made up of Fuc(SO$_3$Na)→Fuc→Fuc→Fuc→Fuc(SO$_3$Na)-OMe, Fuc(SO$_3$Na)→ Fuc(SO$_3$Na)→ Fuc→Fuc→Fuc-OMe and Fuc→Fuc→Fuc→Fuc(SO$_3$Na)→Fuc(SO$_3$Na)-OMe; (5) Singly-charged fragment ions at m/z 243.025 and 389.086 (namely, C1"and C2") arose from the loss of sulfated fucose from the non-reducing terminus, suggesting that MeFuc$_5$(SO$_3$Na)$_2$ might consist of Fuc(SO$_3$Na)→Fuc→Fuc→Fuc→Fuc(SO$_3$Na)-OMe. Thus, it was hypothesized that MeFuc$_5$(SO$_3$Na)$_2$ was made up of Fuc(SO$_3$Na)→Fuc→Fuc→Fuc→Fuc(SO$_3$Na)-OMe, Fuc(SO$_3$Na)→Fuc(SO$_3$Na)→Fuc→Fuc→Fuc-OMe and Fuc→Fuc→Fuc→Fuc(SO$_3$Na)→Fuc(SO$_3$Na)-OMe.

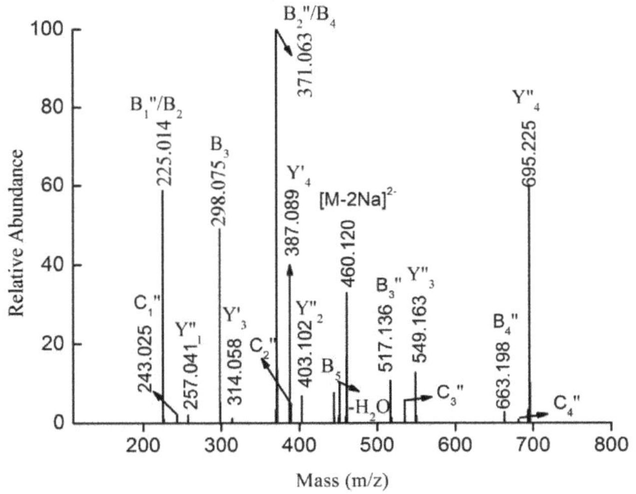

Figure 5. Negative ion mode ESI-CID-MS^2 spectrum of the ion [MeFuc$_5$(SO$_3$Na)$_2$-2Na]$^{2-}$ at m/z 460.120 (-2).

The fragmentation pattern for the doubly charged ion at m/z 526.141 (-2), assigned to the ion [Fuc$_6$(SO$_3$Na)$_2$-2Na]$^{2-}$, was displayed in Figure 6. The ion at m/z 526.141 (-2) was the prototype of the ion [MeFuc$_6$(SO$_3$Na)$_2$-2Na]$^{2-}$, which was confirmed by its similar fragmentation pattern. In other words, both the ion [MeFuc$_6$(SO$_3$Na)$_2$-2Na]$^{2-}$ and the ion [Fuc$_6$(SO$_3$Na)$_2$-2Na]$^{2-}$ were generated under the condition of desulfation. However, the former was methylated, while the latter was not.

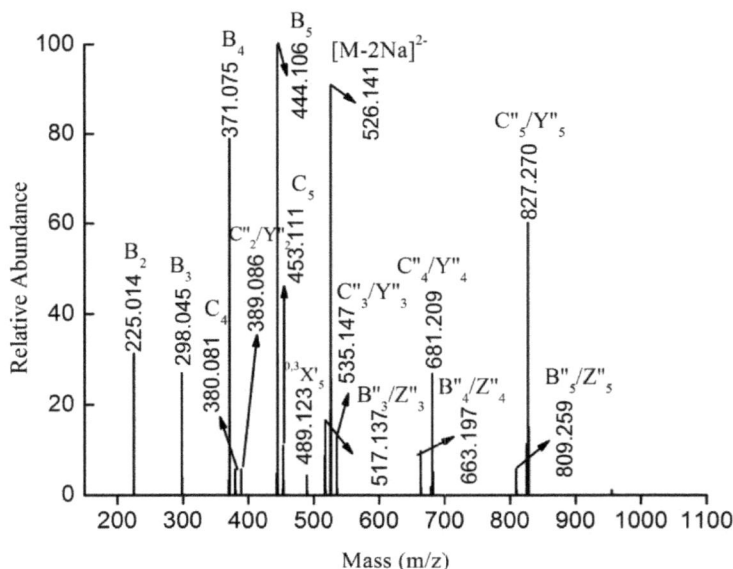

Figure 6. Negative ion mode ESI-MS^2 spectrum of ion [Fuc$_6$(SO$_3$Na)$_2$-2Na]$^{2-}$ at *m/z* 526.141 (−2).

2.4. Anti-Complement Activity

As shown in Figure 7a–d, the effects of the polysaccharides on activation of human complement through the classical pathway (Figure 7a,b) and the alternative pathway (Figure 7c,d) were examined in 1:10-diluted NHS, with heparin used as a reference. The complement group (*i.e.*, positive control) displayed a 93.11% ± 2.96% activation of the classical complement pathway. The activities of KCA, KCW, DKCA, DKCW and heparin were dose-dependent, while ds-DKCA and ds-DKCW showed little or no activity (Figure 7a,b). The activities of KCA and KCW reached a plateau at a concentration of 10 µg/mL, while DKCA plateaued at 50 µg/mL. In addition, the concentration that resulted in 50% inhibition of the classical complement pathway (CH$_{50}$) for DKCW was approximately 218 µg/mL, which was lower than heparin. Therefore, KCA, KCW, and DKCA were more potent than heparin in inhibiting activation of the classical pathway. On the other hand, the concentrations of KCA, KCW, DKCA, DKCW and heparin that resulted in 50% inhibition of the alternative pathway (AP50) were 4.83, 18.60, 24.50, 19.97 and 137.25 µg/mL, respectively (Figure 7c,d). This finding indicated that KCA, KCW, DKCA and DKCW were more potent than heparin in inhibiting activation of the alternative pathway.

KCA and KCW displayed similar activity levels against the two complement pathways, indicating that the extraction methods did not affect the activity levels. The degraded polysaccharides DKCA and DKCW exhibited weaker activity compared to the crude polysaccharides KCA and KCW, which suggested, as others have reported, that the change in molecular weight influenced the anti-complement activity of the two compounds [9,26]. In addition, the finding that ds-DKCA and ds-DKCW showed little or no activity against the two pathways suggested that sulfate was important for anti-complement activity, which was in agreement with other reports [13,27,28].

In sum, the polysaccharides from *Kjellmaniella crassifolia* may be potent drugs that are capable of suppressing complement activation.

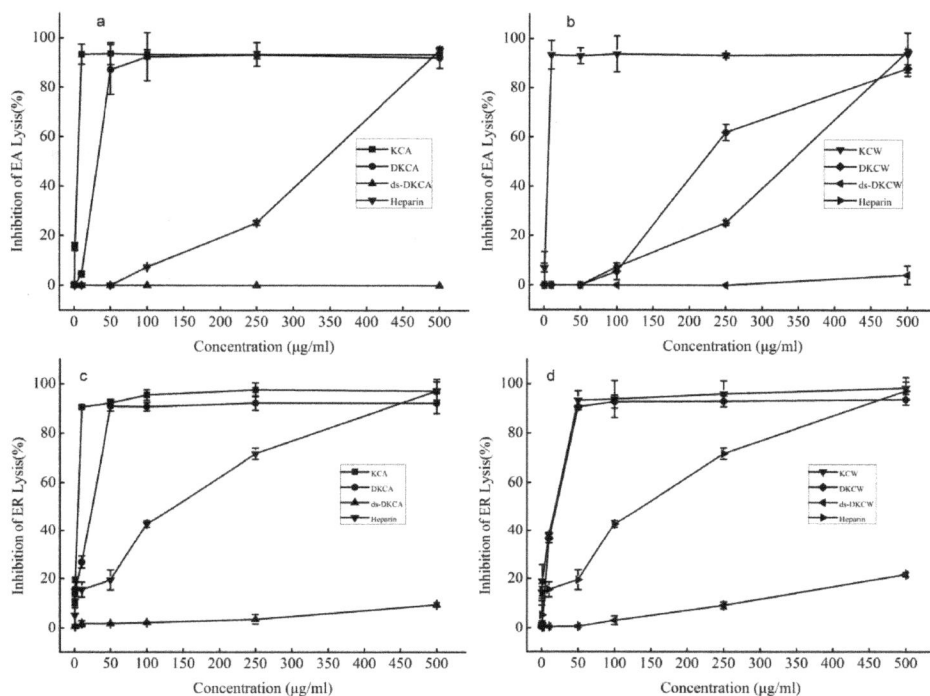

Figure 7. Inhibition of the classical pathway-mediated hemolysis of EA (**a** and **b**) and alternative pathway-mediated hemolysis of ER (**c** and **d**) in 1:10-diluted NHS in the presence of increasing amounts of the polysaccharides. Heparin was used as the reference. The results are expressed as percent inhibition of hemolysis. Data are the means from 3 determinations ± S.E.M.

3. Experimental Section

3.1. Preparation of Polysaccharides

K. crassifolia was collected in Rongcheng, Shandong Province, China, in June of 2013. The polysaccharides were extracted from *K. crassifolia* as previously described [29]. Briefly, the dried algae were cut into pieces, and the polysaccharides were extracted three times in water at room temperature for 2 h. The solution was dialyzed against water and distilled water. Finally, the polysaccharide was concentrated and precipitated with ethanol. The resultant precipitate was named KCW. In an alternative extraction, the dried algae were cut and extracted with 0.1 M HCl at room temperature for 2 h. The solution was neutralized, concentrated, dialyzed and precipitated with ethanol. This precipitate was named KCA.

3.2. Preparation of Low Molecular Weight Fucoidans and Their Desulfated Mixtures

The crude polysaccharides KCA and KCW were degraded using hydrogen dioxide and ascorbic acid to obtain low molecular weight polysaccharides, as previously described [21]. Briefly, crude polysaccharide (1 g) was dissolved in water (100 mL). Ascorbic acid (0.5 g) and hydrogen dioxide (0.3 mL) were then added, and the solution was stirred for 2 h at room temperature. The degraded polysaccharides (e.g., DKCA and DKCW) were obtained after ultrafiltration, concentration and lyophilization.

93

The desulfation of DKCA and DKCW was performed according to the modified method of Nagasawa *et al.* [30]. Desulfation was carried out using its pyridinium salt. Briefly, the sample was dissolved in distilled water (10 mL) and mixed with cationic resin for 3 h. After filtration, the solution was neutralized with pyridinium and lyophilized. The polysaccharide was then dissolved in 20 mL of a 9:1 ratio of dimethyl sulfoxide: methanol (*v:v*) at 80 °C for 5 h. The desulfated solution was dialyzed and lyophilized to give desulfated products (e.g., ds-DKCA and ds-DKCW).

3.3. Composition Methods

Total sugar and fucose content were determined according to the method of Dubois *et al.* [31] and Gibbons [32], using fucose as a standard. The level of sulfation was analyzed by the barium chloride-gelatin method of Kawai *et al.* [33]. The uronic acid (UA) content was estimated with a modified carbazole method using D-glucuronic acid as a standard [34]. The protein level was determined according to the method of Bradford *et al.* [35]. For the determination of sugar composition, the acid-hydrolyzed glycoses were converted into their 1-phenyl-3-methyl-5-pyrazolone derivatives (PMP) and separated by HPLC chromatography [36]. The molecular weight of the samples was assayed by a HP-GPC system on a TSK gel PWxl 3000 column (7 μm, 7.8 × 300 mm) eluted with 0.2 M Na_2SO_4 at a flow rate of 0.5 mL min^{-1} at 30 °C [37].

3.4. Spectroscopic Analysis

Infrared spectra (IR) were recorded from polysaccharide powder in KBr pellets on a Nicolet-360 FTIR spectrometer between 400 and 4000 cm^{-1} (36 scans, at a resolution of 6 cm^{-1}).

MS was performed on a LTQ ORBITRAR XL (Thermo Scientific, Waltham, MA, USA). Samples, dissolved in CH_3CN-H_2O (1:1, *v:v*), were introduced into the MS at a flow rate of 5 μL min^{-1} in the negative ionization mode. The capillary voltage was set to −3000 V, the cone voltage was set to −50 V, the source temperature was set to 80 °C, and the desolvation temperature was set to 150 °C. The collision energy was optimized between 20 and 50 eV. All spectra were analyzed by Xcalibur (Thermo Scientific, Waltham, MA, USA).

3.5. Anti-Complement Activity

The anti-complement activity of the polysaccharides was determined by measuring their ability to inhibit classical [38] and alternative [39] complement-mediated hemolysis. For the classical pathway, 100 μL of various dilutions of tested samples were mixed with 100 μL of 1:10-diluted normal human serum (NHS, which was obtained from healthy, adult donors), 200 μL GVB^{2+} (veronal buffer saline (VBS) containing 0.1% gelatin, 0.5 mM Mg^{2+} and 0.15 mM Ca^{2+}) and 200 μL of sensitized erythrocytes (EA). The mixture was then incubated at 37 °C for 30 min. The following assay controls were incubated under the same conditions: (1) 100% lysis: 200 μL of EA in 400 of μL water; (2) sample control: 100 μL of sample in 500 of μL GVB^{2+}; (3) complement: 100μL of 1:10-diluted NHS and 200 μL EA in 300 μL GVB^{2+}; and (4) blank: 200 μL of EA in 400 μL of GVB^{2+}. After incubation, the mixture was centrifuged (5000 rpm × 10 min) and the erythrocyte lysis was determined at 405 nm. Decreased lysis in the presence of tested polysaccharides indicated anti-complement activity. All of the samples were dissolved in GVB^{2+}. Heparin sodium salt (Beijing Rui Taibio Co., Ltd., Beijing, China), at a concentration of 160 IU/mg, was used as the reference (The molecular weight was 54.1 kDa and sulfate content was 13.60%). The inhibition percentage was calculated using the following equation: inhibition of EA lysis (%) = $(A_{complement} - [A_{sample} - A_{sample\ control}])/A_{complement} \times 100$.

To assay sample inhibition of the alternative complement pathway, 150 μL of various dilutions of tested samples were mixed with 150 μL of 1:10-diluted NHS and 200 μL of rabbit erythrocytes (ER). The mixture was then incubated for 30 min at 37 °C. Cell lysis was determined by the same method as described above for the classical pathway. Controls for 100% lysis, sample control, complement and blank were included. The percent inhibition was calculated using the following equation: inhibition of ER lysis (%) = $(A_{complement} - [A_{sample} - A_{sample\ control}])/A_{complement} \times 100$.

4. Conclusions

In this study, two polysaccharides, KCW and KCA, were extracted from *K. crassifolia* using water and dilute hydrochloric acid, respectively, and their degraded fractions and desulfated fractions were prepared. The chemical compositions indicated that KCA had higher contents of sulfate and Fuc and less of other monosaccharides than KCW. To elucidate the structural features of KCA and KCW, they were degraded and partly desulfated. The desulfated mixtures were determined by ESI-MS. Both were found to contain sulfated fucooligosaccharides, sulfated galactofucooligosaccharides and methyl glycosides of mono-sulfated/multi-sulfated fucooligosaccharides. The major difference between ds-DKCA and ds-DKCW was the intensity of the fragment ions. In addition, the structural features of oligomeric fragments were characterized by ESI-CID-MS^2 and ESI-CID-MS^3. It was shown that the polysaccharides had a backbone of 3-linked Fuc residues sulfated at C-2, C-4 or C-2 and C-4. Some oligomers had the 4-linked Gal residue at the reducing terminus. Some fucooligomers were interspersed by Gal residues. Moreover, activities of the polysaccharides against the classical and alternative complement pathways were measured. The crude polysaccharides KCA and KCW had the highest activity levels, while the desulfated fractions ds-DKCA and ds-DKCW had little or no activity. These data suggested that the change in molecular weight and sulfate content influenced the activity levels. In summary, polysaccharides from *K. crassifolia* may be a good candidate drug for anti-complement therapy.

Acknowledgments: This study was supported by the Ocean Public Welfare Scientific Research Project (201405040), the Shandong Province Science and Technology Development Project (2014GHY115017), the Shandong Provincial Natural Science Foundation (ZR2014DQ024), the Shinan District Science and Technology Development Fund (2013-12-009-SW), the Special Fund for Cooperation between Jilin Province and Chinese Academy of Sciences (2013SYHZ0023), the Natural Science Fundation of China (41406144) and the Applied Research Projects of Nantong Municipal (BK2013011).

Author Contributions: Wenjing Zhang, Weihua Jin and Quanbin Zhang conceived and designed the experiments; Wenjing Zhang performed the experiments; Delin Sun, Luyu Zhao, Jing Wang and Delin Duan contributed reagents, materials and financial support. Wenjing Zhang wrote the paper. All authors have read and approved the final manuscript.

Conflicts of Interest: The authors declare no conflict of interest.

References

1. Yan, H.; Kakuta, S.; Nishihara, M.; Sugi, M.; Adachi, Y.; Ohno, N.; Iwakura, Y.; Tsuji, N.M. *Kjellmaniella crassifolia* miyabe (Gagome) extract modulates intestinal and systemic immune responses. *Biosci. Biotechnol. Biochem.* **2011**, *75*, 2178–2183. [CrossRef] [PubMed]
2. Katayama, S.; Nishio, T.; Kishimura, H.; Saeki, H. Immunomodulatory properties of highly viscous polysaccharide extract from the Gagome alga (*Kjellmaniella crassifolia*). *Plant Foods Hum. Nutr.* **2012**, *67*, 76–81. [CrossRef]
3. Sakai, T.; Kimura, H.; Kojima, K.; Shimanaka, K.; Ikai, K.; Kato, I. Marine bacterial sulfated fucoglucuronomannan (SFGM) lyase digests brown algal SFGM into trisaccharides. *Mar. Biotechnol. (N.Y.)* **2003**, *5*, 70–78. [CrossRef]
4. Wu, J.; Yu, G.; Li, M.; Zhao, X.; Wang, C.; Liu, T.; Gu, C. Extraction, isolation and structural characterization of polysaccharides from *Kjellmanjella crassifolia*. *J. Ocean Univ. China* **2011**, *41*, 127–130.
5. Wang, P.; Zhao, X.; Lv, Y.; Liu, Y.; Lang, Y.; Wu, J.; Liu, X.; Li, M.; Yu, G. Analysis of structural heterogeneity of fucoidan from *Hizikia fusiforme* by ES-CID-MS/MS. *Carbohydr. Polym.* **2012**, *90*, 602–607. [CrossRef] [PubMed]
6. Anastyuk, S.D.; Shevchenko, N.M.; Nazarenko, E.L.; Imbs, T.I.; Gorbach, V.I.; Dmitrenok, P.S.; Zvyagintseva, T.N. Structural analysis of a highly sulfated fucan from the brown alga *Laminaria cichorioides* by tandem MALDI and ESI mass spectrometry. *Carbohydr. Res.* **2010**, *345*, 2206–2212. [CrossRef] [PubMed]
7. Daniel, R.; Chevolot, L.; Carrascal, M.; Tissot, B.; Mourão, P.A.S.; Abian, J. Electrospray ionization mass spectrometry of oligosaccharides derived from fucoidan of *Ascophyllum nodosum*. *Carbohydr. Res.* **2007**, *342*, 826–834. [CrossRef] [PubMed]

8.   Makrides, S.C. Therapeutic inhibition of the complement system. *Pharmacol. Rev.* **1998**, *50*, 59–88. [PubMed]
9.   Blondin, C.; Chaubet, F.; Nardella, A.; Sinquin, C.; Jozefonvicz, J. Relationships between chemical characteristics and anticomplementary activity of fucans. *Biomaterials* **1996**, *17*, 597–603. [CrossRef] [PubMed]
10. Clement, M.J.; Tissot, B.; Chevolot, L.; Adjadj, E.; Du, Y.; Curmi, P.A.; Daniel, R. NMR characterization and molecular modeling of fucoidan showing the importance of oligosaccharide branching in its anticomplementary activity. *Glycobiology* **2010**, *20*, 883–894. [CrossRef] [PubMed]
11. Tissot, B.; Montdargent, B.; Chevolot, L.; Varenne, A.; Descroix, S.; Gareil, P.; Daniel, R. Interaction of fucoidan with the proteins of the complement classical pathway. *Biochim. Biophys. Acta* **2003**, *1651*, 5–16. [CrossRef] [PubMed]
12. Samuelsena, A.B.; Lunda, I.; Djahromia, J.M.; Paulsena, B.S.; Wolda, J.K.; Knutsen, H.S. Structural features and anti-complementary activity of some heteroxylan polysaccharide fractions from the seeds of *Plantago major* L. *Carbohydr. Polym.* **1999**, *38*, 133–143. [CrossRef]
13. Mauzac, M.; Maillet, F.; Jozefonvicz, J.; Kazatchkine, M.D. Anticomplementary activity of dextran derivatives. *Biomaterials* **1985**, *6*, 61–63. [CrossRef] [PubMed]
14. Casu, B. Structure and biological activity of heparin and other glycosaminoglycans. *Pharmacol. Res. Commun.* **1979**, *11*, 1–18. [CrossRef] [PubMed]
15. Foley, S.A.; Mulloy, B.; Tuohy, M.G. An unfractionated fucoidan from *Ascophyllum nodosum*: Extraction, characterization, and apoptotic effects *in vitro*. *J. Nat. Prod.* **2011**, *74*, 1851–1861. [CrossRef] [PubMed]
16. Chen, S.; Xue, C.; Yin, L.A.; Tang, Q.; Yu, G.; Chai, W. Comparison of structures and anticoagulant activities of fucosylated chondroitin sulfates from different sea cucumbers. *Carbohydr. Polym.* **2011**, *83*, 688–696. [CrossRef]
17. Duarte, M.E. R.; Cardoso, M.A.; Noseda, M.D.; Cerezo, A.S. Structural studies on fucoidans from the brown seaweed *Sargassum stenophyllum*. *Carbohydr. Res.* **2001**, *333*, 281–293. [CrossRef] [PubMed]
18. Anastyuk, S.D.; Imbs, T.I.; Shevchenko, N.M.; Dmitrenok, P.S.; Zvyagintseva, T.N. ESIMS analysis of fucoidan preparations from *Costaria costata*, extracted from alga at different life-stages. *Carbohydr. Polym.* **2012**, *90*, 993–1002. [CrossRef] [PubMed]
19. Anastyuk, S.D.; Shevchenko, N.M.; Nazarenko, E.L.; Dmitrenok, P.S.; Zvyagintseva, T.N. Structural analysis of a fucoidan from the brown alga *Fucus evanescens* by MALDI-TOF and tandem ESI mass spectrometry. *Carbohydr. Res.* **2009**, *344*, 779–787. [CrossRef] [PubMed]
20. Tissot, B.; Salpin, J.Y.; Martinez, M.; Gaigeot, M.P.; Daniel, R. Differentiation of the fucoidan sulfated L-fucose isomers constituents by CE-ESIMS and molecular modeling. *Carbohydr. Res.* **2006**, *341*, 598–609. [CrossRef] [PubMed]
21. Jin, W.; Wang, J.; Ren, S.; Song, N.; Zhang, Q. Structural analysis of a heteropolysaccharide from *Saccharina japonica* by electrospray mass spectrometry in tandem with collision-induced dissociation tandem mass spectrometry (ESI-CID-MS/MS). *Mar. Drugs* **2012**, *10*, 2138–2152. [CrossRef] [PubMed]
22. Jin, W.; Zhang, W.; Wang, J.; Ren, S.; Song, N.; Zhang, Q. Structural analysis of heteropolysaccharide from *Saccharina japonica* and its derived oligosaccharides. *Int. J. Biol. Macromol.* **2013**, *62C*, 697–704. [CrossRef]
23. Jin, W.; Zhang, W.; Wang, J.; Ren, S.; Song, N.; Duan, D.; Zhang, Q. Characterization of laminaran and a highly sulfated polysaccharide from *Sargassum fusiforme*. *Carbohydr. Res.* **2014**, *385*, 58–64. [CrossRef] [PubMed]
24. Jin, W.; Guo, Z.; Wang, J.; Zhang, W.; Zhang, Q. Structural analysis of sulfated fucan from *Saccharina japonica* by electrospray ionization tandem mass spectrometry. *Carbohydr. Res.* **2013**, *369*, 63–67. [CrossRef] [PubMed]
25. Saad, O.M.; Leary, J.A. Delineating mechanisms of dissociation for isomeric heparin disaccharides using isotope labeling and ion trap tandem mass spectrometry. *Am. Soc. Mass Spectrom.* **2004**, *15*, 1274–1286. [CrossRef]
26. Crepon, B.; Maillet, F.; Kazatchkine, M.D.; Jozefonvicz, J. Molecular weight dependency of the acquired anticomplementary and anticoagulant activities of specifically substituted dextrans. *Biomaterials* **1987**, *8*, 248–253. [CrossRef] [PubMed]
27. Cofrancsco, E.; Radaelli, F.; Pogliania, E. Correlation of sulfate content and degree of carboxylation of heparin and related glycosaminoglycans with anticomplement activity. Relationships to the anticoagulant and platelet-aggregating activities. *Thromb. Res.* **1979**, *14*, 179–187.

28. Maillet, F.; Maurice Petitou; Jean Choay; Kazatchkine, M.D. Structure-function relationships in the inhibitory effect of heparin on complement activation: Independency of the anti-coagulant and anti-complementary sites on the heparin molecule. *Mol. Immunol.* **1988**, *25*, 917–923. [CrossRef] [PubMed]
29. Jin, W.; Zhang, W.; Wang, J.; Zhang, Q. The neuroprotective activities and antioxidant activities of the polysaccharides from *Saccharina japonica*. *Int. J. Biol. Macromol.* **2013**, *58*, 240–244. [CrossRef] [PubMed]
30. Nagasawa, K.; Inoue, Y.; Tokuyasu, T. An improved method for the preparation of chondroitin by solvolytic desulfation of chondroitin sulfates. *J. Biochem.* **1979**, *86*, 1323–1329. [PubMed]
31. Dubois, M.; Gilles, K.A.; Hamilton, J.K.; Rebers, P.T.; Smith, F. Colorimetric method for determination of sugars and related substances. *Anal. Chem.* **1956**, *28*, 350–356. [CrossRef]
32. Gibbons, M.N. The determination of methylpentoses. *Analyst* **1955**, *80*, 267–276. [CrossRef]
33. Kawai, Y.; Seno, N.; Anno, K. A modified method for chondrosulfatase assay. *Anal. Biochem.* **1969**, *32*, 314–321. [CrossRef] [PubMed]
34. Bitter, T.; Muir, H.M. A modified uronic acid carbazole reaction. *Anal. Biochem.* **1962**, *4*, 330–334. [CrossRef] [PubMed]
35. Bradford, M.M. A rapid and sensitive method for the quantitation of microgram quantities of protein utilizing the principle of protein-dye binding. *Anal. Biochem.* **1976**, *72*, 248–254. [CrossRef] [PubMed]
36. Zhang, J.; Zhang, Q.; Wang, J.; Shi, X.; Zhang, Z. Analysis of the monosaccharide composition of fucoidan by precolumn derivation HPLC. *Chin. J. Oceanol. Limnol.* **2009**, *27*, 1–5. [CrossRef]
37. Zhang, W.; Wang, J.; Jin, W.; Zhang, Q. The antioxidant activities and neuroprotective effect of polysaccharides from the starfish *Asterias rollestoni*. *Carbohydr. Polym.* **2013**, *95*, 9–15. [CrossRef] [PubMed]
38. Xu, H.; Zhang, Y.; Zhang, J.; Chen, D. Isolation and characterization of an anti-complementary polysaccharide D3-S1 from the roots of *Bupleurum smithii*. *Int. Immunopharmacol.* **2007**, *7*, 175–182. [CrossRef] [PubMed]
39. Klerx, J.P.A.M.; Beukelman, C.J.; Dijk, H.V.; Willers, J.M.N. Microassay for colorimetric estimation of complement activity in guinea pig, human and mouse serum. *J. Immunol. Methods* **1983**, *63*, 215–220. [CrossRef] [PubMed]

marine drugs

MDPI

Article

Salt Effect on the Antioxidant Activity of Red Microalgal Sulfated Polysaccharides in Soy-Bean Formula

Ariela Burg [†],* and Levy-Ontman Oshrat [†],*

Department of Chemical Engineering, Sami Shamoon College of Engineering, Basel/Bialik sts., Beer-Sheva 8410001, Israel

* Authors to whom correspondence should be addressed; arielab@sce.ac.il (A.B.); oshrale@sce.ac.il (L.-O.O.); Tel.: +972-8-6475732 (A.B.); +972-8-6475732 (L.-O.O.); Fax: +972-8-6475654 (A.B.); +972-8-6475654 (L.-O.O.).

† These authors contributed equally to this work.

Academic Editor: Paola Laurienzo

Received: 13 August 2015; Accepted: 8 October 2015; Published: 20 October 2015

Abstract: Sulfated polysaccharides produced by microalgae, which are known to exhibit various biological activities, may potentially serve as natural antioxidant sources. To date, only a few studies have examined the antioxidant bioactivity of red microalgal polysaccharides. In this research, the effect of different salts on the antioxidant activities of two red microalgal sulfated polysaccharides derived from *Porphyridium* sp. and *Porphyridium aerugineum* were studied in a soy bean-based infant milk formula. Salt composition and concentration were both shown to affect the polysaccharides' antioxidant activity. It can be postulated that the salt ions intefer with the polysaccharide chains' interactions and alter their structure, leading to a new three-dimensional structure that better exposes antiooxidant sites in comparison to the polysaccharide without salt supplement. Among the cations that were studied, Ca^{2+} had the strongest enhancement effect on antioxidant activities of both polysaccharides. Understanding the effect of salts on polysaccharides' stucture, in addition to furthering knowledge on polysaccharide bioactivities, may also shed light on the position of the antioxidant active sites.

Keywords: antioxidants; calcium; microalgae; polysaccharides; *Porphyridium* sp.; *Porphyridium aerugineum*

1. Introduction

Antioxidants have a positive effect on human health as they can slow down oxidative stress processes caused by reactive oxygen species (ROS), e.g., OH, H_2O_2 and O_2^- [1–3]. These ROS attack macromolecules such as membrane lipids, proteins and DNA [1,4], leading to many health disorders with severe tissue injuries such as cancer, diabetes mellitus, neurodegenerative and inflammatory diseases [5–8]. In the food industry, oxidation of lipids can lead to food product deterioration (e.g., in nutritional value, safety and appearance) due to formation of undesirable secondary lipid peroxidation products. Therefore, many synthetic commercial antioxidants such as butylated hydroxytoluene (BHT), butylated hydroxyanisole (BHA), tert-butylhydroquinone (TBHQ) and propyl gallate (PG) are used to retard the oxidation and peroxidation processes [9]. However, the use of these synthetic antioxidants must be under strict regulation due to potential health hazards [10–12], hence there is an ongoing search for safe, natural alternatives [13,14].

Recently, both the food and pharmaceutical industries have become interested in the development of antioxidants from natural sources. Sulfated polysaccharides derived from marine algae, which are known to exhibit many biological activities, may be a potential natural antioxidant source for these industries [15]. Among the scant studies conducted regarding the antioxidant potential of sulfated polysaccharides, most efforts were focused on those derived from seaweeds [16–22], whereas very little is known regarding red microalgal sulfated polysaccharides [23–25]. The mechanism of these sulfated polysaccharides' antioxidant activity is not certain. It is also known that various residues which could bind to these polysaccharides, such as pigments, flavones, peptides, proteins, and polyphenols can improve their antioxidant activities [26].

Our research is focused on two soluble-sulfated polysaccharide fractions produced by red microalgal species—the seawater *Porphyridium* sp. (hereafter—PS1) and the freshwater *Porphyridium aerugineum* (hereafter—PS2). These microalgae, in contrast to seaweeds, are encapsulated by a thick layer of mucilaginous sulfated polysaccharides that lack the rigid microfibrillar component typical in most of alga species [27]. During growth, the external fraction of these polysaccharides is released to the surrounding aqueous medium, where it accumulates and increases medium viscosity [28–30]. The precise structures of the soluble-sulfated polysaccharide fractions are not fully understood due to their complexity and the lack of known carbohydrolases that degrade them [30–32]. The soluble-sulfated polysaccharide fractions are of high molecular weight (~1–7 $\times$ 10^6 g·mol^{-1}), and due to the presence of glucuronic acid residues and sulfates they are anionic heteropolymers, containing approximately 10 different mono sugars, mainly glucose and galactose [33–36]. In addition, both are built from the same disaccharide building block [37]. PS1 was found to contain several proteins, which are non-covalently bound to it [38]. The most prominent protein detected was a 66-kDa glycoprotein; This protein consists of a polypeptide of approximately 58 kDa and glycan moieties of approximately 8 kDa [38,39]. Sequencing of a cDNA clone encoding the 66-kDa glycoprotein revealed that this is a novel protein, which lacks similarity to any protein in the public domain databases. However, in the SCOP databases, some structural similarities were found between the carbohydrate-binding domain (CBD) and protein superfamilies, such as glycosyltransferases, pectin lyase-like proteins, sialidases, and conA-like lectins/glucanases, indicating a possible role for this 66-kDa glycoprotein in synthesis/modification of the cell-wall polysaccharide [38]. In addition, the glycoprotein was shown to play a role in biorecognition [40].

Physicochemical analysis conducted on these soluble-sulfated polysaccharide fractions indicate that even at low concentrations they create highly viscous polymer solutions, compared to the viscosity of aqueous solutions containing polysaccharides such as xanthans or carrageenans [41]. It is believed that red microalgal polysaccharide layers protect the cells against drought, solar irradiation and maintain the humidity required by the cells [30].

PS1 is stable over a wide range of temperatures (30–160 °C), pH values (2–9), light and salinities [31, 41–49]. Rheological studies have demonstrated that it is comprised of an oriented, single two-fold helical structure with a pitch of 1.6 nm, *i.e.*, a single chain helix with a regular chemical repeat [49]. The stiffness of PS1 was reported to be in the same rigid helical range as that of xanthan gum and DNA [43]. A conformational transition state in the PS1 chain was observed at low ionic strength (<0.01 M), most likely reflecting a contraction of the polymer chain from a highly stretched to a stiff, wormlike chain [43]. Furthermore, rheological studies have also indicated that the nature of PS1 is affected by ionic strength, valence, type and concentration of cations, pH, polymer concentration, and temperature [43]. For example, in a salt-free solution, PS1 favors a stretched chain conformation, the result of long-range electrostatic effects, but the addition of an electrolyte such as NaCl, masks the electrostatic interactions and allows a more flexible configuration resulting in lower intrinsic viscosities [43].

To our knowledge, only a few reports have evaluated the antioxidant activity of PS1 and PS2 [23, 24]. These fractions were shown to exhibit dose-dependent antioxidant activity, and were found to serve as better inhibitors of linoleic acid auto-oxidation compared to carrageenan and cellulose [24]. In addition, PS1 inhibited the oxidative damage to 3T3 cells caused by FeSO$_4$ in a dose-dependent manner [24]. A search for the active antioxidant component was also conducted; PS1 polysaccharide fragments created by microwave and sonication exerted an inhibitory effect on oxidative damage in dependence on the fraction received following fragmentation [23]. Thus, it seems that a significant part of the polysaccharide is not required for its antioxidant activity. The 66-kDa glycoprotein within PS1 may also play role in the polysaccharide's antioxidant activity [24]. It was suggested that the antioxidant activity helps the cells cope with the ROS to which they are exposed under stress such as drought, extreme temperature, solar irradiation, *etc.* [24]. Taking together these findings, together with the fact that they are safe to use in food, are stable over a wide range of temperatures/environmental conditions, and have favorable rheological and chemical properties, the PS1 and PS2 polysaccharides may be used as antioxidant additives in food or/and biomedical products. Despite this potential of PS1 and PS2, knowledge on their antioxidant activity is still lacking, and the effect of salt on this activity has not been studied to date.

Herein, the effect of salt on the antioxidant activity of PS1 and PS2 in the presence of soy bean-based infant milk formula was studied. Infant milk formula based on soy bean was chosen as a good platform for oxidative stress because of its composition; it is rich in polyunsaturated fatty acids (PUFA) (Table S1 in supplementary information) which are sensitive to oxidative stress. The results of this study contribute to our understanding of red microalgal polysaccharide structure and function, especially as antioxidants, under various salt conditions.

2. Results and Discussion

The antioxidant activity of red microalgal soluble-sulfated polysaccharide fractions in infant milk formula was evaluated by measuring the Malondialdehyde (MDA)-oxidant product after adding KO_2 as an oxidative reagent. MDA levels were significantly reduced in oxidant-damaged formulas containing either of the polysaccharides, in comparison to the control, see Figure 1. MDA levels were also measured in pure PS solutions with and without KO_2; in both cases MDA levels were negligible; thus it can be assumed that the polysaccharides are probably not affected by KO_2—induced oxidant damage.

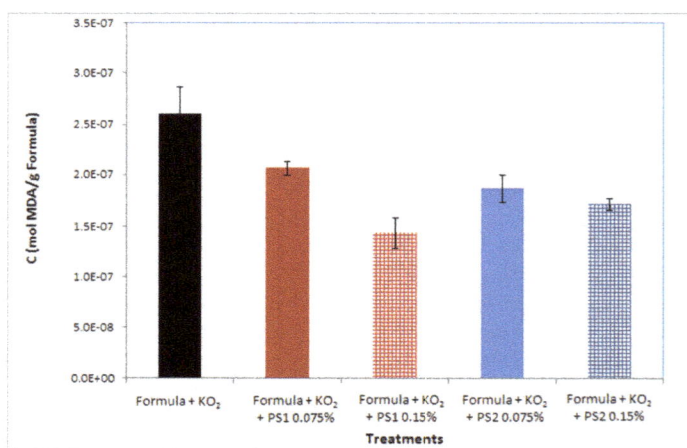

Figure 1. Antioxidant activity of PS1 and PS2 in soy bean milk formula following addition of KO_2. C_{MDA} was measured by thiobarbitoric acid (TBA) method. All treatments contained 0.5 g formula, 0.35 mM KO_2, with or without 0.075/0.15% w/v of either PS1 or PS2 polysaccharides. Values are expressed as mean $\pm$ SD, $n = 3$.

The effective antioxidant activity of polysaccharides derived from red microalgae that was demonstrated in infant milk formula (Figure 1) is in accordance with previous reports demonstrating the antioxidant activity of red microalgal polysaccharides [23–25]. The antioxidant activity of both soluble-sulfated polysaccharide fractions seems to be similar with a slight advantage to PS1. The increase in the polysaccharide concentration (PS1/PS2) also improved their antioxidant activity, but it seems that PS1 has a better dose-concentration effect than that of PS2 (Figure 1). The similarity in the antioxidant capability of these polysaccharides is not surprising since it is known that the structure and the physical behavior of the polysaccharides are similar; they are composed of the same building block with the same predominant monosaccharides and are both negatively charged. The relative advantage of PS1 may be attributed to differences between the polysaccharides, such as sulfate content, ratio of monosaccharides, *etc.* [41].

In order to expand our knowledge on the antioxidant activity of the polysaccharides, it is important to study their activities under different conditions. Ionic strength and cation type were found to influence the intrinsic viscosity of PS1, probably by varying the electrostatic forces and leading to conformational changes [43]. Herein, the effect of potassium, sodium and divalent cation salts with different anions on the antioxidant ability of polysaccharides was tested in an oxidant-damaged infant formula. Salts were added to the polysaccharides prior to their exposure to the oxidant-damaged infant formula, and MDA concentrations were measured as will be explained in the experimental section, to assess the effect of salts on the antioxidant activity of the polysaccharides. All results are displayed as MDA reduction percentages, relative to each experimental setup without the salt.

All types of salts tested, regardless of their anion/cation composition and their concentrations, positively induced the antioxidant activity of both polysaccharides in comparison to the native polysaccharide (control-without salt supplement). We were thus interested in elucidating the effect of different cations and anions, as well as different salt concentrations, on this induction.

Monovalent cation type effect: In order to elucidate the effect of different monovalent cations on the polysaccharides' antioxidant bioactivity, a comparison was performed between NaCl and KCl, between Na_2SO_4 and K_2SO_4 and between Na_3PO_4 and K_3PO_4. Generally, regardless of the anion present, the K+ cations were found to have a stronger influence on the antioxidant activity of the polysaccharides, inducing a marked improvement in MDA level reduction, in comparison to the Na+—formula treatments, as seen by comparing Figures 2 and 3.

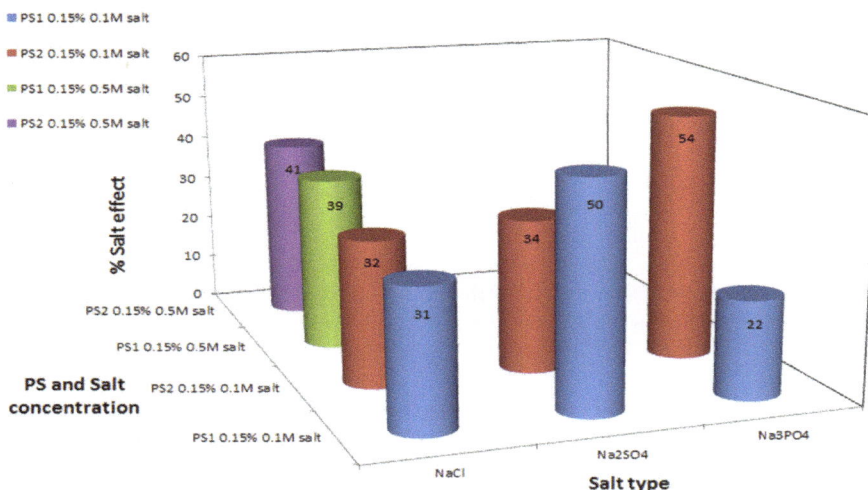

Figure 2. Polysaccharide antioxidant activity induction by sodium salts. C_{MDA} was measured by TBA method. Experimental setup included 0.5 g formula with 0.35 mM KO_2, 0.15% *w/v* polysaccharide and various sodium salts at two different concentrations, in a final volume of 10 mL. Values represent the mean of at least 3 repeats; Maximum standard deviation equals 5%.

The cation antioxidant enhancement effect can be explained: cations derived from the salts bind to negatively charged residues in the polysaccharide, such as sulfate and glucuronic acid residues. This interaction with the polysaccharide, depending on the cation type, may change the polysaccharide's three-dimensional structure due to conformational or structural changes in the solution, by masking the electrostatic repulsions between the polysaccharide chain molecules (which decreases association between the polysaccharide chains). The newly formed structure probably better exposes the potential antioxidant sites in comparison to a reduced-salt native polysaccharide. This postulation is based on a former viscometry study, which indicated that salts can change the polysaccharide conformational structure [43]. In our case, the potassium ions probably induced a change in the intermolecular bonds in the polysaccharide—leading to better exposure of the antioxidant sites in comparison to the three dimensional structure that was formed in the presence of sodium salts. It was expected that an incease in cation concentration will lead to stronger reduction of MDA levels. However, this is not always the case, as seen by comparing the results obtained for Na_2SO_4 with those obtained for Na_3PO_4 (Figure 2), and K_2SO_4 with those obtained for K_3PO_4 (Figure 3)—where an increase in the cation concentration did not lead to an increase in MDA reduction rates by the polysaccharide. This phenomenon can be explained by the anion effect detailed below.

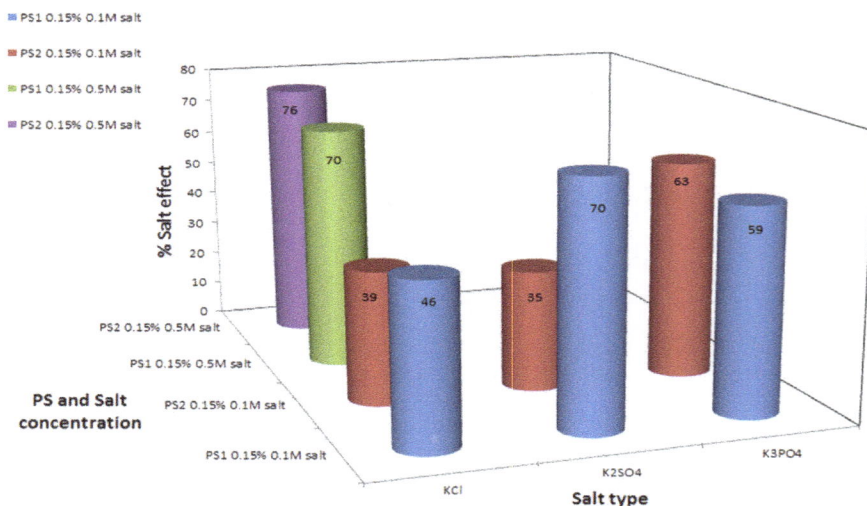

Figure 3. Polysaccharide antioxidant activity induction by potassium salts. C_{MDA} was measured by TBA method. Experimental setup included 0.5 g formula with 0.35 mM KO_2, 0.15% w/v polysaccharide and various potassium salts at two different concentrations, in a final volume of 10 mL. Values represent the mean of at least 3 repeats; Maximum standard deviation equals 8%.

Based on former reports which suggested that proteoglycans and proteins which are arrayed with carbohydrate polymers contribute to induction of polysaccharide antioxidant activities and therefore are probably involved in the polysccharide antioxidative protective activities [24,50–54], another possible explanation for the cation antioxidant effect can be postulated; a specific interaction of cations with the cell-wall proteins, which are part of the polysaccharides, probably through specific amino acid residues (most likely negatively charged) may occur. The interaction may cause conformational or structural changes, leading to better exposure of the antioxidant sites.

Divalent cation type effect: Comparing the divalent cation treatments indicats that Ca^{2+} had the strongest influence on the reduction of MDA levels: 0.5M $CaCl_2$ reduced MDA level by 75% and 81% in PS1 and PS2 formula-treatments, respectively (Figure 4). This finding is not surprising, since

Ca^{2+} ions are known to be effective molecules that stabilize polysaccharides by creating a cross-linked network [55]; for example, Ca^{2+} ions are used for alginate gel formation, which is commonly explained by the egg-box model [56]. According to this model, the structural features of the glucoronic acid allow strong complexation with Ca^{2+} ions, which are embodied in cavities of stiff alginate chains like eggs in a cardboard egg box. Since PS1 and PS2 also contain anionic glucuronic acid blocks and sulfate groups, Ca^{2+} ions probably create a strong complexation with the anions, stabilizing the interactions between polysaccharide chains. The Ca^{2+} ions induced a significantly stronger enhancement in the antioxidant activity of both polysaccharides, in comparison to other divalent ions (Mg^{2+}/Sr^{2+}), see Figure 4. This enhancement is attributed to the conformational change of the polysaccharide which occurs due to the masking of its negative charges by the cation. It is likely that the cation's charge density and radius play an important role in this sense—thus, we suggest that, of the three divalent cations tested, Ca^{2+} is best suited for rendering stability to the polysaccharide, due to the suitability of its charge density and radius.

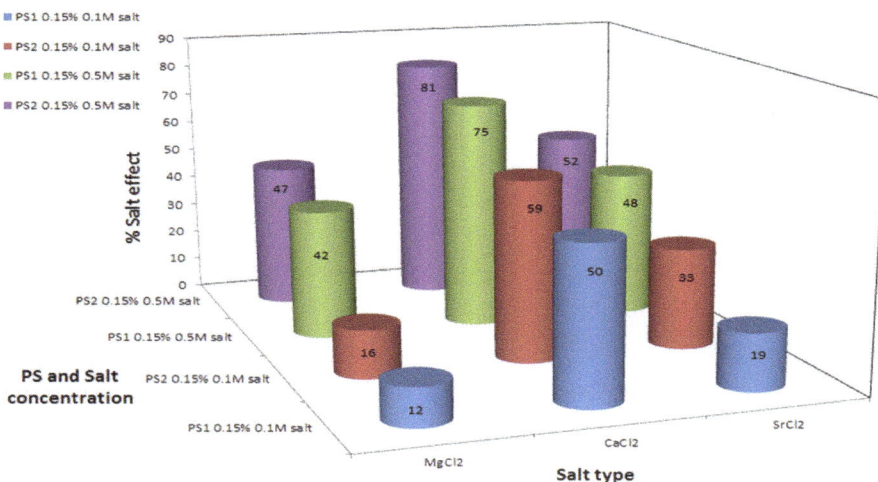

Figure 4. Polysaccharide antioxidant activity induction by divalent cation salts. C_{MDA} was measured by TBA test. Experimental setup included 0.5 g formula with 0.35 mM KO_2, 0.15% w/v polysaccharide, and various divalent salts at two different concentrations, in a final volume of 10 mL. Values represent the mean of at least 3 repeats. Maximum standard deviation equals 8%.

The weakest antioxidant effect among the divalent ion treatments was observed in the low concentration $MgCl_2$-polysaccharide treatments: 0.1 M $MgCl_2$ reduced the MDA level by 12% and 16% in the PS1- and PS2-formula treatments respectively, compared to the same treatment without salt (Figure 4). This is also not suprising, since alginate hydrogels have been obtained with various divalent ions except for Mg^{2+} [57,58].

Anion type effect: It can be postulated that anion type also influence the polysaccharide antioxidant activity. A comparison between the effects of potassium and sodium sulfate salts (Na_2SO_4 and K_2SO_4), and between the potassium and sodium phosphate salts (Na_3PO_4 and K_3PO_4) point out that the type of anion does not only play a role in the induction of the polysaccharides' antioxidant activity, but also differs between the two polysaccharides examined. For PS1 % salt effect values were much higher in the presence of $SO_4{}^{2-}$ in comparison to $PO_4{}^{3-}$ ions, the opposite effect was apparent in PS2 system, % salt effect values were higher in the presence of phosphate ions in comparison to sulfate ions, see Figures 2 and 3.

The counter anion (e.g., PO_4^{3-} in the presence of PS1-formula or SO_4^{-2} in the presence of PS2-formula), can cause the salt and/or the polysacchride to aggregate. In the case of salt aggregation the actual concentration of the free cations which could bind to the PS is not the analytical concentration. In the case of polysaccharide aggregation, the concentration of available polysaccharide chains is reduced, leading to less interactions between the cations and the negatively charged residues within the polysaccharide chains. The different effect that the anions have on the two polysaccharides is probably due to their chemical structure difference.

Salt concentration effect: Salt concentration affects the antioxidant activity of both polysaccharides. High concentration of each salt (including $NaCl/KCl/MgCl_2/CaCl_2/SrCl_2$), 0.5 M, was resulted in higher % salt effect, indicating that the polysaccharides act as better antioxidants in comparison to the low salt concentrations, 0.1 M (Figures 2–4). The antioxidant activity is not linearly dose-dependent and differs in correspondence to the salt type, e.g., formula-polysaccharide treatments that contained 0.5 M KCl reduced the MDA levels up to two fold in comparison to treatments that contained 0.1 M KCl, see Figure 3. Increasing the salt concentration in the PS1-$MgCl_2/SrCl_2$ formula-treatments, Figure 4, also led to an increase in the antioxidant activity (up to three fold of MDA reduction). Noticeably, increasing the NaCl concentration had only a slight effect on the MDA levels in comparison to the other salts studied (Figure 2). The nonlinear effect of the salt concentration could be explained by the same explanation as was explained above in the anions effect. High concentration could cause to salt aggregation which reduced the actual free cation concentration in the solution. As a result less cation-polysaccharide interactions could be formed.

3. Experimental Section

3.1. Materials

All solutions were prepared from analytical-grade chemicals and double distilled water (DDW) that was passed through a Millipore setup at a final resistivity that was above 10 MΩ/cm.

All reagents and solvents were of analytical grade. Trichloroacetic acid (TCA), thiobarbituric acid (TBA), and butylated hydroxyl toluene (BHT), KO_2, KCl, NaCl, K_2SO_4, Na_2SO_4, K_3PO_4, Na_3PO_4, $MgCl_2$, $CaCl_2$, $SrCl_2$ were purchased from Sigma-Aldrich (Rechovot, Israel). Milk sample was a commercial newborn formula based on soy bean; its composition is given in Table S1 in the supplementary information.

3.2. Algae and Growth Conditions

P. sp. (UTEX 637) was obtained from Culture Collection of Texas University, Austin, Texas, USA. *P. arg.* (B111.79) was obtained from the culture collection of the University of Göttingen, Göttingen, Germany. Both species were cultivated in 250 mL Erlenmeyer flasks, each containing 100 mL of growth medium—*P.* sp. in artificial sea water (ASW) [59], and *P. arg.* in freshwater [45]. Algae were grown under shaking speed of 100 rpm; temperature of (25 ± 3) °C; illumination was supplied continuously from above with fluorescent cool-white lamps at a photon flux density of 90 μmol photons $m^{-2} \cdot s^{-1}$; aeration was provided through bubbling air containing 2%–3% CO_2 into the shaker. All cultures were inoculated with 48-h-old cells. Initial cell concentration was adjusted to 2×10^6 cells/mL.

3.3. Isolation and Quantitative Analysis of Soluble-Sulfated Polysaccharide Fraction

Cultures at stationary phase of growth were centrifuged (3000 *g*, 10 min, 4 °C) and the supernatants containing the dissolved polysaccharides were dialyzed (MW cutoff 8000 Da) against DDW at 4 °C until the conductivity of the water reached 300 μs × cm^{-1}. The dialyzed polysaccharide was then freeze-dried and resuspended in DDW to reach 1% *w*/*v*, comprising the PS stock. The polysaccharide concentration was determined by a known procedure [60], with D-galactose (Sigma G-0625, St. Louis, MO, USA) as the standard, in a concentration range of 0.1–1 μg/mL.

3.4. Sample Preparation

The sample preparation procedure included three steps: (1) Preparation of the PS-salt solution samples: The diluted PS solution (0.15% w/v in DDW) was incubated for 60 min with an appropriate concentration of one of the following salts: KCl, NaCl, K_2SO_4, Na_2SO_4, K_3PO_4, Na_3PO_4, $MgCl_2$, $CaCl_2$, $SrCl_2$ at 0.1 M, or KCl, NaCl, $MgCl_2$, $CaCl_2$, $SrCl_2$ at 0.5 M. The solubility of K_2SO_4, Na_2SO_4, K_3PO_4, Na_3PO_4 in aqueous solutions is lower than 0.5 M, therefore the only concentration used was 0.1 M; (2) Preparation of the PS-formula systems: The PS-salt solution samples were added to 0.5 g milk powder with 35 mM KO_2, at a total volume of 10 mL; (3) Malondialdehyde (MDA) levels were determined according to the thiobarbitoric acid (TBA) method described below. Control experiments were performed separately in order to study the oxidation of the polysaccharides and the formula by KO_2 (without salts) using the same procedure detailed above. To extend our knowledge regarding the dose-dependency of the polysaccharide effect on the milk powder without salt supplement, two polysaccharide concentrations (0.075/0.15% w/v) were studied for PS1 and PS2.

3.5. Production of Free Radicals

The free radicals were formed by addition of KO_2 to the samples. As mentioned above O_2^- is known as an oxidative agent [4,61].

The KO_2 solution was prepared fresh for each experiment and added to the sample immediately after its preparation.

3.6. Malondialdehyde (MDA) Measurements

MDA levels were measured spectroscopically according to Fenaille *et al.* [62] with minor modifications. The assay is based on a reaction between oxidized lipids and TBA under acidic conditions to yield a pink chromogen with a maximum absorbance at 532 nm. After sample preparation an aliquot of the slurry (2 mL) was transferred to a 5-mL tube, to which the following were added: 1.6 mL DDW, 0.8 mL TBA 0.67%, 0.8 mL TCA 5% and 0.0128 g of BHT. The mixture was homogenized and centrifuged at 3500 rpm for 30 min. The supernatant was incubated in a 70 °C water bath for 60 min, and cooled to room temperature before measuring absorbance. The color produced by the chemical reaction was read at 532 nm in a Cary 100 Bio UV-Visible spectrophotometer, and the amount of MDA formed was determined by using the molar extinction coefficient $\varepsilon(532\,\text{nm}) = 1.56 \times 10^{-5}$ cm·nmol^{-1} [62].

3.7. Calculation of % Salt Effect

The induction levels of the PS antioxidant activity under different salt environments was calculated according to the following formula: % salt effect $= (A_0 - A)/A_0 \times 100$, where A_0 is the absorbance of the control sample without salt, and A is the absorbance of the test sample with the salt.

3.8. Statistical Analyses

All data were expressed as means $\pm$ standard deviation (SD) of at least three replications, and the ANOVA test was used for statistical analysis. Each experiment was repeated at least three times.

4. Conclusions

Red microalgae polysaccharides hold great potential to be used as agents in various products for human benefit, for example, as natural antioxidants in food or cosmetic products. However, there are very few reports on sulfated red micoalgae polysaccharide antioxidant bioactivities and there are still many open questions to be answered regarding the mechanisms through which the polysaccharide copes with the ROS. The results of this study show that both red microalgae soluble-sulfated polysaccharide fractions (PS1 and PS2) exhibit significant radical scavenging abilities in soy bean formula in the presence of KO_2.

It was demonstrated that salt composition (cation and anion type) and concentration affect the antioxidant activity of these red microalgal polysaccharides. Thus we can postulate that the salt ions intefere with the polysaccharide chain interactions by changing the intermolecular bonds and causing changes in the three-dimensional structure of the polysaccharide, leading to a newly formed structure that better exposes the antioxidant sites. Among the cations which were tested, Ca^{2+} had the strongest enhancement effect on antioxidant activities of both polysaccharides. Since proteins are also part of the cell-wall complex polysaccharides [26,38] they can be involved in the newly-formed three-dimensional structure. It seems, however, that the activity of the polysaccharides does not result from one single component but from synergism of its oligomeric components. Due to the polysaccharide complexity and the lack of knowledge regarding its structure, the mechanism of the salt effect cannot be explained in depth.

Here we report for the first time the potential of the polysaccharides to act as an antioxidant ingredient in infant food formula, which can be used as a model for food products, as well as for other disciplines. We further demonstrate the enhancing effect of salts on this antioxidant activity and propose a hypothesis which can explain the differential effect of the various cations and anions on the two polysaccharides. The physicochemical properties which enable the polysaccharides to be used as a stabilizer or thickener, the relative ease in which they can be produced by controlling algal growth conditions and their uniqueness as a natural material which is dissolved in aqueous systems, reinforce their potential to be used as antioxidants in industrial systems. Our findings have a contribution in furthering knowledge on the polysaccharides' antioxidant behavior under different environmental conditions.

Author Contributions: All authors contributed equally to the manuscript preparation.

Conflicts of Interest: The authors declare no conflict of interest.

References

1. Angele-Martinez, C.; Goodman, C.; Brumaghim, J. Metal-mediated DNA damage and cell death: Mechanisms, detection methods, and cellular consequences. *Metallomics* **2014**, *6*, 1358–1381. [CrossRef] [PubMed]
2. Bregano, J.W.; Dichi, I. Inflammatory bowel disease. In *Role of Oxidative Stress Chronic Diseases*; Dichi, I., Breganó, J.W, Simão, A.N.C., Cecchini, R., Eds.; CRC Press, Taylor & Francis Group: Boca Raton, FL, USA, 2014; pp. 501–520.
3. Furlan, A.L.; Jobin, M.-L.; Buchoux, S.; Grelard, A.; Dufourc, E.J.; Gean, J. Membrane lipids protected from oxidation by red wine tannins: A proton NMR study. *Biochimie* **2014**, *107*, 82–90. [CrossRef] [PubMed]
4. Harrison, I.P.; Selemidis, S. Understanding the biology of reactive oxygen species and their link to cancer: NADPH oxidases as novel pharmacological targets. *Clin. Exp. Pharmacol. Physiol.* **2014**, *41*, 533–542. [CrossRef] [PubMed]
5. Butterfield, D.A. Amyloid β-peptide (1-42)-induced oxidative stress and neurotoxicity: Implications for neurodegeneration in Alzheimer's disease brain. *Free Radic. Res.* **2002**, *36*, 1307–1313. [CrossRef] [PubMed]
6. Halliwell, B.; Aruoma, O.I. DNA damage by oxygen-derived species. Its mechanism and measurement in mammalian systems. *FEBS Lett.* **1991**, *281*, 9–19. [CrossRef]
7. Froelich, L.; Riederer, P. Free radical mechanisms in dementia of Alzheimer type and the potential for antioxidative treatment. *Arzneimittelforschung* **1995**, *45*, 443–446.
8. Yang, C.S.; Landau, J.M.; Huang, M.-T.; Newmark, H.L. Inhibition of carcinogenesis by dietary polyphenolic compounds. *Annu. Rev. Nutr.* **2001**, *21*, 381–406. [CrossRef] [PubMed]
9. Ramsaha, S.; Aumjaud, B.E.; Neergheen-Bhujun, V.S.; Bahorun, T. Polyphenolic rich traditional plants and teas improve lipid stability in food test systems. *J. Food Sci. Technol.* **2015**, *52*, 773–782. [CrossRef] [PubMed]
10. Park, P.-J.; Jung, W.-K.; Nam, K.-S.; Shahidi, F.; Kim, S.-K. Purification and characterization of antioxidative peptides from protein hydrolyzate of lecithin-free egg yolk. *J. Am. Oil Chem. Soc.* **2001**, *78*, 651–656. [CrossRef]

11. Kazi, A.; Deshmukh, S.; Mirza, R.; Murtadak, S. A comprehensive review on flavonoids. *World J. Pharm. Res.* **2015**, *4*, 560–572.

12. Mao, G.; Zhang, S.; Song, H.; Ding, S.; Zhu, P.; Wang, X.; Liang, C. Synthesis, biological activities and therapeutic properties of esculetin and its derivatives. *J. Chem. Pharm. Res.* **2015**, *7*, 122–130.

13. Pena-Ramos, E.A.; Xiong, Y.L. Antioxidative activity of whey protein hydrolysates in a liposomal system. *J. Dairy Sci.* **2001**, *84*, 2577–2583. [CrossRef]

14. Sharma, R.; Rani, D.; Jain, N.; Kantwa, S.M.; Jaitawat, A. Cinnamon—A natural replacement for synthetic drugs: A review. *Int. J. Curr. Res. Biosci. Plant Biol.* **2015**, *2*, 69–78.

15. Jiao, G.; Yu, G.; Zhang, J.; Ewart, H.S. Chemical structures and bioactivities of sulfated polysaccharides from marine algae. *Mar. Drugs* **2011**, *9*, 196–223. [CrossRef] [PubMed]

16. Ruperez, P.; Ahrazem, O.; Leal, J.A. Potential Antioxidant Capacity of Sulfated Polysaccharides from the Edible Marine Brown Seaweed Fucus vesiculosus. *J. Agric. Food Chem.* **2002**, *50*, 840–845. [CrossRef] [PubMed]

17. Rocha de Souza, M.C.; Marques, C.T.; Guerra Dore, C.M.; Ferreira da Silva, F.R.; Oliveira Rocha, H.A.; Leite, E.L. Antioxidant activities of sulfated polysaccharides from brown and red seaweeds. *J. Appl. Phycol.* **2007**, *19*, 153–160. [CrossRef] [PubMed]

18. Wang, J.; Zhang, Q.; Zhang, Z.; Song, H.; Li, P. Potential antioxidant and anticoagulant capacity of low molecular weight fucoidan fractions extracted from *Laminaria japonica*. *Int. J. Biol. Macromol.* **2010**, *46*, 6–12. [CrossRef] [PubMed]

19. Dore, C.M.; das C Faustino Alves, M.G.; Will, L.S.; Costa, T.G.; Sabry, D.A.; de Souza Rêgo, L.A.; Accardo, C.M.; Rocha, H.A.; Filgueira, L.G.; Leite, E.L. A sulfated polysaccharide, fucans, isolated from brown algae Sargassum vulgare with anticoagulant, antithrombotic, antioxidant and anti-inflammatory effects. *Carbohydr. Polym.* **2013**, *91*, 467–475. [CrossRef] [PubMed]

20. Qi, H.; Zhang, Q.; Zhao, T.; Hu, R.; Zhang, K.; Li, Z. In vitro antioxidant activity of acetylated and benzoylated derivatives of polysaccharide extracted from *Ulva pertusa* (Chlorophyta). *Bioorg. Med. Chem. Lett.* **2006**, *19*, 2441–2445. [CrossRef] [PubMed]

21. Xue, Z.; Xue, C.; Cai, Y.; Wang, D.; Fang, Y. The study of antioxidant activities of fucoidan from *Laminaria japonica*. *High Technol. Lett.* **2005**, *11*, 91–94.

22. Costa, L.S.; Fidelis, G.P.; Cordeiro, S.L.; Oliveira, R.M.; Sabry, D.A.; Camara, R.B.; Nobre, L.T.; Costa, M.S.; Almeida-Lima, J.; Farias, E.H.; *et al.* Biological activities of sulfated polysaccharides from tropical seaweeds. *Biomed. Pharmacother.* **2010**, *64*, 21–28. [CrossRef] [PubMed]

23. Sun, L.; Wang, C.; Shi, Q.; Ma, C. Preparation of different molecular weight polysaccharides from *Porphyridium cruentum* and their antioxidant activities. *Int. J. Biol. Macromol.* **2009**, *45*, 42–47. [CrossRef] [PubMed]

24. Tannin-Spitz, T.; Bergman, M.; van-Moppes, D.; Grossman, S.; Arad, S. Antioxidant activity of the polysaccharide of the red microalga Porphyridium sp. *J. Appl. Phycol.* **2005**, *17*, 215–222. [CrossRef]

25. Chen, B.; You, W.; Huang, J.; Yu, Y.; Chen, W. Isolation and antioxidant property of the extracellular polysaccharide from *Rhodella reticulata*. *World J. Microbiol. Biotechnol.* **2010**, *26*, 833–840. [CrossRef]

26. Wang, J.; Hu, S.; Nie, S.; Yu, Q.; Xie, M. Reviews on Mechanisms of in Vitro Antioxidant Activity of Polysaccharides. *Oxid. Medi. Cell. Longev.* **2015**; in press.

27. Bold, H.C.; Wynne, M.J. *Introduction to the Algae: Structure and Reproduction*, 2nd ed.; Prentice Hall Inc.: Englewood Cliiffs, NJ, USA, 1985.

28. Ramus, J. Rhodophyte unicells, biopolymer, physiology and production. In *Algal Biomass Technologies—An Interdisciplinary Perspective*; Barclay, W.R., McIntosh, R.P., Eds.; Cramer: Berlin, Germany, 1986; pp. 51–55.

29. Arad, M.S. Production of sulfated polysaccharides from red unicellular algae. In *Algal Biotechnology—An Interdisciplinary Perspective*; Stadler, T., Mollion, J., Verduset, M.C., Eds.; Elsevier Applied Science: London, UK, 1988; pp. 65–87.

30. Arad, S.; Levy-Ontman, O. Red microalgal cell wall polysaccharides: Biotechnological aspects. *Curr. Opin. Biotechnol.* **2010**, *21*, 358–364. [CrossRef] [PubMed]

31. Ucko, M.; Cohen, E.; Gordin, H.; Arad, S. Relationship between the unicellular red alga *Porphyridium* sp. and its predator, the dinoflagellate *Gymnodinium* sp. *Appl. Environ. Microbiol.* **1989**, *55*, 2990–2994. [PubMed]

32. Arad, S.; Keristovesky, G.; Simon, B.; Barak, Z.E.; Geresh, S. Biodegradation of the sulfated polysaccharide of *Porphyridium* by soil bacteria. *Phytochemistry* **1993**, *32*, 287–290. [CrossRef]

33. Percival, E.; Foyle, R.A.J. The extracellular polysaccharides of *Porphyridium cruentum* and *Porphyridium aerugineum. Carbohydr. Res.* **1979**, *72*, 165–176. [CrossRef]

34. Geresh, S.; Arad, S. The extracellular polysaccharides of the red microalgae: Chemistry and rheology. *Bioresour. Technol.* **1991**, *38*, 195–201. [CrossRef]

35. Heaney-Kieras, J.; Chapman, D.J. Structural studies on the extracellular polysaccharide of the red alga, *Porphyridium cruentum. Carbohydr. Res.* **1976**, *52*, 169–177. [CrossRef]

36. Geresh, S.; Adin, I.; Yarmolinsky, E.; Karpasas, M. Characterization of the extracellular polysaccharide of *Porphyridium* sp.: Molecular weight determination and rheological properties. *Carbohydr. Polym.* **2002**, *50*, 183–189. [CrossRef]

37. Geresh, S.; Dubinsky, O.; Arad, M.S.; Christiaen, D.; Glaser, R. Structure of 3-*O*-(alpha-D-glucopyranosyluronic acid)-L-galactopyranose, an aldobiouronic acid isolated from the polysaccharides of various unicellular red algae. *Carbohydr. Res.* **1990**, *208*, 301–305. [CrossRef]

38. Shrestha, R.P.; Weinstein, Y.; Bar-Zui, D.; Arad, S. A glycoprotein noncovalently associated with cell-wall polysaccharide of the red microalga *Porphyridium* sp. (Rhodophyta). *J. Phycol.* **2004**, *40*, 568–580. [CrossRef]

39. Levy-Ontman, O.; Arad, S.; Harvey, D.J.; Parsons, T.B.; Fairbanks, A.; Tekoah, Y. Unique N-glycan moieties of the 66-kDa cell wall glycoprotein from the red microalga *Porphyridium* sp. *J. Biol. Chem.* **2011**, *286*, 21340–21352. [CrossRef] [PubMed]

40. Ucko, M.; Shrestha, R.P.; Mesika, P.; Bar-Zvi, D.; Arad, S. Glycoprotein moiety in the cell wall of the red microalga *Porphyridium* sp. (Rhodophyta) as the biorecognition site for the *Crypthecodinium cohnii*-like dinoflagellate. *J. Phycol.* **1999**, *35*, 1276–1281. [CrossRef]

41. Arad, S.; Levy-Ontman, O. Sulfated polysaccharides in the cell wall of red microalgae. In *Handbook of Biopolymer-Based Materials: From Blends and Composites to Gels and Complex*; Sabu, T., Dominique, D., Christophe, C., Jyotishkumar, P., Eds.; Wiley-VCH Verlag: Berlin, Germany, 2013; Volume 2, pp. 351–370.

42. Yaron, A.; Cohen, E.; Arad, S.M. Stabilization of aloe vera gel by interaction with sulfated polysaccharides from red microalgae and with xanthan gum. *J. Agric. Food Chem.* **1992**, *40*, 1316–1320. [CrossRef]

43. Eteshola, E.; Gottlieb, M.; Arad, S. Dilute solution viscosity of red microalga exopolysaccharide. *Chem. Eng. Sci.* **1996**, *51*, 1487–1494. [CrossRef]

44. Ginzberg, A.; Korin, E.; Arad, S.M. Effect of drying on the biological activities of a red microalgal polysaccharide. *Biotechnol. Bioeng.* **2008**, *99*, 411–420. [CrossRef] [PubMed]

45. Savins, J.G. Oil Recovery Process Employing Thickened Aqueous Driving Fluid. U.S. Patent 4,079,544, 21 March 1978.

46. Arad, S.; Weinstein, Y. Novel lubricants from red microalgae: Interplay between genes and products. *Biomedic* **2003**, *1*, 32–37.

47. Arad, M.S.; Richmond, A. Industrial production of microalgal cell-mass and secondary products—Species of high potential: *Porphyridium* sp. In *Handbook of Microalgal Culture: Biotechnology and Applied Phycology*; Richmond, A., Ed.; Blackwell Publishing Ltd.: Oxford, UK, 2004; pp. 289–299.

48. Arad, M.S.; Rapoport, L.; Moshkovich, A.; van-Moppes, D.; Karpasas, M.; Golan, R.; Golan, Y. Superior biolubricant from a species of red microalga. *Langmuir* **2006**, *22*, 7313–7317. [CrossRef] [PubMed]

49. Eteshola, E.; Karpasas, M.; Arad, S.; Gottlieb, M. Red microalga exo polysaccharides. Part 2. Study of the rheology, morphology, and thermal gelation of aqueous preparations. *Acta Polym.* **1998**, *49*, 549–556. [CrossRef]

50. Liu, F.; Ooi, V.E.C.; Chang, S.T. Free radical scavenging activities of mushroom polysaccharide extracts. *Life Sci.* **1997**, *60*, 763–771. [CrossRef]

51. Liu, D.; Sheng, J.; Li, Z.; Qi, H.; Sun, Y.; Duan, Y.; Zhang, W. Antioxidant activity of polysaccharide fractions extracted from *Athyrium multidentatum* (Doll.) Ching. *Int. J. Biol. Macromol.* **2013**, *56*, 1–5. [CrossRef] [PubMed]

52. Leung, P.H.; Zhao, S.; Ho, K.P.; Wu, J.Y. Chemical properties and antioxidant activity of exopolysaccharides from mycelial culture of *Cordyceps sinensis* fungus Cs-HK1. *Food Chem.* **2009**, *114*, 1251–1256. [CrossRef]

53. Cheung, Y.-C.; Siu, K.-C.; Liu, Y.-S.; Wu, J.-Y. Molecular properties and antioxidant activities of polysaccharide-protein complexes from selected mushrooms by ultrasound-assisted extraction. *Process Biochem.* **2012**, *47*, 892–895. [CrossRef]

54. Huang, Q.-L.; Siu, K.-C.; Wang, W.-Q.; Cheung, Y.-C.; Wu, J.-Y. Fractionation, characterization and antioxidant activity of exopolysaccharides from fermentation broth of a *Cordyceps sinensis* fungus. *Process Biochem.* **2013**, *48*, 380–386. [CrossRef]

55. Dumitriu, S.; Vidal, P.F.; Chornet, E. Hydrogels based on polysaccharides. In *Polysaccharides in medical application*; Dumitriu, S., Ed.; Marcel Dekker, Inc.: New York, NY, USA, 1996; pp. 125–241.

56. Shao, F.; Ankur, T.; Diana, M.S.; Riccardo, L.B.; Ira, S.B.; Sachin, V.; Eric, J.M.; Lawrence, H.B. Relevance of Rheological Properties of Sodium Alginate in Solution to Calcium Alginate Gel Properties. *Res. Artic.* **2011**, *12*, 453–460.

57. Seely, G.R.; Hart, R.L. Binding of alkaline earth metal ions to alginate. *Macromolecules* **1974**, *7*, 706–710. [CrossRef] [PubMed]

58. Florian-Algarin, V.; Acevedo, A. Rheology and Thermotropic Gelation of Aqueous Sodium Alginate Solutions. *J. Pharm. Innov.* **2010**, *5*, 37–44. [CrossRef]

59. Jones, R.H.; Speer, H.L.; Kury, W. Studies on the growth of the red alga *Porphyridium cruentum*. *Physiol. Plant.* **1963**, *16*, 636–643. [CrossRef]

60. Dubois, M.; Gilles, K.A.; Hamilton, J.K.; Rebers, P.A.; Smith, F. Colorimetric method for determination of sugars and related substances. *Anal. Chem.* **1956**, *28*, 350–356. [CrossRef]

61. Halliwell, B. Free radicals and antioxidants—Quo vadis? *Trends Pharmacol. Sci.* **2011**, *32*, 125–130. [CrossRef]

62. Fenaille, F.; Mottier, P.; Turesky, R.J.; Ali, S.; Guy, P.A. Comparison of analytical techniques to quantify malondialdehyde in milk powders. *J. Chromatogr. A* **2001**, *921*, 237–245. [CrossRef]

marine drugs

MDPI

Article

Preliminary Characterization, Antioxidant Properties and Production of Chrysolaminarin from Marine Diatom *Odontella aurita*

Song Xia [1,2,†], **Baoyan Gao** [1,†], **Aifen Li** [1], **Jihai Xiong** [2], **Ziqiang Ao** [2] and **Chengwu Zhang** [1,*]

[1]   Institute of Hydrobiology, Jinan University, Guangzhou 510632, China; xiasongsummer212@163.com (S.X.); gaobaoyan1211@126.com (B.G.); tiger@jnu.edu.cn (A.L.)

[2]   Institute of Energy Research, Jiangxi Academy of Sciences, Nanchang 330096, China; xjh6110@vip.sina.com (J.X.); aoziqiang628@163.com (Z.A.)

*   Author to whom correspondence should be addressed; tzhangcw@jnu.edu.cn; Tel./Fax: +86-20-85224366.

†   These authors contributed equally to this work.

Received: 26 June 2014; in revised form: 9 September 2014; Accepted: 9 September 2014; Published: 23 September 2014

Abstract: A new chrysolaminarin, named CL2, with a molecular mass of 7.75 kDa, was purified from the marine diatom, *Odontella aurita*, using DEAE-52 cellulose anion-exchange chromatography and Sephadex G-200 gel-filtration chromatography. The monosaccharide and structural analysis revealed that CL2 was a glucan mainly composed of glucose, which was linked by the β-D-(1→3) (main chain) and β-D-(1→6) (side chain) glycosidic bond, demonstrated by infrared spectroscopy (IR) and nuclear magnetic resonance (NMR). The antioxidant activity tests revealed that the CL2 presented stronger hydroxyl radical scavenging activity with increasing concentrations, but less was effective on reducing power analysis and scavenging 1,1-diphenyl-2-picrylhydrazyl (DPPH) radical. The influences of nitrogen concentration and light intensity on chrysolaminarin production of *O. aurita* were further investigated in a glass column photobioreactor, and a record high chrysolaminarin productivity of 306 mg L^{-1} day^{-1} was achieved. In conclusion, the chrysolaminarin CL2 from *O. aurita* may be explored as a natural antioxidant agent for application in aquaculture, food and pharmaceutical areas.

Keywords: *Odontella aurita*; chrysolaminarin; structural characteristics; antioxidant activity; productivity

---

1. Introduction

Oxidative stress causes lots of damage to biological macromolecules, such as nucleic acids, proteins, lipids and carbohydrates, which may lead to the development of chronic and degenerative ailments [1]. Although a variety of synthetic chemicals, such as phenolic compounds, are found to be effective radical scavengers, they usually have side effects. Thus, many efforts have been spent on searching for compounds with antioxidant activity and low cytotoxicity from natural materials.

Recently, accumulated evidence has demonstrated that natural polysaccharides are effective antioxidants for scavenging reactive oxygen species (ROS) [2–4]. Marine diatoms accumulate β-D-1,3-glucans, also called chrysolaminarin, as an energy storage carbohydrate, especially upon nutrient-depletion conditions [5]. The chrysolaminarin from several diatoms have been characterized and found to be a β-1,3-glucan with a degree of polymerization (DP) in the range of 5–60 and a degree of branching (DB) of 0–0.2 at Position 6 [6]. A large number of studies indicated that β-glucans obtained from various organisms, such as plants, algae and microorganisms, are effective agents in scavenging ROS, stimulating immunity in fish, as well as treating diseases, like cancer, infection, inflammation and influenza [7–10]. However, little attention has been devoted to the production and biological activities of chrysolaminarin from marine diatoms.

The marine diatom, *Odontella aurita*, has been industrially cultured in raceways and used as a dietary supplement rich in ω-3 polyunsaturated fatty acids (PUFAs) for several years [11,12]. Some other bioactive compounds contained in this microalga, such as fucoxanthin and phytosterols, have been isolated, structurally elucidated and proved to be beneficial to human health [13–15]. No studies have been reported, to our knowledge, on the structural characteristics and bioactivities of chrysolaminarin in *O. aurita*. In this study, a new storage chrysolaminarin was purified from *O. aurita*. Its monosaccharide composition, chemical structure and antioxidant activity were characterized. The production of chrysolaminarin from *O. aurita* was also investigated in glass column photobioreactors. This research is aimed at characterizing the structure and antioxidant capacity of a new chrysolaminarin purified from the marine diatom, *O. aurita*, and evaluating its production potential in glass column photobioreactor.

2. Results and Discussion

2.1. Isolation, Purification and Characterization of Chrysolaminarin

2.1.1. Purification and Homogeneity of Chrysolaminarin

The crude polysaccharide was isolated from freeze-dried *O. aurita* and then chromatographed on a DEAE-52 cellulose column (Pharmacia, Uppsala, Sweden). After being gradient eluted with an aqueous solution of NaCl (0.1, 0.3 and 0.5 M), one major peak, named CL1, was obtained from the 0.1 M NaCl eluate (Figure 1). Next, the fraction CL1 was applied to Sephadex G-200 gel-filtration column chromatography (Pharmaci, Uppsala, Sweden) for further purification, yielding a single, symmetric and sharp peak, which indicated that the obtained polysaccharide, named CL2 (chrysolaminarin 2), was homogeneous (Figure 2). The UV spectrum showed no significant absorbance at 260 nm or 280 nm, indicating that CL2 contained no protein or nucleic acid (data not shown).

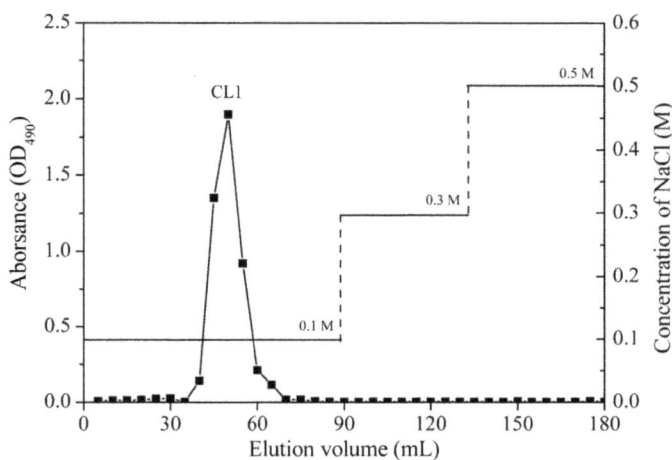

Figure 1. DEAE-cellulose column elution profile of crude polysaccharide from *O. aurita*.

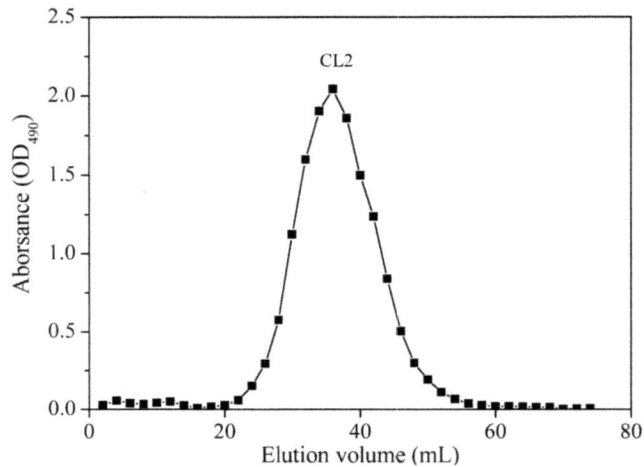

Figure 2. Sephadex G-200 gel-filtration chromatogram of the fraction CL1 (chrysolaminarin 1) obtained from DEAE-cellulose column elution.

2.1.2. Monosaccharide Composition and Molecular Weight of CL2

Chrysolaminarins from various diatoms span a great number of different molecular weights, from ~1 to 40 kDa [6]. The average molecular weight of CL2 was found to be 7.75 kDa based on the results of gel-filtration chromatography. The monosaccharide composition of CL2 was analyzed by complete acid hydrolysis and the GC-MS analysis methods. The results shown in Table 1 indicate that the sugar components of CL2 are mainly composed of glucose (82.23%) with smaller amounts of mannose (13.27%) and traces of ribose, arabinose, xylose and galactose.

Table 1. Monosaccharide composition of CL2 from *O. aurita*.

Sugar Components (Total Sugar [a] %)					
Glucose	Mannose	Ribose	Arabinose	Xylose	Galactose
82.23	13.27	0.46	3.62	0.26	0.16

[a] Total sugar: the sum of the six monosaccharides.

2.1.3. FTIR Spectra Analysis

As shown in Figure 3, the FTIR spectra of CL2 showed a significant, strong broad characteristic peak at around 3431 cm^{-1}, corresponding to the stretching vibration of O-H groups, as well as a C-H band at 2923 cm^{-1} [16,17]. Two prominent absorption bands between 1200 and 1000 cm^{-1} are dominated by ring vibrations overlapped with C-O glycosidic band vibration [10,18,19]. The stretching peak at 1639 cm^{-1} and a weak stretching peak appeared at 1377 cm^{-1} were due to the presence of carboxyl groups [20]. The characteristic peak of β-glycosidic linkage at 889 cm^{-1} demonstrated that the obtained polysaccharide CL2 was a β-type polysaccharide [20].

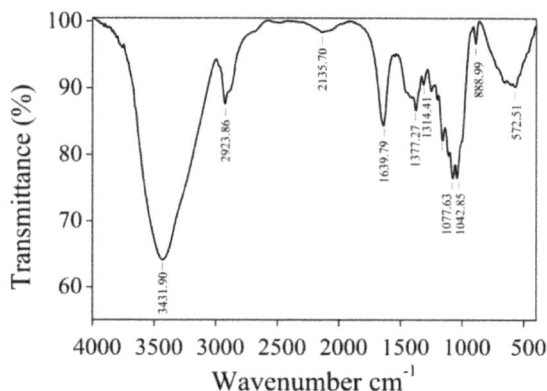

Figure 3. The FTIR spectra of CL2 from *O. aurita*.

2.1.4. NMR Spectra Analysis

The ^{1}H-NMR and ^{13}C-NMR spectra of CL2 are presented in Figure 4. The anomeric proton single at δ 4.39 and 4.08 ppm in the ^{1}H NMR were assigned to H-1 of the β-1,3-linkage and the β-1,6-linkage, respectively [21,22], which agree with the presence of an IR band at 889 cm^{-1}. Based on the respective peak areas at 4.39 and 4.08 ppm on the ^{1}H-NMR spectrum, the ratio of the β-1,3- to β-1,6-linkage was estimated to be 4:1. The ^{13}C-NMR spectrum showed major signals at δ 104.2–104.6, 86.5–87.3, 78.1–78.4, 77.6–78.1, 75.1–75.4, 71.5, 69.1–70.6 and 62.3–63.1 (Figure 4b). These signals are in agreement with the results from previous analysis of microalgae *Chaetoceros muelleri* and *Pleurochrysis haptonemofera* [21,23], which indicated that the CL2 obtained from *O. aurita* has a β-D-(1→3)- (main chain) and β-D-(1→6) (branch chain)-linked glucopyranan structure. The integrated analysis of structural information demonstrated that the obtained polysaccharide CL2 from *O. aurita* is a medium molecular weight chrysolaminarin, mainly composed of a β-D-(1→3)- (main chain) and β-D-(1→6) (side chain)-linked glucose.

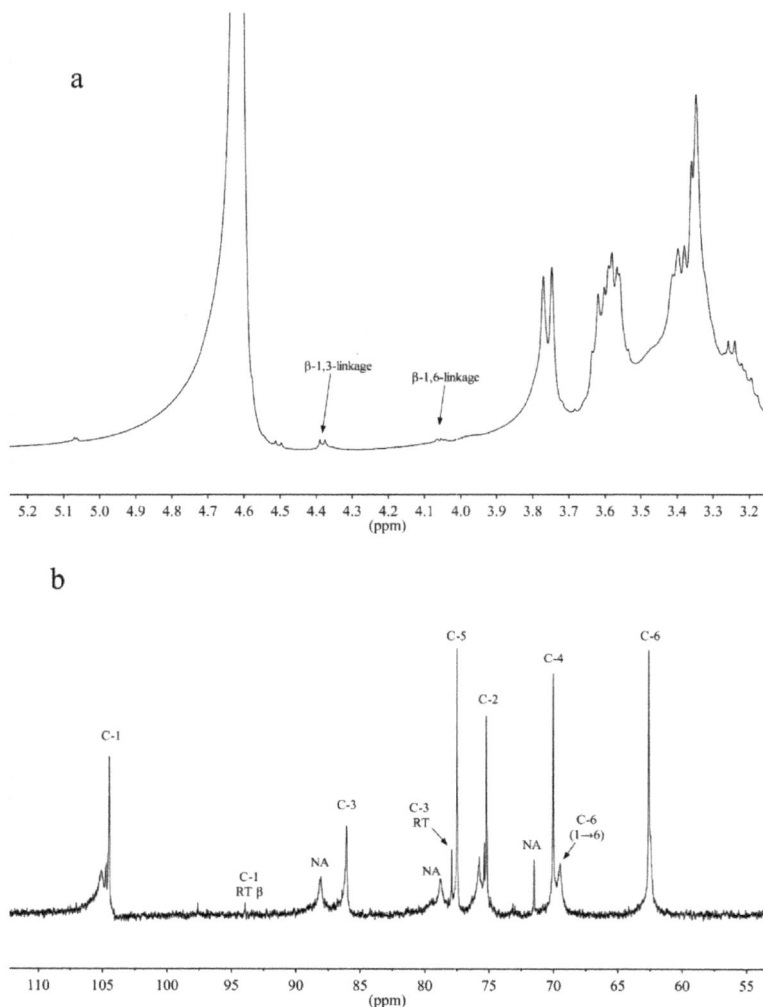

Figure 4. (a) ^{1}H-NMR and (b) ^{13}C-NMR spectrum of CL2 from *O. aurita* (NA: not assigned).

2.2. Assay for Antioxidant Activity

2.2.1. Reducing Power

For the assessment of the reducing power, the Fe^{3+}–Fe^{2+} transformation of CL2 was investigated using the potassium ferricyanide reduction method (Figure 5a). The results revealed that the reducing power of CL2 was weak. At 2 mg mL^{-1}, CL2 showed a reducing power of 0.046 ± 0.015 abs, which gradually increased to 0.554 ± 0.139 abs at 100 mg mL^{-1}; whereas the reducing power of ascorbic acid (VC) reached a plateau of 2.508 ± 0.13 abs at 2 mg mL^{-1}. Kozarski *et al.* [24] declared that starch exhibited no reducing power in their study. Lo *et al.* [25] reported a weak relationship between reducing power and monosaccharide composition, and they also found that the reducing power of polysaccharides was much lower than that of ascorbic acid. Kanmani *et al.* [26] found that the exopolysaccharide from *Streptococcus phocae* exhibited a rather weak reducing power (0.2 abs at 2 mg mL^{-1}), which was much lower than that of ascorbic acid (reaching its plateau of 2.5 abs at 0.4 mg mL^{-1}).

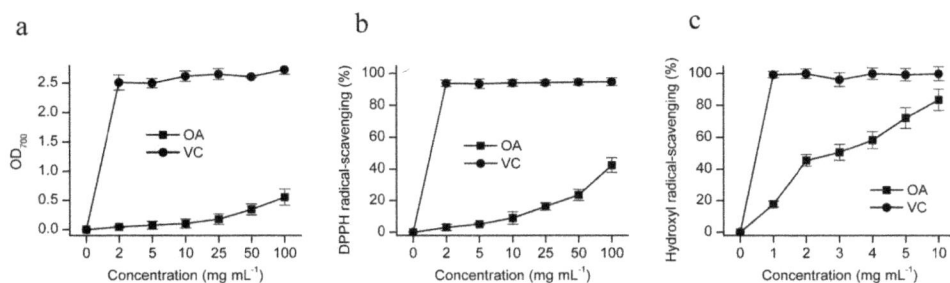

Figure 5. Antioxidant assays for the chrysolaminarin CL2 from *O. aurita*. (**a**) Reducing power; (**b**) scavenging of DPPH radicals; (**c**) scavenging of hydroxyl radicals. Values are the means ± SD (*n* = 3). When error bars cannot be seen, the error is less than the size of the symbol.

2.2.2. DPPH Radical Scavenging Activity

The DPPH free radical has been widely used as a tool to evaluate the antioxidant activity. The results shown in Figure 5b indicated that the scavenging activity of CL2 increased slowly with the increase of dosage in a concentration-dependent manner. When the concentration of CL2 was at 100 mg mL^{-1}, its scavenging activity reached 42.455% ± 4.671%. Ascorbic acid expressed a much higher scavenging ability and reached a plateau of 93.774% ± 2.089% at 2 mg mL^{-1}. The structural characteristics of polysaccharide, such as molecular weight, monosaccharide composition, availability of hydroxyl group and conformation of side chains, were reported to be responsible for the scavenging ability of polysaccharide [2,25,27].

2.2.3. Hydroxyl Radical Scavenging Activity

Hydroxyl radical is believed to be the most harmful free radical in the reactive oxygen species, as it could induce severe damage to adjacent biomolecules [28]. The hydroxyl radical scavenging activity of CL2 is shown in Figure 5c. The CL2 exhibited high scavenging activity on hydroxyl radical, and the scavenging effect of polysaccharide enhanced with increasing dosage. At 1 mg mL^{-1}, CL2 showed hydroxyl radical scavenging activity of 17.72% ± 1.98%, which gradually increased to 83.54% ± 6.71% at 10 mg mL^{-1}. However, the scavenging activity of CL2 was much lower than that of ascorbic acid, as it reached a plateau of 99.38% ± 2.089% at 1 mg mL^{-1}. The potential antioxidant activity of chrysolaminarin from *O. aurita* may lead to the development of a novel natural antioxidant agent. The results were in accord with Kanmani *et al.* [26], who found that the purified exopolysaccharide exhibited low reducing power, but relatively strong hydroxyl radical scavenging activity; at 1.2 mg mL^{-1}, the hydroxyl radical scavenging activity of purified exopolysaccharide reached almost 20%, which increased to 40%–45% at 2.4 mg mL^{-1}. In all experiments, the control ascorbic acid showed much better antioxidant activity than exopolysaccharide produced from *S. phocae*.

There are a number of reports on the evaluation of antioxidant activity in diatoms (*Phaeodactylum tricornutum* and *Chaetoceros calcitrans*) or other microalgae (*Botryococcus braunii*, *Porphyridium cruentum* and *Scenedesmus obliquus*) [29,30]. These studies concluded that several microalgal genera contain potent antioxidants. However, the experiments were carried out with lipophilic and hydrophilic extracts, not a purified compound. Kanmani *et al.* [26] also found that crude exopolysaccharide extract showed higher reducing power and hydroxyl radical scavenging activity than purified exopolysaccharide, which may be due to the antioxidant components, such as proteins, amino acids, organic acids and other microelements, in crude exopolysaccharide.

2.3. Production of Chrysolaminarin from O. aurita in a Column Photobioreactor

To investigate the accumulation pattern of chrysolaminarin in *O. aurita*, the biomass concentration and chrysolaminarin content of *O. aurita* cultivated in the glass column photobioreactor were studied (Figure 6). The biomass concentration of *O. aurita* had been illustrated in our previous paper [15], which concluded that high light intensity (300 μmol photons $m^{-2} s^{-1}$) and nitrogen-replete condition (18 mM) were favorable to obtain high biomass accumulation, and the maximum biomass concentration of 6.36 g L^{-1} was achieved on Day 10 (Figure 6b). Under low light, the chrysolaminarin content in the low nitrogen cultures increased from 15.09% of dry weight (DW) to 61.34% DW during a 12-day cultivation. In the high nitrogen cultures, chrysolaminarin content remained stable in the first eight days and then gradually increased to 39.67% DW at the end of the culture period (Figure 6c). Under high light, the changes in chrysolaminarin content followed the similar trends of their counterparts under low light with the difference being that the chrysolaminarin accumulated earlier. A maximum chrysolaminarin content of 64.86% DW was obtained in the low nitrogen cultures (Figure 6d). These results indicated that chrysolaminarin acts as storage products in *O. aurita*, especially in nitrogen-depleted condition. Myklestad [31] also found that the chrysolaminarin contents of marine diatoms *Chaetoceros affinis* and *Skeletonema costatum* were usually low in the exponential phase of growth, but they increased very rapidly when nutrients were exhausted in the stationary phase. The light intensity also affects the content of glucan; at nutrient saturation, high light led to the higher level of chrysolaminarin, which was consistent with the results in this study.

However, nutrient depletion also limited the growth of *O. aurita*, leading to the decline of biomass concentration (Figure 6a,b). Chrysolaminarin productivity, as a combined effect of biomass concentration and chrysolaminarin content, was a more suitable evaluation index for chrysolaminarin production. The chrysolaminarin volumetric productivity of *O. aurita* cultivated under different conditions was compared (Table 2). High light (HL) was demonstrated to be beneficial for maximizing chrysolaminarin productivity, as the biomass concentration and chrysolaminarin content of *O. aurita* under HL were significantly higher than their counterparts under low light (LL). Under HL, because of the significant enhancement in biomass concentration, the obtained chrysolaminarin productivity in high nitrogen (HN) was 14.18% higher than that in low nitrogen (LN), resulting in a record high chrysolaminarin productivity of 306 mg L^{-1} day^{-1}. The results indicated that the microalga, *O. aurita*, may be a promising natural source for the production of antioxidative chrysolaminarin.

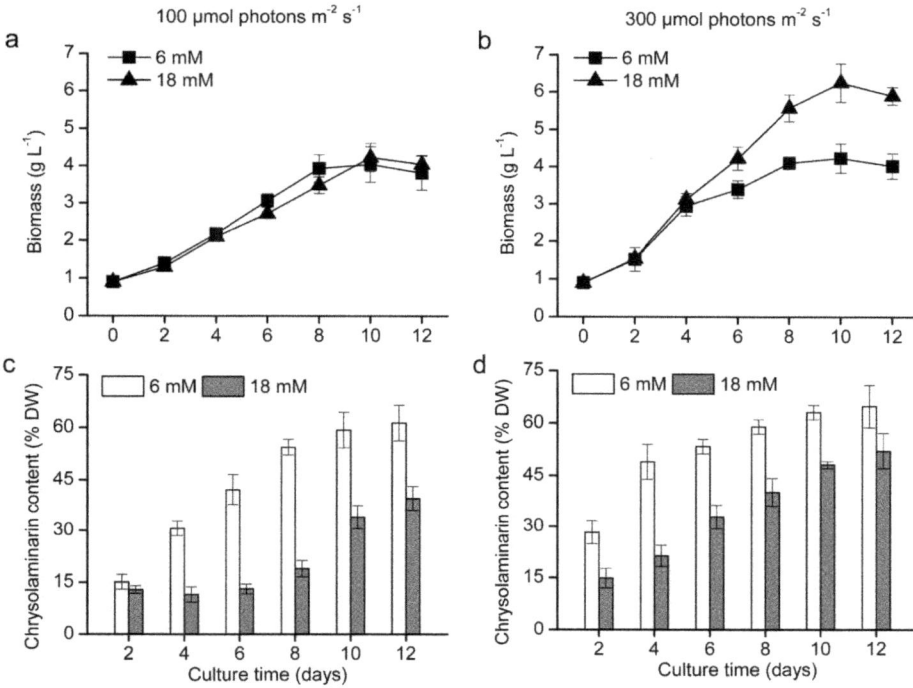

Figure 6. The biomass (**a,b**) and chrysolaminarin content (**c,d**) of *O. aurita* cultivated in the column photobioreactor under 100 (**a,c**) and 300 (**b,d**) μmol photons m^{-2} s^{-1} with a replete (18 mM) and deficient (6 mM) nitrate supply. Values are the means $\pm$ SD (n = 3). When error bars cannot be seen, the error is less than the size of the symbol.

Table 2. The chrysolaminarin volumetric productivity of *O. aurita* cultivated under different conditions in a column photobioreactor at Day 10. Data represent the mean of three replicates.

Culture Condition [a]	Biomass Concentration (g L^{-1})	Chrysolaminarin Content (% Dry Weight)	Chrysolaminarin Productivity (mg L^{-1} Day^{-1})
LL + LN	4.04	59.33	240
LL + HN	4.23	34.05	144
HL + LN	4.24	63.11	268
HL + HN	6.36	48.16	306

[a] LL: low light (100 μmol photons m^{-2} s^{-1}); HL: high light (300 μmol photons m^{-2} s^{-1}); LN: low nitrogen (6 mM); HN: high nitrogen (18 mM).

3. Experimental Section

3.1. Organism and Culture Conditions

The diatom, *Odontella aurita* K-1251, was obtained from the Scandinavian Culture Collection of Algae and Protozoa (SCCAP) at the University of Copenhagen (Copenhagen, Denmark) and deposited in our laboratory with modified L1 medium. The influences of nitrogen concentration and light intensity on chrysolaminarin accumulation were conducted with column photobioreactors (60 cm length, 3 cm diameter). Cultures were aerated with air supplemented with 1% CO_2 through a hollow glass rod and maintained at 25 $\pm$ 2 °C in an air-conditioned room. The growth medium recipe and cultural systems have been illustrated in our previous papers [5,15]. Two light intensities

(100 and 300 μmol photons m^{-2} s^{-1}) and two nitrogen concentrations (6 and 18 mM) were designed for investigating the accumulation pattern of chrysolaminarin in *O. aurita*. The paste of algal cells collected by centrifugation was inoculated into different treatments at roughly the same starting cell concentration based on the optical density at a 750-nm wavelength. The cultures were harvested every two days, and the corresponding biomass concentration and chrysolaminarin content were determined.

3.2. Biomass Measurement

Briefly, 10-mL cultures were filtered onto a pre-weighed GF/B filter paper and dried at 105 °C. The biomass concentration was determined by the difference in weight [15].

3.3. Determination of Chrysolaminarin Content

Chrysolaminarin was extracted from *O. aurita* according to Granum and Myklestad [32] with minor modification. Briefly, freeze-dried algal powder (50 mg) was extracted with 5 mL of 50 mM sulfuric acid at 60 °C for 30 min. The extract was assayed quantitatively for chrysolaminarin content using the phenol-sulfuric acid method [33]. Briefly, 1 mL of chrysolaminarin extract was mixed with 0.5 mL 6% (w/v) phenol solution and 5 mL concentrated sulfuric acid. After standing for 30 min, the absorbance of the mixture at 490 nm was measured, and the chrysolaminarin content was determined by comparison to a calibration curve prepared with glucose.

3.4. Preparation of Chrysolaminarin

The isolation and purification procedure of chrysolaminarin from *O. aurita* was performed according to the flowchart shown in Figure 7. Freeze-dried microalgal powder (10 g) was extracted twice with 500 mL sulfuric acid (50 mM) in a 60 °C water bath for 30 min. The supernatants were collected by centrifugation and precipitated with four volumes of 95% ethanol at 4 °C. The precipitate was recovered by centrifugation (5000 rpm, 5 min) and washed twice with ethanol and acetone, then freeze-dried. The freeze-dried extracts were dissolved in deionized water, and the proteins were removed by the mixture of chloroform: *n*-butyl alcohol (4:1, v/v) according to the Sevag method [34]. The deproteinized solution was dialyzed against deionized water for 48 h, and the crude chrysolaminarin was obtained under freeze drying. Then, 3 mL of crude chrysolaminarin solution (10 g L^{-1}) were applied to a DEAE-52 cellulose chromatography column (2 × 30 cm) and gradient eluted with sodium chloride (0.1 M, 0.3 M and 0.5 M). Each 5 mL of eluate was collected at a flow rate of 0.5 mL/min and monitored for the presence of polysaccharides using the phenol-sulfuric acid method [33]. The collected fraction (named CL1) was dialyzed, concentrated and loaded on a Sephadex G-200 column (Pharmacia, Uppsala, Sweden) with 0.1 M sodium chloride as the mobile phase. Each 2 mL of eluate was collected at a flow rate of 0.2 mL/min. The corresponding chrysolaminarin fraction was collected, dialyzed against distilled water for 48 h and concentrated with air flow. Freeze drying of the resulting solution obtained the chrysolaminarin as a faintly white powder (named CL2) and gave the chrysolaminarin in a 10% yield, which was used for gas chromatography-mass spectrometry (GC-MS), infrared spectroscopy (IR) and nuclear magnetic resonance spectroscopy (NMR).

Algal powder of *O. aurita*

↓ ←——— Extracted with 50 mM H_2SO_4, 30 min at 60 °C

Supernatant

↓ ←——— Added × 4 V ethanol (95%), stand for 12 h at 4 °C

Precipitate

↓   Washed with ethanol and acetone

↓ ←——— Freeze-dried

Freeze-dried extracts

↓   Dissolved in deionized water

↓ ←——— Removing proteins and nucleic acids by Sevag's method

↓   Dialyzed against deionized water for 48 h, then freeze-dried

Crude glucans

↓   Separated by DEAE-52 and Sephadex G-200

↓ ←——— Dialyzed against deionized water for 48 h, then freeze-dried

Chrysolaminarin

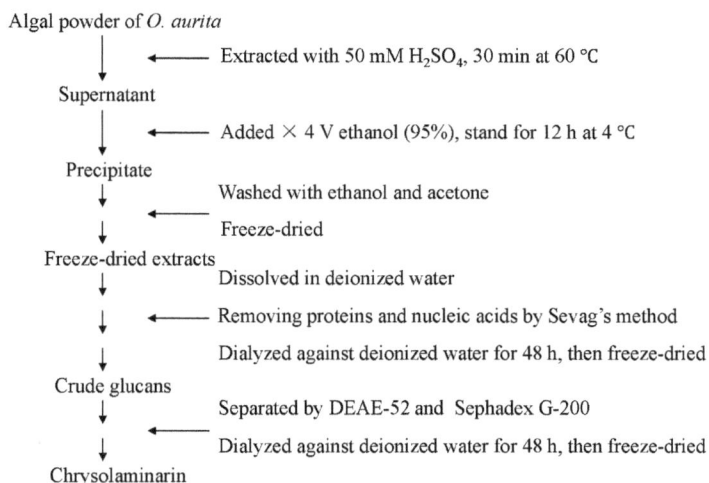

Figure 7. Isolation and purification procedure of chrysolaminarin from *O. aurita.*

3.5. Structural Analysis

3.5.1. Molecular Weight

The average molecular weight of chrysolaminarin CL2 was determined by comparison to a calibration curve prepared with the T-series Dextran standards (Sigma-Aldrich, ST, Louis, MO, USA) as molecular mass markers using gel-filtration chromatography (GPC) [35].

3.5.2. Monosaccharide Composition

GC-MS (TRACE, Thermo Finnigan, Waltham, MA, USA) was used for analysis of the monosaccharide components. The chrysolaminarin, CL2 (10 mg), was hydrolyzed in 10 mL 3 M trifluoroacetic acid (TFA) at 105 °C for 6 h. The product was reduced with $NaBH_4$ for 2 h at room temperature, acetic anhydride at 100 °C for 2 h and then analyzed by gas chromatography at a temperature program of 150–220 °C with a rate of 4 °C/min using N_2 as the carrier. A standard curve was set up with standard monosaccharides derivatized and measured under the same procedure [10].

3.5.3. IR Spectroscopy

The chrysolaminarin, CL2, was ground with dry KBr powder and pressed for Fourier transform infrared (FTIR) measurement using an EQUINOX55 spectrometer (Bruker, Bremen, Germany) at the frequency range of 4000–400 cm^{-1}.

3.5.4. NMR Analysis

Dried chrysolaminarin CL2 (15 mg) was dissolved in D_2O for NMR measurements. NMR spectra were obtained on a Bruker AVANCE III 500 spectrometer (Bruker Biospin, Rheinstetten, Germany) and recorded at 500.26 MHz for ^{1}H and 125.8 for ^{13}C nuclei, using a 5-mm broadband probe head. Spectra were obtained at 298 K in D_2O, with DSS as the internal reference standard. The signals were assigned regarding Størseth *et al.* [6].

3.6. Antioxidant Activity Assessment

3.6.1. Reducing Power

The reducing power of chrysolaminarin CL2 was determined based on Deng *et al.* [10] with minor modification. Briefly, 1 mL of chrysolaminarin CL2 solution was mixed with 0.2 mL 2 M sodium phosphate buffer (pH 6.6) and 0.5 mL 1% (w/v) aqueous potassium ferricyanide. The mixture was incubated at 50 °C for 20 min in a water bath. Then, 2.5 mL 10% (w/v) of trichloroacetic acid were added to the mixture. The resultant mixture was centrifuged at 3500 rpm for 10 min. Two milliliters of the supernatant were diluted with 3 mL distilled water and then mixed with 0.5 mL 0.3% (w/v) ferric chloride. The absorbance was measured at 700 nm against distilled water. The increase in absorbance indicated an increase in reducing power.

3.6.2. DPPH Radical Scavenging Activity

The scavenging activity of 1,1-diphenyl-2-picrylhydrazyl (DPPH) radical was carried out according to Sachindra *et al.* [36]. Briefly, 2 mL 0.16 mM ethanolic DPPH solution was added to 2 mL of related solution. The mixture was shaken vigorously and left to stand for 30 min at room temperature in the dark, and then, the absorbance was measured at 517 nm. The inhibition of DPPH radicals by the samples was calculated as follows: DPPH radical scavenging activity (%) = [1 − (absorbance of sample − absorbance of blank)/absorbance of control)] × 100%.

3.6.3. Hydroxyl Radical Scavenging Activity

Hydroxyl radical scavenging activity was determined based on Yang *et al.* [2] with minor modification. Briefly, 0.2 mL of chrysolaminarin CL2 solution were mixed with 2 mL EDTA-Fe solution (0.15 mM) and 0.8 mL salicylic acid (2 mM). Afterwards, 2 mL H_2O_2 (6 mM) were added to the reaction mixture and incubated for 30 min at 37 °C. Absorbance was measured at 510 nm. The capability of hydroxyl radical scavenging by the samples was calculated as follows: hydroxyl radical scavenging activity (%) = [1 − (absorbance of sample − absorbance of blank)/Abs. of control)] × 100%.

4. Conclusions

The new chrysolaminarin, CL2, was purified from marine diatom, *O. aurita*, and structurally determined as a glucan linked by the β-D-(1→3) (main chain) and β-D-(1→6) (side chain) glycosidic bond. The results of antioxidant experiments indicate that CL2 possesses potent antioxidant, especially for scavenging hydroxyl radicals. The high volumetric productivity of chrysolaminarin in the column photobioreactor suggests that the microalga *O. aurita* could be developed as a new natural source of antioxidant or as a food supplement.

Acknowledgments: The research was supported by the following funding: the Special Program for Low-Carbon, Reform and Development Commission of Guangdong Province; the Natural Science Foundation of China (No. 31170337 and No. 41176105); the National High Technology Research and Development Program of China (863 Program) (No. 2013AA065805); the National Basic Research Program of China (973 Program) (No. 2011CB2009001); the Key Science and Technology Innovation Project of Jiangxi Provincial Department of Science and Technology (No. 20124ACB01200); the Knowledge Innovation Team Project of Jiangxi Province (20113BCB24017); the National Natural Science Foundation of China (No. 41263006).

Author Contributions: Conceived of and designed the experiments: Chengwu Zhang and Aifen Li. Performed the experiments: Song Xia and Baoyan Gao. Analyzed the data: Song Xia, Jihai Xiong and Ziqiang Ao. Wrote and revised the paper: Song Xia, Baoyan Gao and Chengwu Zhang.

Conflicts of Interest: The authors declare no conflict of interest.

References

1. Finkel, T.; Holbrook, N.J. Oxidants, oxidative stress and the biology of aging. *Nature* **2000**, *408*, 239–247. [CrossRef]

2. Yang, X.; Wang, R.; Zhang, S.; Zhu, W.; Tang, J.; Liu, J.; Chen, P.; Zhang, D.; Ye, W.; Zheng, Y.; *et al.* Polysaccharides from *Panax japonicus* C.A. Meyer and their antioxidant activities. *Carbohydr. Polym.* **2014**, *101*, 386–391.

3. Zhang, Z.S.; Wang, X.M.; Mo, X.F.; Qi, H.M. Degradation and the antioxidant activity of polysaccharide from *Enteromorpha linza*. *Carbohydr. Polym.* **2013**, *92*, 2084–2087. [CrossRef] [PubMed]

4. Sarkar, R.; Nandan, C.K.; Bhunia, S.K.; Maiti, S.; Maiti, T.K.; Sikdar, S.R.; Islam, S.S. Glucans from alkaline extract of a hybrid mushroom (backcross mating between *PfloVv12* and *Volvariella volvacea*): Structural characterization and study of immunoenhancing and antioxidant properties. *Carbohydr. Res.* **2012**, *347*, 107–113. [CrossRef]

5. Xia, S.; Li, A.F.; Wan, L.L.; Sang, M.; Zhang, C.W. Effects of nutrients and light intensity on the growth and biochemical composition of a marine microalga *Odontella aurita*. *Chin. J. Oceanol. Limnol.* **2013**, *31*, 1163–1173. [CrossRef]

6. Størseth, T.R.; Kirkvold, S.; Skjermo, J.; Reitan, K.I. A branched β-D-(1→3,1→6)-glucan from the marine diatom *Chaetoceros debilis* (Bacillariophyceae) characterized by NMR. *Carbohydr. Res.* **2006**, *341*, 2108–2114. [PubMed]

7. Muramatsu, D.; Iwai, A.; Aoki, S.; Uchiyama, H.; Kawata, K.; Nakayama, Y.; Nikawa, Y.; Kusano, K.; Okabe, M.; Miyazaki, T.; *et al.* β-glucan derived from *Aureobasidium pullulans* is effective for the prevention of influenza in mice. *PLoS One* **2012**, *7*, e41399. [CrossRef] [PubMed]

8. Ahmad, A.; Anjum, F.M.; Zahoor, T.; Nawaz, H.; Dilshad, S.M. Beta glucan: A valuable functional ingredient in foods. *Crit. Rev. Food Sci. Nutr.* **2012**, *52*, 201–212. [CrossRef] [PubMed]

9. Bhanja, S.K.; Nandan, C.K.; Mandal, S.; Bhunia, B.; Maiti, T.K.; Mondal, S.; Islam, S.S. Isolation and characterization of the immunostimulating β-glucans of an edible mushroom *Termitomyces robustus* var. *Carbohydr. Res.* **2012**, *357*, 83–89. [CrossRef] [PubMed]

10. Deng, C.; Hu, Z.; Fu, H.; Hu, M.; Xu, X.; Chen, J. Chemical analysis and antioxidant activity *in vitro* of a β-D-glucan isolated from *Dictyophora indusiata*. *Int. J. Biol. Macromol.* **2012**, *51*, 70–75. [CrossRef] [PubMed]

11. Braud, J.P. Simultaneous culture in pilot tanks of the macroalga *Chondrus crispus* (Gigartinaceae) and the microalga *Odontella aurita* (Eupodiscaceae) producing EPA. In *Marine Microorganisms for Industry*; Le Gal, Y., Muller-Feuga, A., Eds.; Ifremer Edition: Plouzane, France, 1998; pp. 39–47.

12. Mimouni, V.; Ulmann, L.; Pasquet, V.; Mathieu, M.; Picot, L.; Bougaran, G.; Cadoret, J.P.; Morant-Manceau, A.; Schoefs, B. The potential of microalgae for the production of bioactive molecules of pharmaceutical interest. *Curr. Pharm. Biotechnol.* **2012**, *13*, 2733–2750. [CrossRef] [PubMed]

13. Toume, K.; Ishibashi, M. 5α,8α-Epidioxysterol sulfate from a diatom *Odontella aurita*. *Phytochemistry* **2002**, *61*, 359–360. [CrossRef] [PubMed]

14. Moreau, D.; Tomasoni, C.; Cathrine, J.; Kaas, R.; Guedes, L.; Cadoret, J.P.; Muller-Feuga, A.; Kontiza, I.; Vagias, C.; Roussis, V.; *et al.* Cultivated microalgae and the carotenoid fucoxanthin from *Odontella aurita* as potent anti-proliferative agents in bronchopulmonary and epithelial cell lines. *Environ. Toxicol. Pharmacol.* **2006**, *22*, 97–103. [CrossRef] [PubMed]

15. Xia, S.; Wang, K.; Wan, L.L.; Li, A.F.; Hu, Q.; Zhang, C.W. Production, characterization, and antioxidant activity of fucoxanthin from the marine diatom *Odontella aurita*. *Mar. Drugs* **2013**, *11*, 2667–2681. [CrossRef] [PubMed]

16. Wang, R.; Chen, P.; Jia, F.; Tang, J.; Ma, F. Optimization of polysaccharides from *Panax japonicus* C.A. Meyer by RSM and its anti-oxidant activity. *Int. J. Biol. Macromol.* 2012; 50, 331–336.

17. You, Q.H.; Yin, X.L.; Zhang, S.N.; Jiang, Z.H. Extraction, purification, and antioxidant activities of polysaccharides from *Tricholoma mongolicum Imai*. *Carbohydr. Polym.* **2014**, *99*, 1–10. [CrossRef] [PubMed]

18. Zou, C.; Du, Y.; Li, Y.; Yang, J.; Zhang, L. Preparation and *in vitro* antioxidant activity of lacquer polysaccharides with low molecular weights and their sulfated derivatives. *Int. J. Biol. Macromol.* **2010**, *46*, 140–144. [CrossRef] [PubMed]

19. Ding, X.; Tang, J.; Cao, M.; Guo, C.X.; Zhang, X.; Zhong, J.; Zhang, J.; Sun, Q.; Feng, S.; Yang, Z.R.; *et al.* Structure elucidation and antioxidant activity of a novel polysaccharide isolated from *Tricholoma matsutake*. *Int. J. Biol. Macromol.* 2010, *47*, 271–275.

20. Zhang, Y.X.; Dai, L.; Kong, X.W.; Chen, L.W. Characterization and *in vitro* antioxidant activities of polysaccharides from *Pleurotus ostreatus*. *Int. J. Biol. Macromol.* **2012**, *51*, 259–265. [CrossRef] [PubMed]

21. Hirokawa, Y.; Fujiwara, S.; Suzuki, M.; Akiyama, T.; Sakamoto, M.; Kobayashi, S.; Tsuzuki, M. Structural and physiological studies on the storage β-polyglucan of haptophyte *Pleurochrysis haptonemofera*. *Planta* **2008**, *227*, 589–599. [CrossRef] [PubMed]

22. Choi, J.I.; Kim, H.J.; Lee, J.W. Structural feature and antioxidant activity of low molecular weight laminarin degraded by gamma irradiation. *Food Chem.* **2011**, *129*, 520–523. [CrossRef]

23. Størseth, T.R.; Hansen, K.; Skjermo, J.; Krane, J. Characterization of a β-D-(1→3)-glucan from the marine diatom *Chaetoceros mülleri* by high-resolution magic-angle spinning NMR spectroscopy on whole algal cells. *Carbohydr. Res.* **2004**, *339*, 421–424. [CrossRef] [PubMed]

24. Kozarski, M.; Klaus, A.; Niksic, M.; Vrvic, M.M.; Todorovic, N.; Jakovljevic, D.; van Griensven, L.J.L.D. Antioxidative activities and chemical characterization of polysaccharide extracts from the widely used mushrooms *Ganoderma applanatum*, *Ganoderma lucidum*, *Lentinus edodes* and *Trametes versicolor*. *J. Food Compos. Anal.* **2012**, *26*, 144–153. [CrossRef]

25. Lo, T.C.-T.; Chang, C.A.; Chiuc, K.-H.; Tsayd, P.-K.; Jena, J.-F. Correlation evaluation of antioxidant properties on the monosaccharide components and glycosyl linkages of polysaccharide with different measuring methods. *Carbohydr. Polym.* **2011**, *86*, 320–327. [CrossRef]

26. Kanmani, P.; Kumar, R.S.; Yuvaraj, N.; Paari, K.A.; Pattukumar, V.; Arul, V. Production and purification of a novel exopolysaccharide from lactic acid bacterium *Streptococcus phocae* PI80 and its functinoal characteristics activity invitro. *Bioresour. Technol.* **2011**, *102*, 4827–4833. [CrossRef] [PubMed]

27. Ker, Y.-B.; Chen, K.-C.; Chyau, C.-C.; Chen, C.-C.; Guo, J.-H.; Hsien, C.-L.; Wang, H.-E.; Peng, C.-C.; Chang, C.-H.; Peng, R.-P.; *et al.* Antioxidant capability of polysaccharides fractionated from submerge-cultured *Agaricus blazei* Mycelia. *J. Agric. Food Chem.* 2005; 53, 7052–7058.

28. Li, J.; Liu, Y.; Fan, L.; Ai, L.; Shan, L. Antioxidant activities of polysaccharides from the fruiting bodies of *Zizyphus jujuba* cv. Jinsixiaozao. *Carbohydr. Polym.* **2011**, *84*, 390–394. [CrossRef]

29. Goiris, K.; Muylaert, K.; Fraeye, I.; Foubert, I.; Brabanter, J.D.; Cooman, L.D. Antioxidant potential of microalgae in relation to their phenolic and carotenoid content. *J. Appl. Phycol.* **2012**, *24*, 1477–1486. [CrossRef]

30. Guedes, A.C.; Giao, M.S.; Seabra, R.; Ferreira, A.C.S.; Tamagnini, P.; Moradas-Ferreira, P.; Malcata, F.X. Evaluation of the antioxidant activity of cell extracts from microalgae. *Mar. Drugs* **2013**, *11*, 1256–1270. [CrossRef] [PubMed]

31. Myklestad, S.M. Production, chemical structure, metabolism, and biological function of the (1→3)-linked, β-D-glucans in diatoms. *Biol. Oceanogr.* **1989**, *6*, 313–326.

32. Granum, E.; Myklestad, S.M. A simple combined method for determination of β-1,3-glucan and cell wall polysaccharides in diatoms. *Hydrobiologia* **2002**, *477*, 155–161. [CrossRef]

33. Dubois, M.; Gillies, K.A.; Hamilton, J.K.; Rebers, P.A.; Smith, F. Colorimetric method for the determination of sugars and related substances. *Anal. Chem.* **1956**, *28*, 350–356. [CrossRef]

34. Sevag, M.G.; Lackman, D.B.; Smolens, J. The isolation of the components of streptococcal nucleoproteins in serologically active form. *J. Biol. Chem.* **1938**, *124*, 425–436.

35. Ye, M.; Chen, W.X.; Qiu, T.; Yuan, R.Y.; Ye, Y.W.; Cai, J.M. Structural characterization and anti-ageing activity of extracellular polysaccharide from a strain of *Lachnum* sp. *Food Chem.* **2012**, *132*, 338–343. [CrossRef]

36. Sachindra, N.M.; Sato, E.; Maeda, H.; Hosokawa, M.; Niwano, Y.; Kohno, M.; Miyashita, K. Radical scavenging and singlet oxygen quenching activity of marine carotenoid fucoxanthin and its metabolites. *J. Agric. Food Chem.* **2007**, *55*, 8516–8522. [CrossRef] [PubMed]

marine drugs

MDPI

Article

Structural Analysis and Anticoagulant Activities of the Novel Sulfated Fucan Possessing a Regular Well-Defined Repeating Unit from Sea Cucumber

Mingyi Wu [1], Li Xu [1,2], Longyan Zhao [1,2], Chuang Xiao [1,2], Na Gao [1,2], Lan Luo [1], Lian Yang [1], Zi Li [1], Lingyun Chen [3,*] and Jinhua Zhao [1,*]

[1] State Key Laboratory of Phytochemistry and Plant Resources in West China, Kunming Institute of Botany, Chinese Academy of Sciences, Kunming 650201, China; wumingyi@mail.kib.ac.cn (M.W.); xulib@mail.kib.ac.cn (L.X.); zhaolongyan@mail.kib.ac.cn (L.Z.); xiaochuang@mail.kib.ac.cn (C.X.); gaona@mail.kib.ac.cn (N.G.); luolanjya@sina.com (L.L.); yanglian@mail.kib.ac.cn (L.Y.); lizi@mail.kib.ac.cn (Z.L.)

[2] School of Life Sciences, University of Chinese Academy of Sciences, Beijing 100000, China

[3] Pharmacy Department, Yunnan University of TCM, Kunming 650200, China

* Authors to whom correspondence should be addressed; chenlingyun@ynutcm.edu.cn (L.C.); Zhaojinhua@mail.kib.ac.cn (J.Z.); Tel./Fax: +86-871-6591-8230 (L.C.); +86-871-6522-6278 (J.Z.).

Academic Editor: Paola Laurienzo

Received: 26 January 2015; Accepted: 26 March 2015; Published: 13 April 2015

Abstract: Sulfated fucans, the complex polysaccharides, exhibit various biological activities. Herein, we purified two fucans from the sea cucumbers *Holothuria edulis* and *Ludwigothurea grisea*. Their structures were verified by means of HPGPC, FT-IR, GC–MS and NMR. As a result, a novel structural motif for this type of polymers is reported. The fucans have a unique structure composed of a central core of regular (1→2) and (1→3)-linked tetrasaccharide repeating units. Approximately 50% of the units from *L. grisea* (100% for *H. edulis* fucan) contain sides of oligosaccharides formed by nonsulfated fucose units linked to the O-4 position of the central core. Anticoagulant activity assays indicate that the sea cucumber fucans strongly inhibit human blood clotting through the intrinsic pathways of the coagulation cascade. Moreover, the mechanism of anticoagulant action of the fucans is selective inhibition of thrombin activity by heparin cofactor II. The distinctive tetrasaccharide repeating units contribute to the anticoagulant action. Additionally, unlike the fucans from marine alga, although the sea cucumber fucans have great molecular weights and affluent sulfates, they do not induce platelet aggregation. Overall, our results may be helpful in understanding the structure-function relationships of the well-defined polysaccharides from invertebrate as new types of safer anticoagulants.

Keywords: polysaccharide; sulfated fucan; chemical structure; anticoagulant

1. Introduction

Thromboembolic diseases continue to be the leading cause of death throughout the world [1]. Most thromboembolic processes require anticoagulant therapy. Thus, the current efforts are to develop specific and potent anticoagulant agents.

Unfractionated heparin (UFH) and low-molecular-weight heparins (LMWHs) have been cornerstones of antithrombotic treatment and prophylaxis for the last 70 years, which are the only sulfated polysaccharides currently used as anticoagulant drugs. However, these compounds have several side effects such as hemorrhagic effects, development of thrombocytopenia, ineffectiveness in congenital or acquired antithrombin deficiencies, incapacity to inhibit thrombin bound to fibrin, and so on [1–3]. In addition, the commercial sources of heparins are mainly pig intestinal mucosa or bovine lung, where they occur in low concentrations. The possibility that prions and viruses could be carried by these molecules in addition to the increasing needs for antithrombotic therapies indicate the necessity to look for alternative sources of anticoagulant agents [3,4].

Marine invertebrate and alga are abundant sources of anticoagulant polysaccharides, such as a variety of sulfated fucans (also called as fucoidan from brown alga) [3–11]. The proposed mechanisms of action of these compounds are predominantly related to the inhibition of factors Xa and thrombin (IIa) mediated by antithrombin (AT) and heparin cofactor II (HCII) [6–8]. Besides the anticoagulant and antithrombotic activities, some sulfated fucans also possess other important biological activities such as inducing the sperm acrosome reaction, gastroprotective activities and inhibition of osteoclastogenesis [12–14].

In spite of the high level of interest shown in functional aspects of sulfated fucans, their structural properties have been relatively little studied. Thus, the structure-activity relationships remain to be elucidated. Most of the difficulties for these studies arise from the fact that these compounds are very heterogeneous polysaccharides or various sulfate substituted homogeneous ones which give complex NMR spectra with broad signals hampering resolution [4,11,15]. It is not always possible to define whether these polysaccharides from invertebrates have repetitive units. For example, a study on the structure of the sulfated fucan from sea cucumber, containing regular 1→3-linked units, was reported [16,17]. Furthermore, the structure of sulfated fucans may vary according to the species of invertebrates, as it is the case for heparan sulfates in vertebrates [18]. Thus, each new sulfated polysaccharide purified from a sea cucumber may be a new compound with unique structures and, consequently, with potential novel biological activities.

Recently, during the process of searching for new anticoagulant sulfated polysaccharide, we obtained the sulfated fucans from two species of sea cucumbers *Holothuria edulis* and *Ludwigothurea grisea*, and preliminarily described physicochemical characteristics of the *H. edulis* fucan [9]. Here we report structural characterization and pharmacological activities of two new sulfated fucans in detail. These two polysaccharides have a similar unique structure composed of a central core of regular α(1→3)- and α(1→2)-linked tetrasaccharide repeating units. Approximately 50% of the units from *L. grisea* (100% for *H. edulis* fucan) contain branches of oligosaccharides formed by nonsulfated fucose units linked to the O-4 position of the central core. Of particular significance was the finding that the type of sulfated fucan exhibits selectively antithrombin activity by heparin cofactor II and shows potent anticoagulant activity without inducing platelet aggregation.

2. Results and Discussion

2.1. Physicochemical Characteristics

Sulfated fucans were extracted from the body wall of two species of sea cucumbers *H. edulis* and *L. grisea*. Purification was achieved by Sephadex G-100 and anion exchange chromatography on a DEAE-Sepharose FF column according to our previous method [9]. Analysis of sulfated fucans by anion exchange chromatography on a DEAE Sepharose FF column confirmed the high negative charge densities of the two polysaccharides [9]. The purities of these polysaccharides were confirmed by gel filtration chromatography on a Shodex OH-pak SB-804 HQ column. The results show that they each migrate as a single homogeneous peak [9] and do not have any ultraviolet absorption near 260 or 280 nm by the measurement of an UV-detector, indicating no contaminants such as protein and peptide.

These polysaccharides were obtained as water-soluble white powder after lyophilization. These polysaccharide fractions have been characterized by different analytical techniques to compare their physicochemical properties, as shown in Table 1. The chemical analysis of purified sulfated fucan revealed fucose as the only sugar with a high content of sulfate ester with a ratio ~1:(0.80–0.90). The molecular masses of *H. edulis* fucan and *L. grisea* fucan are 616 and 554 kDa, respectively, as determined by the high-performance gel permeation chromatography. Their FT-IR spectra display the existence of sugar backbone (1130–1170 and 1000 cm^{-1}) (Figure 1) [19]. These spectra show several bands corresponding to sulfate ester: the peaks at 1266 and 854 cm^{-1} are derived from the stretching vibration of S=O of sulfate and the bending vibration of C–O–S of sulfate in axial position, respectively. The signals at 3442 and 1031 cm^{-1} are from the stretching vibration of O–H and C–O, respectively. Additionally, the strongly negative specific rotation of the sulfated fucans is compatible with residues of L-fucopyranose [20]. Thus, partial physicochemical characteristics of these polysaccharides from the body walls of sea cucumbers *H. edulis* and *L. grisea* confirm that they are sulfated fucans. As shown in Table 1, comparison analysis of the sulfated fucans shows that their physicochemical characteristics vary according to the species of invertebrates, possible reflecting other important structural differences.

Table 1. Chemical composition and physicochemical properties of the sulfated fucans from the body wall of two sea cucumbers *H. edulis* and *L. grisea*.

Source	Species	Chemical Composition (Molar Ratios)		Average Molecular Weight (kDa)	Specific Rotation	Ref.
		Fuc	Sulfate/Monosaccharide			
Sea cucumber	*Holothuria edulis*	1.0	0.80 [a]	616 [b]	−181°	[9]
	Ludwigothurea grisea	1.0	0.89 [a]	554 [b]	−178°	This work
	Apostichopus japonicas	1.0	0.57 [c]	420 [b]	−182°	[9]
	Stichopus japonicus	1.0	0.79 [c]	32 [b]	ND [e]	[14]
Sea urchin	*Strongylocentrotus pallidus*	1.0	1.0 [a]	100 [d]	ND	[7]
	Strongylocentrotus purpuratus	1.0	1.3 [a]	100 [d]	ND	[7]

[a] Based on interpretation of the ^{1}H-NMR spectrum; [b] Determined by high-performance gel permeation chromatography; [c] Based on chemical analysis; [d] Determined by polyacrylamide gel electrophoresis; [e] ND, not determined.

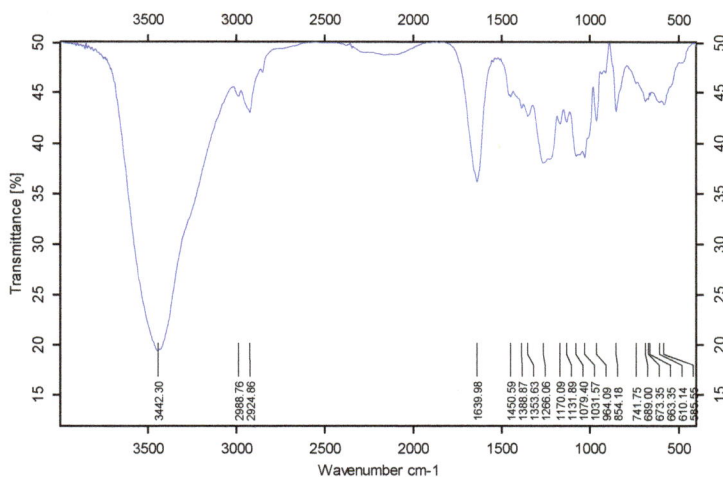

Figure 1. FT-IR spectrum of the sulfated fucan from sea cucumber.

2.2. Methylation Analysis

The position of the glycosidic linkages as well as the position of the sulfate ester in the polysaccharide was determined by methylation analysis (Table 2). Methylation analysis confirms the occurrence of (1→2) and (1→3) linkages in the sulfated L-fucans: ~40% of 4-methylfucose, 25% of 2-methylfucose, ~20% of 2,3,4-tri-*O*-methylfucose in the *H. edulis* sulfated L-fucan (~10% of 2,3,4-tri-*O*-methylfucose in the *L. grisea* sulfated L-fucan) were formed from the native polysaccharide. Although the proportions of the methylated derivatives are not exactly as expected, they are consistent with a polysaccharide composed of 3-linked and 2-linked fucose residues, sulfated at the O-2 position, O-4 and unsulfated units. These structures could be confirmed and further detailed by NMR analysis.

Table 2. Partically methylated alditol acetates derived from the sulfated fucans.

Derivative	Positions of Substitution	Primary Mass Fragments (*m/e*)	Composition (w%, mol) H. edulis	L. grisea
1,5-di-*O*-acetyl-2,3,4-tri-*O*-methyl-L-fucitol		89, 101, 117,131, 161,175	4.02% (0.047)	4.86% (0.057)
1,3,5-tri-*O*-acetyl-2,4-di-*O*-methyl-L-fucitol	3	89,101,117, 131, 233, 247	30.44% (0.325)	26.39% (0.283)
1,2,3,5-tetra-*O*-acetyl-4-*O*-methyl-L-fucitol	2,3	89, 131, 201, 261	34.14% (0.335)	40.27% (0.397)
1,2,3,4,5-penta-*O*-acetyl-L-fucitol	2,3,4	128, 170, 231, 289	22.05% (0.200)	23.02% (0.201)

2.3. NMR Analysis

For the NMR analysis of the sulfated fucans, they give overlapping spectra with broad signals hampering resolution, since line widths are several Hz, as expected for polysaccharides of high molecular mass (616 and 554 kDa) [9]. Attempts to record two-dimensional NMR spectra for these polysaccharides gave no useful result in this study. To further elucidate the structures in detail, their depolymerized products (~15 kDa) were prepared according to previous reported methods on the sulfated polysaccharide, which keep almost similar to chemical compositions and structures of native compounds during the free radical depolymerization [21,22].

The ^{1}H, ^{13}C one- and two-dimensional spectra of the sulfated fucans from *H. edulis and L. grisea* are shown in Figures 2–5. The chemical shifts in Table 3 are based on the interpretations of ^{1}H/^{1}H correlated spectroscopy (COSY), total correlation spectroscopy (TOCSY), and ^{1}H/^{13}C heteronuclear single-quantum coherence (HSQC) spectra (Figures 3–5). As shown in Figure 2A, the signals at about 1.10–1.40 ppm could be readily assigned to the methyl protons of fucose residues (CH$_3$) [9,12]. In addition, the chemical shifts of the envelope of anomeric signals at 4.9–5.6 ppm were consistent with the existence of major five types of α-L-fucose units (designated by A–E in Figure 2A). Integration of regions of the ^{1}H NMR spectrum indicated that the five types of residues were present in equal proportions (Figure 2B). Similarly, for the *L. grisea* sulfated fucan, it has five types of residues with a ratio ~1:1:1:1:0.5.

Table 3. ^{1}H and ^{13}C chemical shifts from identified 2D NMR spectra of the sulfated fucans from two sea cucumbers.

Sugar Residues			Chemical Shifts [a]					
			1	2	3	4	5	6
Holothuria edulis								
A	→3)-α-L-Fucp-(2SO$_3^-$)-(1→	H	5.40	**4.58** [b]	*4.18* [c]	4.12	4.46	1.30
		C	94.16	**73.75**	*73.61*	68.85	67.70	15.92
B	→3)-α-L-Fucp-(2SO$_3^-$)-(1→	H	5.43	**4.60**	*4.39*	4.11	4.47	1.32
		C	98.89	**75.12**	*73.07*	68.89	67.71	16.17
C	→3)-α-L-Fucp-(2,4SO$_3^-$)-(1→	H	5.47	**4.63**	*4.48*	**4.98**	4.51	1.33
		C	94.31	**75.36**	*74.67*	**81.24**	67.08	16.17
D	→2,4)-α-L-Fucp-(1→	H	5.13	*4.03*	*4.10*	*4.12*	4.36	1.29
		C	96.64	*75.93*	*67.27*	*69.80*	67.47	15.80
E	α-L-Fucp-(1→	H	5.16	4.01	4.08	3.97	4.36	1.29
		C	95.93	67.14	67.11	68.91	67.54	15.80
Ludwigothurea grisea								
A	→3)-α-L-Fucp-(2SO$_3^-$)-(1→	H	5.36	**4.55**	*4.14*	4.08	4.40	1.26
		C	94.29	**73.78**	*73.60*	69.25	68.01	15.97
B	→3)-α-L-Fucp-(2SO$_3^-$)-(1→	H	5.39	**4.57**	*4.36*	4.08	4.32	1.26
		C	98.89	**75.18**	*73.08*	69.25	68.01	15.97
C	→3)-α-L-Fucp-(2,4SO$_3^-$)-(1→	H	5.44	**4.60**	*4.39*	**4.94**	4.47	1.29
		C	94.29	**75.40**	*74.70*	**81.25**	67.11	16.24
D (D')	→2,4)-α-L-Fucp-(1→ or →2,-α-L-Fucp-(1→	H	5.10	*3.98*	*4.06*	*4.13 (4.09)*	4.48	1.29
		C	96.77	*75.97*	66.85 (67.11)	69.79 (68.92)	67.47	16.24
E	α-L-Fucp-(1→	H	5.07	3.99	4.01	4.06	4.41	1.26
		C	96.77	66.85	67.11	68.92	67.47	15.97

[a] Data were recorded on a Bruker Advance DRX 500 spectrometer; chemical shifts are given in ppm with reference to trimethylsilyl-propionic acid (TSP) d4; [b] Values in boldface indicate positions bearing sulfate groups; [c] Values in italic type indicate glycosylated positions.

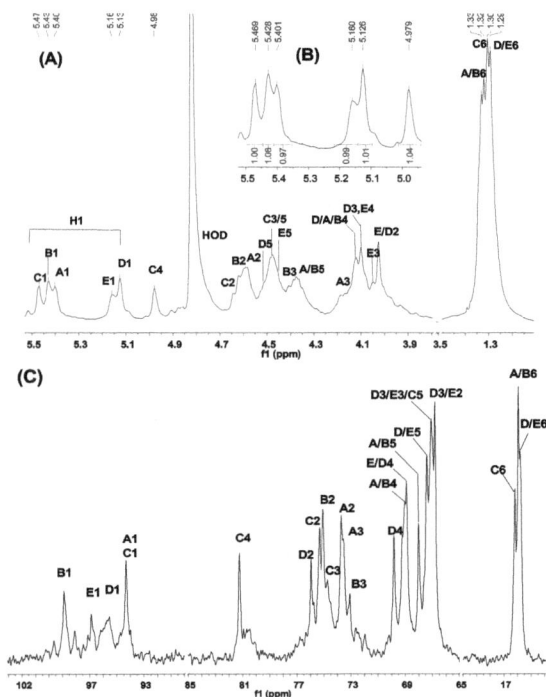

Figure 2. ^{1}H (**A,B**) and ^{13}C (**C**) one-dimensional NMR spectra at 500 MHz of the sulfated fucan from *H. edulis*. The spectra were recorded at 300 K for samples in D_2O solution. Chemical shifts are relative to external trimethylsilylpropionic acid at 0 ppm. The residual water has been suppressed by pre-saturation. The anomeric signals assigned by ^{1}H/^{13}C HSQC (see Figure 5) are labeled A–E in the sulfated fucan. Expansion of the 4.9–5.6 ppm region of the ^{1}H spectrum is shown in the *inset* in (**A**). The integrals were listed under the anomeric signals (**B**).

Furthermore, ^{1}H/^{1}H COSY and TOCSY spectra of the sulfated fucans from *H. edulis* and *L. grisea* can be assigned to five spin systems C, B, A, D (D') and E (Figures 3 and 4), allowing the assignments of almost all proton signals of five fucose residues as shown in Table 3, each consistent with fucose [17]. The contributions to the ^{1}H NMR spectrum of residues A, B and C varied most significantly in their H-2 and H-4 shifts. These differences may be attributed to sulfation shift. The values of δ_{H-2} for them, at 4.55–4.60 ppm, were shifted ~0.6 ppm downfield of δ_{H-2} for D (D') and E, indicating that residues A, B and C were 2-*O*-sulfated. The values of δ_{H-4} of residue C was ~0.9 ppm downfield of δ_{H-4} for other four fucose residues, indicating that only residue C was 4-*O*-sulfated. Therefore, residue C bears two sulfate groups, at position 4 and 2; A and B are 2-*O*-sulfated residues, and residues D and E are unsulfated.

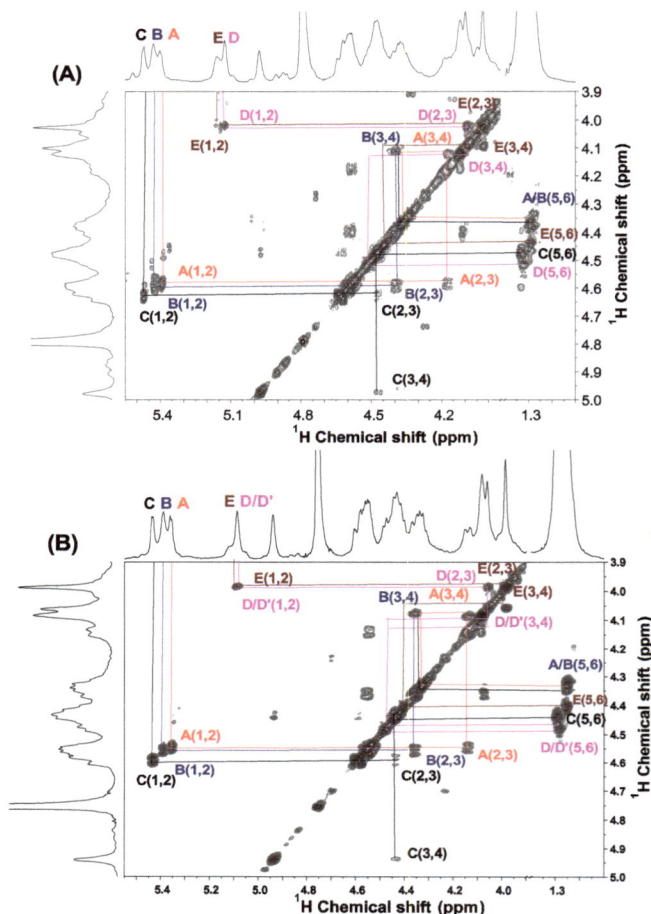

Figure 3. $^1H/^1H$ COSY spectra of the sulfated fucans from *H. edulis* (**A**) and *L. grisea* (**B**). The spectra were recorded at 300 K for samples in D_2O solution. Chemical shifts are relative to external trimethylsilylpropionic acid at 0 ppm. The residual water has been suppressed by pre-saturation. The anomeric signals assigned by $^1H/^{13}C$ HSQC (see Figure 5) are labeled A–E in the sulfated fucans.

In addition, carbon chemical shifts for sulfated and unsulfated fucose residues were also observed by the ^{13}C (Figure 2C) and $^1H/^{13}C$ HSQC experiment (Figure 5A,B). Strong downshifts of C-2 of residues A, B and C (~8 ppm) and C-4 of residues C (~13 ppm) (Figure 5A,B and Table 3) relative to those of residues D and E indicate that three of the five residues are 2-*O*-sulfated, that one are 2, 4-dis-*O*-sulfated and that the other two are non-*O*-sulfated. Additionally, the anomeric carbon signals labeled A–E in the sulfated fucan were also easily assigned by $^1H/^{13}C$ HSQC, giving the chemical shifts ~95–99 ppm as shown in Table 3. Notably, strong downshifts of C-3 of residues A, B and C (~7 ppm), C-2 (~9 ppm) and C-4 (~2 ppm) of residues D may display existence of 3-linked, 2-linked and 4-linked glycosidic linkages.

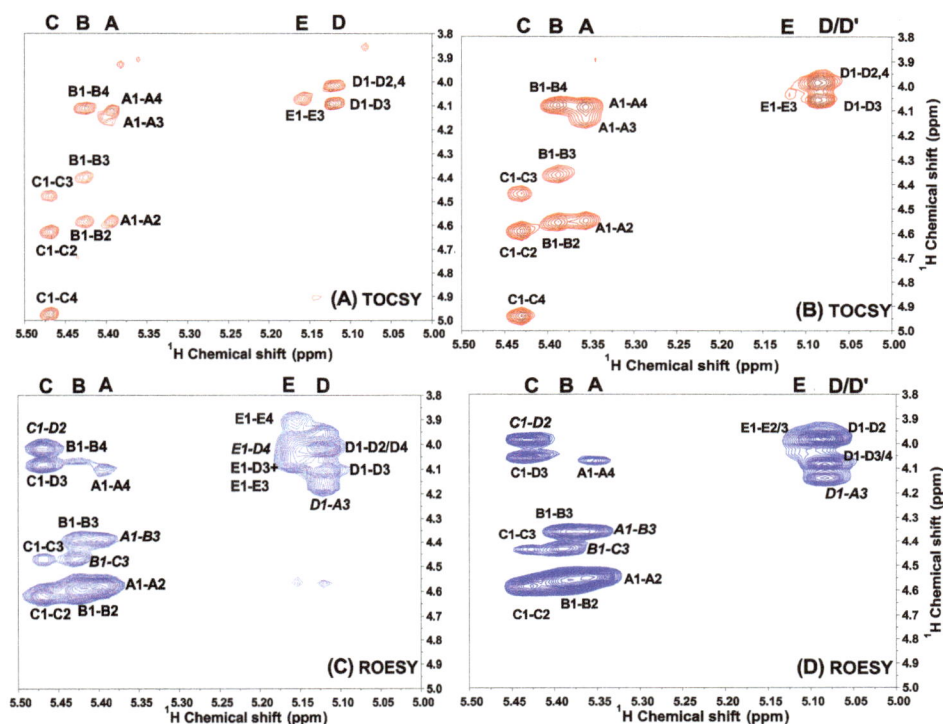

Figure 4. Expansions of the TOCSY (**A,B**) and ROESY (**C,D**) spectra of two sulfated fucans from *H. edulis* and *L. grisea* (**B,D**).The TOCSY spectra (**A,B**)show some cross-peaks used in the assignment of the fucose residue, especially positions bearing sulfate esters. The ROESY spectra (**B,D**) show ROEs, the sequence-defining A1–B3, B1–C3, C1–D2, D1–A3 and E1–D4. The five fucose residues in the repeating unit are marked A–E as described in the legend of Figure 6.

The order of the five residues can be easily deduced. The only possible array is one 2, 4-dis-*O*-sulfated residue followed by two consecutive 2-*O*-sulfated residues and one unsulfated residues with an unsulfated residues as a side chain. Our proposition was confirmed by the ROESY and HMBC spectra (Figures 4 and 5). As in the ROESY spectra of other fucans from echinoderms [12,17], ROEs between protons of different units can be seen, and they were used to reveal the sequence (besides, of course, ROEs on other protons in the same residue). In the two sulfated fucans, H-1 of residue A shows cross-peaks to H-3 of residue B; H-1 of residue B shows cross-peaks to H-3 of residue C; H-1 of residue C shows cross-peaks to H-2 of residue D; H-1 of residue D shows cross-peaks to H-3 of residue A; and notably, H-1 of residue E shows cross-peaks to H-4 of residue D. Similarly, the HMBC spectrum (Figure 5C,D) also show the sequence-defining A1–B3, B1–C3, C1–D2, D1–A3 and E1–D4. These evidences indicate the sequence and linkage -3-A-1→3-B-1→3-C-1→2-D-1→ as the linear tetrasaccharide backbone and occurrence of linkage E-1→4-D as a side chain, as shown in Figure 6.

To further estimate the configurations at the glycosidic linkages, the direct coupling constants ($^1J_{C-H}$) of C-1 of each sugar were also obtained from HMBC spectrum (Figure 5).The large values of 170–175 Hz for these fucose residues indicate the protons are equatorial [23]. Taking account of the vicinal coupling constant ($^3J_{1H-2H}$) of 3 Hz for fucose residues, the configurations at C-1 of these sugars are determined to α-L-fucose, consistent with the strongly negative specific rotation (−181° and −178°

for *H. edulis* and *L. grisea* fucan, respectively) and the proton shifts of anomeric signals of the sulfated fucans (Table 3).

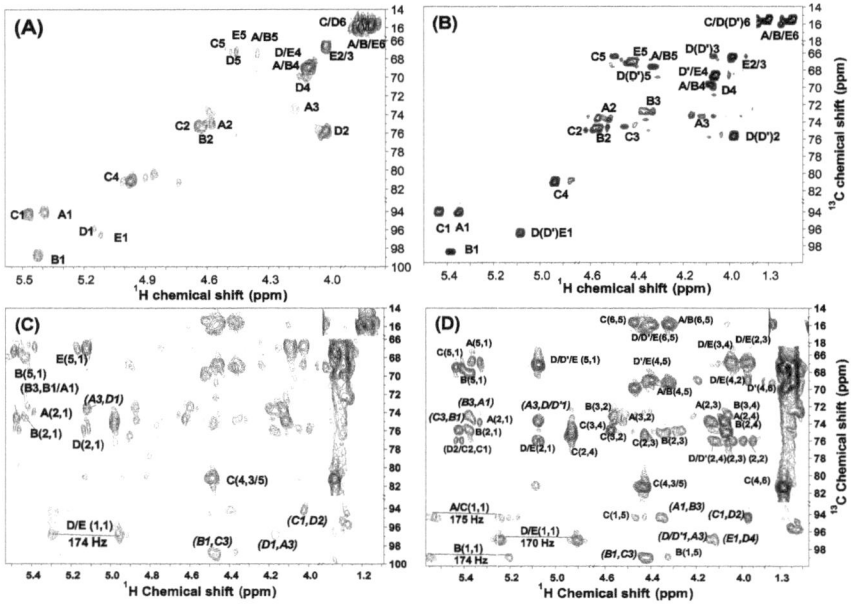

Figure 5. ^{1}H/^{13}C HSQC (**A,B**) and HMBC (**C,D**) spectra of two sulfated fucans from two sea cucumbers *H. edulis* (**A,C**) and *L. grisea* (**B,D**). The assignments were based on TOCSY and COSY spectra. The anomeric signals were identified by the characteristic carbon chemical shifts and are marked A–E. The HMBC spectra (**C,D**) also show the sequence-defining A1–B3, B1–C3, C1–D2, D/D'1–A3 and E1–D4. The five fucose residues in the repeating unit are marked A–E as described in the legends of Figure 6.

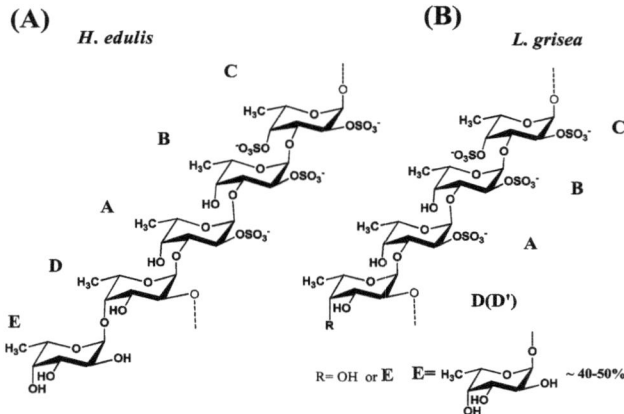

Figure 6. Proposed regular repeating units of sulfated fucan isolated from two sea cucumbers *H. edulis* (**A**) and *L. grisea* (**B**). The five fucose residues in the repeating unit are marked A–E as described in the legends.

Thus, the sulfated fucans from *H. edulis* and *L. grisea* are mostly a tetrasaccharide repeating backbone unit consisting of 2- and 3-linked α-L-fucose residues and 4-linked unsulfated α-L-fucose residue as a side chain.

Overall, the combination of chemical analysis, specific optical rotation, methylation experiments and NMR spectroscopy has allowed us to determine the fine structure of the sulfated polysaccharides isolated from the body walls of two species of sea cucumbers. The sulfated α-L-fucans from *H. edulis* and *L. grisea* are essentially linear polymers, composed of a regular tetrasaccharide repeat backbone unit defined by the pattern of *O*-sulfation with an unsulfated fucose residue as a side chain (Figure 6). The proportion of unsulfated fucose residue in the side chain varies in the two species. The α-L-fucan from *H. edulis* consists of total fucosylated at the *O*-4 position of the tetrasaccharide repeat backbone (Figure 6A), while the sulfated α-L-fucan from *L. grisea* consists of about half fucosylated at the backbone (Figure 6B).

A variety of sulfated fucoidans from marine algae have been described [11,17,24]. These compounds are among the most abundant and widely studied of all sulfated polysaccharides of non-mammalian origin [12]. The algal fucans have complex heterogeneous structures [11,24]. Their regular repeating sequences are not easily deduced; even high-field NMR is at the limit of its resolution, and complete description of their structure is much difficult at present [4,12,15,17,25]. In contrast to the algal fucans, these sea cucumber polysaccharides have novel regular structures composed of well-defined repeating units of oligosaccharides as shown in Figure 6. To our best knowledge, two new sulfated polysaccharides composed of a 3-linked and 2-linked tetrasaccharide repeat backbone unit defined by the pattern of *O*-sulfation with an unsulfated fucose residue as a side chain had not been reported. Especially, the position (1→2) of the glycosidic linkage may be firstly found in the marine invertebrate fucans, though it had been reported the occurrence of (1→2) linkage in the fucoidans from a variety of brown seaweeds [26,27].

2.4. Anticoagulant Activities

Anticoagulant activities of the polysaccharides from the two sea cucumbers were assessed by measuring the activated partial thromboplastin time (APTT), prothrombin time (PT) and thrombin time (TT), and compared with the same activities of unfractionated heparin and dermatan sulfate from mammalian sources. The APTT, PT, and TT are used to determine the ability to inhibit blood clotting through the intrinsic, extrinsic and common pathways of the coagulation cascade, respectively [9,28]. As shown in Table 4, the assays indicated that the sulfated fucans from *H. edulis* and *L. grisea* had similar APTT-prolonging activity (~10 heparin U/mg), but did not affect PT and TT in human plasma at the concentrations tested (1–100 μg/mL). Its APTT-prolonging activity is two-fold greater than that of mammalian dermatan sulfate, and is similar to that of another two sulfated fucans from marine sea urchin in the literature [7]. The results indicated that the type of sulfated fucans may affect the intrinsic but not the extrinsic and common coagulation process.

To further investigate mechanism of anticoagulant action, anti-thrombin activity in the presence of heparin cofactor II, anti-factor Xa and anti-thrombin activities mediated by AT were also examined with chromogenic substrates (Figures 7–9). Increasing concentrations of the sulfated fucans resulted in essentially complete inhibition of thrombin activation by heparin cofactor II (Figure 7), corresponding with a noncompetitive inhibition pattern [29]. As shown in Figures 7–9 and Table 4, the sea cucumber polysaccharides displayed significantly weaker anti-factor Xa and anti-thrombin activities mediated by AT than heparin and LMWH as positive anticoagulant drugs, suggesting that their anticoagulant mechanisms are different from those of heparin-like drugs. Like DS as a positive reference, in the presence of heparin cofactor II, the sulfated fucans showed very strong inhibition of thrombin (IC$_{50}$ 0.5–0.7 μg/mL, ~1 nM). In contrast, their thrombin and factor Xa inhibition activities mediated by AT were much weaker (Figures 8 and 9, Table 4). As shown in Table 4, HCII-dependent anti-thrombin activity of the type of the sulfated fucan is above 1000-fold higher than anti-factor Xa activity and 100-fold higher than anti-thrombin activity in the presence of AT. These results suggested that the sulfated polysaccharides selectively inhibit thrombin activity in the presence of heparin cofactor II. Thus, the structural requirements for interaction of these polysaccharides with coagulation cofactors (HCII and AT) and their target proteases may be macromolecularity such as conformation and length of repetitive units, *etc.*

Interestingly, the sulfated fucans from two species of sea cucumber, which possess the same tetrasaccharide repeat backbone unit but various proportions of fucosylated side chains, show essentially the same anticoagulant activities such as APTT-prolonging activity and HCII-dependent anti-thrombin activity (Figure 7 and Table 4). These results implied that anticoagulant action of the sulfated α-L-fucan from sea cucumber might be assigned to the tetrasaccharide repeat backbone unit with the specific pattern of sulfation and the novel position of the glycosidic linkage.

Table 4. Activated partial thromboplastin (APTT) and IC_{50} of the sulfated fucans for thrombin or factor Xa inhibition in the presence of antithrombin or heparin cofactor II.

Source (Polysaccharides)	Structure	APTT (U/mg)	IC_{50} (µg/mL)			Reference
			Thrombin/Antithrombin	Thrombin/HCII	Factor Xa/Antithrombin	
Holothuria edulis (sulfated fucan)	Figure 6A	9.4 ± 0.47 [a]	78.7 ± 6.2	0.7 ± 0.02	>1000	This work
Ludwigothurea grisea (sulfated fucan)	Figure 6B	13 ± 0.4 [a]	66.5 ± 11	0.5 ± 0.03	>1500	This work
Strongylocentrotus purpuratus (sulfated fucan)	→3)-α-L-Fucp-(2,4SO$_3^-$)-(1→3)-α-L-Fucp-(4SO$_3^-$)-(1→3)-α-L-Fucp-(4SO$_3^-$)-(1→	10 [b]	0.9 [c]	2	ND	[7]
Strongylocentrotus pallidus (sulfated fucan)	→3)-α-L-Fucp-(2SO$_3^-$)-(1→3)-α-L-Fucp-(2SO$_3$)-(1→3)-α-L-Fucp-(4SO$_3^-$)-(1→3)-α-L-Fucp-(4SO$_3^-$)-(1→	18 [b]	>500	3	25	[7]
pig intestinal mucosa (Heparin [c])		212	0.015 ± 0.001	0.2 ± 0.01	0.03 ± 0.003	This work
pig intestinal mucosa (Dermatan sulfate)	HexA [(either β-D-GlcA or α-L-IdoA)-]→3-β-D-GalNAc] disaccharides joined by 1 →4 linkages	5 ± 0.2 [a]	0.95 ± 0.06	0.07 ±0.002	2.4 ± 0.28	This work

[a] The activity of the polysaccharides to prolong APTT is expressed as USP units/mg (U/mg) using a parallel standard curve based on the Heparin 212 units/mg from Sigma (St. Louis, MO, USA); [b] The activity is expressed as international units/mg using a parallel standard curve based on the International Heparin Standard (193 U/mg); [c] Only 80% thrombin inhibition was observed with this sulfated fucan. ND, not determined.

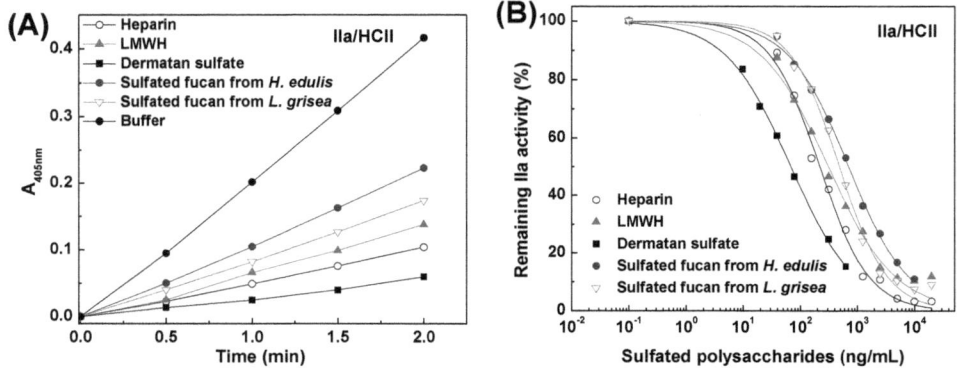

Figure 7. Inhibitory effects of the sulfated fucans, heparin, low molecular weight heparin (LMWH) and dermatan sulfate (DS) on thrombin mediated by heparin cofactor II. (**A**) Shows the time course of thrombin inhibition. HCII (~1 μM) was incubated with thrombin (20 NIH/mL) in the presence of 30 μL (625 ng/mL) samples at 37 °C. After 2 min, 30 μL of 4.5 mM CS-01 (38) was added, the residual thrombin activity was recorded by absorbance at 405 nm; (**B**) Shows the dependence on the sulfated polysaccharide concentration for thrombin inactivation in the presence of HCII. The reaction mixtures were as described in (**A**), except that different concentrations of sulfated polysaccharides were used. Results are shown as means of duplicates. See Table 4 for IC_{50} values.

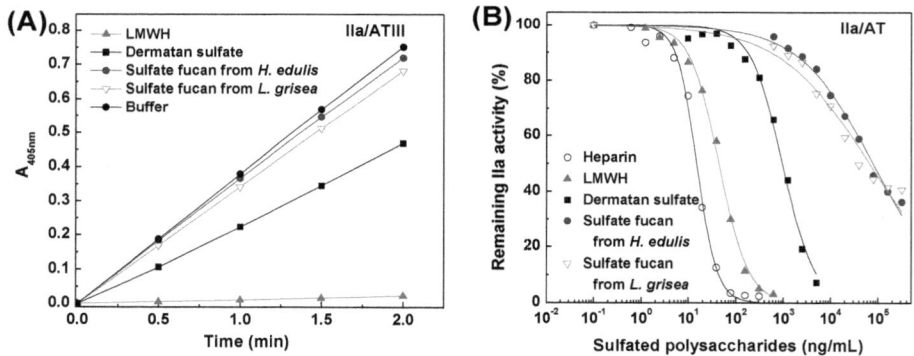

Figure 8. Inhibitory effects of the sulfated fucans, heparin, LMWH and DS on thrombin in the presence of antithrombin. (**A**) Shows the time course of thrombin inhibition. Mixed samples of 30 μL of polysaccharides (625 ng/mL) and 30 μL of 0.25 IU/mL AT were incubated at 37 °C for 2 min, and 30 μL of 24 NIH/mL IIa was then added. After incubation for 2 min, 30 μL of 1.25 mM CS-01 (38) was added, the residual factor IIa activity was recorded by absorbance at 405 nm; (**B**) Shows the dependence on the sulfated polysaccharide concentration for thrombin inactivation mediated by AT. The reaction mixtures were as described in (**A**), except that different concentrations of sulfated polysaccharides were used. Results are shown as means of duplicates. See Table 4 for IC_{50} values.

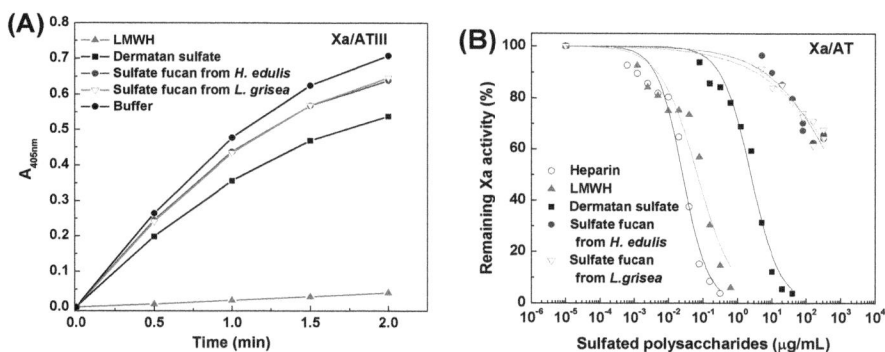

Figure 9. Inhibitory effects of the sulfated fucans, heparin, LMWH and DS on factor Xa in the presence of antithrombin. (**A**) Shows the time course of Xa inhibition. Mixed samples of 30 μL of polysaccharides (625 ng/mL) and 30 μL of 1 IU/mL AT were incubated at 37 °C for 2 min, and 30 μL of 8 μg/mL bovine Xa was then added. After incubation for 1 min, 30 μL of 1.20 mM CS-11(65) was added, the residual Xa activity was recorded by absorbance at 405 nm; (**B**) Shows the dependence on the sulfated polysaccharide concentration for Xa inactivation in the presence of AT. The reaction mixtures were as described in (**A**), except that different concentrations of sulfated polysaccharides were used. Results are shown as means of duplicates. See Table 4 for IC$_{50}$ values.

There are few polysaccharides which provide a suitable comparison for the anticoagulant activity of the sulfated fucans from sea cucumber. Polymers with sulfated fucose as their primary constituent, the fucans from the egg jelly of sea urchin, also have anticoagulant properties [6–8]. A study of the fucan from *Strongylocentrotus pallidus* established that the major anti-thrombin activity of this preparation was mediated by heparin cofactor II, with low ability to potentiate anti-thrombin; however, another species, *Strongylocentrotus purpuratus*, has yielded a fucan in which the balance of antithrombin and heparin cofactor II mediated activities is more in favor of the former [7]. Our results suggested that the type of novel fucans from sea cucumber possesses selectively HCII-dependent thrombin inhibition. The structures of fucans vary from species to species [12,17,30], so this may give rise to variation in the detailed mechanisms of anticoagulant action.

To investigate the structure-activity relationship, it may be valuable to compare anticoagulant action of the fucans from different resources with each other, although differences in experimental techniques between laboratories make direct comparisons between the results of different studies difficult. A study of the fucans from sea urchin suggested that, the occurrence of 2,4-di-O-sulfated units is an amplifying motif for 3-linked L-fucan enhanced thrombin inhibition by antithrombin; the major structural requirement for anti-thrombin activity by heparin cofactor II becomes single 4-O-sulfated fucose units; the presence of 2-O-sulfated fucose residues always had a deleterious effect on anticoagulant activity [7]. However, in this work, the sulfated fucans from sea cucumber without single 4-O-sulfated fucose units possess strong thrombin inhibition mediated by heparin cofactor II (Table 4). These sulfated L-fucans from sea urchins and sea cucumbers have similar charge density and sulfate content (Table 1), but possess the specific pattern of sulfation and the position of the glycosidic linkage as shown in Table 4 and Figure 6. Therefore, these specific structural characters of the sea cucumber polysaccharides, especially the position of the glycosidic linkage, not merely a consequence of their charge density and sulfate content, may contribute to the anticoagulant action.

Overall, our results demonstrated that combining structural analysis of sulfated polysaccharides with specific biological assays is a useful tool to investigate anticoagulant activity. These studies may help elucidate a closer relationship between structure and biological activity of sulfated polysaccharides. New compounds with obvious practical applications may be found.

2.5. The Sea Cucumber Sulfated Fucans do not Induce Human Platelet Aggregation

The effects of the sea cucumber fucans on platelet aggregation were studied via a conventional turbidimetric assay [31,32]. Aggregation was quantified by measuring the maximum extent of the increase in light transmittance 5 min after the addition of the agonist. As shown in Figure 10, the results showed that the fucans did not significantly cause platelets to aggregate in citrated human platelet rich plasma at several concentrations (7.5–30 μg/mL), similar to saline as a negative control. However, a study of fucoidans from marine alga established that they demonstrated dose-dependent irreversible platelet aggregation [33]. Although the absence of a more detailed description on the structure of the polysaccharide used in the study makes it difficult to compare with our results, various structures of the fucans from different resources may account for their different effects on platelet aggregation. In contrast, the aggregation response of another polysaccharide, the fucosylated glycosaminoglycan from the sea cucumber *H. edulis* [9], was dose-dependent and was detected immediately after the addition of the agonist. We observed a considerable effect of the glycosaminoglycan at 30 μg/mL, which was ~75% aggregation as the same as arachididonic acid (AA) at 0.5 mM as a positive control (Figure 10B). The platelet aggregation induced by the sea cucumber glycosaminoglycan had been shown [28,34]. Therefore, the different structural polysaccharides from sea cucumber, in spite of the same species, may demonstrate various effects on platelet aggregation.

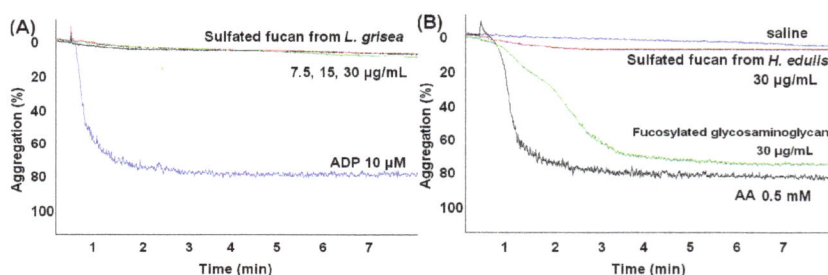

Figure 10. Profile of the platelet aggregation induced by the sea cucumber polysaccharides: the sulfated fucan from *L. grisea* (**A**); the sulfated fucan and fucosylated glycosaminoglycan from *H. edulis* (**B**). The profile showed that the sulfated fucans from sea cucumber do not cause platelets to aggregate at several concentrations.

It is known that a sequence of responses such as further platelet aggregation and fibrin formation result in thrombosis [35]. Thus, the platelet aggregation of the compound as an anticoagulant agent may be its undesirable effect. In this work, the sea cucumber fucans do not cause platelets to aggregate in human plasma at the concentration of 100% thrombin inhibition medicated by HCII. Therefore, our data support the further investigation of the sulfated fucans as novel anticoagulant agents without inducing platelet aggregation.

3. Experimental Section

3.1. Materials

The sea cucumber *H. edulis* were obtained commercially (Qingdao, China), and the sea cucumber *L. grisea* was kindly donated by Mourão (Instituto de BioquímicaMédica, Universidade Federal do Rio de Janeiro, Rio de Janeiro, Brazil). The monosaccharides including glucuronic acid (GlcA), glucose (Glc) and galactose (Gal) were purchased from Alfa Aesar. The *N*-acetylgalactosamine (GalNAc) was purchased from TCI and fucose (Fuc) was from sigma Chemical Co. (Shanghai, China). The 1-phenyl-3-methyl-5-pyrazolone (PMP, 99%) was purchased from ACROS organics. Heparin (212 USP U/mg) and dermatan sulfate (DS) (~41,400 Da) were purchased from Sigma (St. Louis, MO, USA).

LMWH (Enoxaparin, 0.4 mL × 4000 AXaIU) was purchased from Sanofi-Aventis (Paris, France). The activated partial thromboplastin time (APTT), prothrombin time (PT), thrombin time (TT) reagents and standard human plasma were from Teco Medical (Neufahrn N.B., Germany). Biophen Heparin Anti-IIa kits, Biophen Antithrombin 2.5 kits, Human HCII, ATIII, thrombin, thrombin chromogenic substrate CS-01(38) and factor Xa chromogenic substrate SXa-11 were all from Hyphen Biomed (Paris, France). Arachidonic acid (AA), adenosine diphosphate (ADP) and collagen were purchased from Chronolog Corporation, (Havertown, PA, USA). All of other chemicals used were of reagent grade and were obtained commercially.

3.2. Purification of the Sulfated Fucans and Preparation of Their Degraded Products

The polysaccharides were extracted from the body wall of sea cucumbers as the previous description [9,16]. The tissue of the dried body wall was digested by 0.5 M sodium hydroxide for 2 h, and then core protein combined with polysaccharides was released by the papain (EC 3.4.22.2) for 6 h. The crude polysaccharides were purified by gel filtration with a Sephadex G-100 (2 cm × 100 cm, GE Healthcare Technology, Uppsala, Sweden) and ion-exchange chromatography with a DEAE-sephadex FF column (3 cm × 7 cm, Amersham Biosciences, Uppsala, Sweden). The purity of preparation was assayed by high-performance gel permeation chromatography (HPGPC) using a Agilent technologies1200 series (Agilent Co., Santa Clara, CA, USA) apparatus with RID (G1362A) and DAD (G1315D) detectors, equipped with a Shodex OH-pak SB-804 HQ column (8 mm × 300 mm). Chromatographic conditions and procedures were performed according to the previous method [19,36].

To further elucidate the structures in detail, their depolymerized products (~15 kDa) were prepared according to previous reported methods on the sulfated polysaccharide [19]. The native fucan (5.0 g) and 160 mg of copper (II) acetate monohydrate were dissolved in 180 mL of 6% sodium acetate and sodium chloride solution at 35 °C. A 10% H_2O_2 solution was added. The pH of the solution was maintained at 7.3~7.5 by addition of 1 M NaOH solution. The reaction was stopped at different time and after the reaction, 0.5 g of disodium ethylenediamine tetra-acetate dihydrate was added to remove contaminating copper from the product. Depolymerized product was precipitated with ethanol (1:4 (v/v) reaction mixture/ethanol). The crude product was collected by centrifugation (4000× *g* for 20 min) and washed with ethanol. The precipitate was dissolved in water, dialyzed with a molecular weight cut-off of 3 kDa (Spectrum Laboratories Inc., Piscataway, NJ, USA) and lyophilized.

3.3. Monosaccharide Composition Analysis

The monosaccharide components of the sulfated fucans were analyzed by reverse-phase HPLC according to PMP derivatization procedures [9]. Briefly, standard monosaccharides or hydrolyzed sample were dissolved in 0.6 M NaOH (50 μL) and a 0.5 M PMP (50 μL) solution before the derivatization, and incubated at 70 °C for 30 min in a heating block. Then the mixture was neutralized by 50 μL of 0.3 M HCl solution and filtered through 0.22 μm membrane (Millipore, MA, USA). 10 μL of the resulting solution was injected into the RP-C_{18} column (Agilent Eclipse XDB C18, 150 mm × 4.6 mm). The flow rate was 1 mL/min, and UV absorbance of the effluent was monitored at 250 nm. Buffers A and B were 0.1 M ammonium acetate, pH 5.5, containing 22% acetonitrile, respectively. Sugar identification was achieved by comparison with reference sugars.

3.4. Methylation Analysis

Methylation of the sulfated fucan was carried out according to the literature with minor modification [37–40]. Briefly, 5 mg of the sample was placed in a vacuum oven at 40 °C overnight in the presence of phosphorus pentoxide. Then it was dissolved in 2 mL of anhyd DMSO and sonicated completely. Afterwards, 0.6 mL NaOH–DMSO solution under nitrogen was added to the mixture, which was sonicated for at least 30 min. The derivatization was triggered by loading 1 mL of cold CH_3I dropwise until it was fully cooled. The resulting solution was allowed to react for 30 min in the ultrasonic bath and kept for more 30 min. The methylated polysaccharides were extracted with 4 mL of chloroform and dried at low pressure on a rotary evaporator. After hydrolysis with 10 mL of 2 M trifluoroacetic acid, the fucanhydrolysates were dissolved in 2 mL of NaOH aqueous solution. A total of 20 mg of NaBH$_4$ were added to reduce the hemiacetal group. After incubation at 25 °C for 2 h, 100 μL of glacial acetic acid were used to terminate the reduction. The sample was dried under low pressure, and then acetylated by adding 2 mL of acetic anhydride and 2 mL of pyridine. The reaction was kept at 100 °C for 1 h. A total of 2 mL of distilled water were used to decompose the remained acetic anhydride. The acetylated derivatives were extracted with 4 mL of methylene chloride. A gas chromatography/ mass spectrometer (GCMS-QP 2010, Shimadzu, Kyoto, Japan) was used to analyze the glycosidic linkage. The acetylated derivatives were loaded into a RTX-5 capillary column. The temperature program was set as follows: the initial temperature of column was 150 °C, increased to 180 °C at 10 °C/min, then to 220 °C at 2 °C/min, increasing to 240 °C at 5 °C/min, holding for 5 min; injection temperature: 230 °C. The ion source of mass spectrometer was set at 240 °C.

3.5. NMR Analysis

The structure analysis were performed by NMR analyses at 300 K in D$_2$O with a Bruker Avance spectrometer of 500 MHz equipped with a ^{13}C/^{1}H dual probe in FT mode, as previously described [12]. All samples were previously dissolved in deuterium (D$_2$O, 99.9% D) and lyophilized thrice to replace exchangeable protons with deuterium. The lyophilized samples were then dissolved in D$_2$O at a 20–30 g/L concentration. All spectra were recorded with HOD suppression by presaturation. The interpretations of ^{1}H/^{1}H correlated spectroscopy (COSY), total correlation spectroscopy (TOCSY), rotating frame overhauser effect spectroscopy (ROESY), and ^{1}H/^{13}C heteronuclear single-quantum coherence (HSQC), heteronuclear multiple bond coherence (HMBC) spectra were recorded using state-time proportion phase incrementation for quadrature detection in the indirect dimension. All chemical shifts were relative to internal 3-trimethylsilyl-(2,2,3,3-^{2}H4)-propionic acid (TSP, δ_H and δ_C = 0.00).

3.6. Determination of the Anticoagulant Activities

The activated partial thromboplastin time (APTT), prothrombin time (PT), and thrombin time (TT) were determined with a coagulometer (TECO MC-4000, Teco Medical Inc., Neufahrn N.B., Germany) by using APTT, PT, TT reagents and standard human plasma as previously described [41,42].

3.7. Inhibition of Thrombin by Heparin Cofactor II

Inhibition of thrombin by HCII was measured with thrombin chromogenic substrate CS-01(38) using previously described method with modifications [28,43]. A mixture containing 30 μL of HCII (1 μM) in 20 mM Tris-HCl (pH 7.4) and 0.1% PEG-8000, and 30 μL of various concentrations of each glycosaminoglycan in 20 mM Tris-HCl (pH 7.4) was incubated at 37 °C for 1 min. A 30 μL aliquot of 20 NIH/mL thrombin in 20 mM Tris-HCl (pH 7.4) containing 0.1% PEG-8000 was then added. After incubation at 37 °C for 1 min, 30 μL of 4.5 mM thrombin chromogenic substrate CS-01(38) solution was added and the thrombin activity was measured. The absorbance at 405 nm was measured on a Bio-Tek Microplate Reader (ELx 808, BioTek Instruments, Inc., Winooski, VT, USA).

3.8. Inhibition of Thrombin and Factor Xa by Antithrombin

The antithrombin and anti-factor Xa activities in the presence of AT were measured with Biophen Heparin Anti-IIa kits and Biophen Heparin Anti-Xa kits [28]. A mixture of 30 μL samples and 30 μL 0.25 IU/mL AT (or 1 IU/mL AT) was incubated at 37 °C for 2 min; 30 μL of 24 NIH /mL thrombin (or 8 μg/mL bovine factor Xa) was added. After incubation for 2 min (or 1 min for factor Xa), the residual thrombin or factor Xa activity was measured by the addition of 30 μL of 1.25 mM thrombin chromogenic substrate CS-01(38) or 1.20 mM factor Xa chromogenic substrate CS-11(65). The absorbance of the reaction mixture was read at 405 nm on a Bio-Tek Microplate Reader.

Mar. Drugs **2015**, *13*, 2063–2084

3.9. Platelet Aggregation Assays

Turbidometric measurements of platelet aggregation of the polysaccharides were performed in a Chronolog Model 700 Aggregometer (Chronolog Corporation, Havertown, PA, USA) according to Born's method [28,31,32]. Venousblood from a young volunteer (31 years old, male, 72 kg) were anticoagulated with 3.8% sodium citrate (9:1, v/v). Platelet-rich plasma (PRP) and platelet-poor plasma (PPP) were prepared shortly after blood collection by spinning the sample at 180 g for 10 min at 22 °C. The PRP was carefully removed and the remaining blood centrifuged at 2400 g for 10 min to obtain PPP. The centrifuge temperature was maintained at 4 °C. Platelet counts were adjusted by the addition of PPP to the PRP to achieve a count of 250×10^9 L^{-1}. Platelet aggregation studies were completed within 3 h of preparation of PRP. Immediately after preparation of PRP, 250 μL was transferred into each of the three test tubes, with 250 μL PPP as a control. After 5 min of warming, PRP and PPP were put in testing places and were warmed for a further 5 min. Final concentrations of agonists were: AA 0.5 mM and ADP 10 μM as positive references, respectively. The change of optical density as a result of platelet aggregation was recorded. The present study was approved by the Research Ethics Committee of Kunming Institute of Botany (SYXK-K2013-0004), Chinese Academy of Sciences. The study subject provided written informed consent (20 September 2013).

4. Conclusions

In conclusion, structural characterization and pharmacological activities of two new sulfated fucans from sea cucumber have been elucidated. As described above, the polysaccharides are composed of a central core of regular $\alpha(1\rightarrow3)$- and $\alpha(1\rightarrow2)$-linked tetrasaccharide repeating units together with an unsulfated fucose residue as a side chain. Investigation of the structure-activity relationship suggested that novel tetrasaccharide repeating units of the sea cucumber polysaccharides may contribute to the anticoagulant action. The anticoagulant and platelet aggregation assays further indicated that the type of fucans possesses selectively anti-thrombin activity by heparin cofactor II without causing platelets to aggregate, whose anticoagulant mechanisms are different from those of heparin-like drugs. Further work on *in vivo* antithrombotic activity of the well-defined sulfated fucan as a potential anticoagulant therapeutics is worthy to be conducted.

Acknowledgments: This work was supported by Grants from the Yunnan Provincial Science and Technology Department in China (2010CI116 and 2013FA046), the National Natural Science Foundation of China (81102372 and 81373292), Outstanding Technical Talent Foundation of the Chinese Academy of Sciences and the West Light Foundation of the Chinese Academy of Sciences. We thank Mourão (Instituto de Bioquímica Médica, Universidade Federal do Rio de Janeiro, Rio de Janeiro, Brazil) for kindly donation of the sea cucumber *L. grisea*. We also thank Rong Huang for performing the NMR experiments.

Author Contributions: Jinhua Zhao, Mingyi Wu and Lingyun Chen conceived the idea of discovery of novel sulfated fucan, designed the experiments, supervised students to carry out the experiments, drew and summarized the figures, and finalized the preparation of the manuscript. Li Xu and Chuang Xiao carried out the anticoagulant activities of the sulfated fucans and their effects on coagulation factors and cofactors. Longyan Zhao prepared, analyzed the sulfated fucan. Lian Yang carried out the activity assays of platelet aggregation and performed statistical analyses. Na Gao, Zi Li and Lan Luo extracted, purified and carried out the physicochemical analysis of the sulfated fucans. Mingyi Wu wrote the manuscript. Li Xu and Chuang Xiao participated in the discussion and revised the manuscript.

Conflicts of Interest: The authors declare no conflicts of interest.

References

1. Streiff, M.B.; Bockenstedt, P.L.; Cataland, S.R.; Chesney, C.; Eby, C.; Fanikos, J.; Fogarty, P.F.; Gao, S.W.; Grcia-Aguilara, J.; Goldhaber, S.Z.; *et al.* Venous thromboembolic disease. *J. Natl. Compr. Cancer Netw.* **2011**, *9*, 714–777.
2. Warkentin, T.E.; Levine, M.N.; Hirsh, J.; Horsewood, P.; Roberts, R.S.; Gent, M.; Kelton, J.G. Heparin-induced thrombocytopenia in patients treated with low-molecular weight heparin or unfractionated heparin. *N. Engl. J. Med.* **1995**, *332*, 1330–1335. [CrossRef] [PubMed]

3. Mourão, P.A.S.; Pereira, M. Searching for alternatives to heparin: Sulfated fucans from marine invertebrates. *Trends Cardiovasc. Med.* **1999**, *9*, 225–232. [CrossRef] [PubMed]

4. Rocha, H.A.O.; Moraes, F.A.; Trindade, E.S.; Franco, C.R.C.; Torquato, R.J.S.; Veiga, S.S.; Valente, A.P.; Mourão, P.A.S.; Leite, E.L.; Nader, H.B.; *et al.* Structural and hemostatic activities of a sulfated galactofucan from the brown alga *Spatoglossum schroederi*: An ideal antithrombotic agent? *J. Biol. Chem.* **2005**, *280*, 41278–41288.

5. Bordbar, S.; Anwar, F.; Saari, N. High-value components and bioactives from sea cucumbers for functional foods—A review. *Mar. Drugs* **2011**, *9*, 1761–1805. [CrossRef] [PubMed]

6. Mourão, P.A.S. Use of sulfated fucans as anticoagulant and antithrombotic agents: Future perspectives. *Curr. Pharm. Des.* **2004**, *10*, 967–981. [CrossRef] [PubMed]

7. Pereira, M.S.; Melo, F.R.; Mourão, P.A.S. Is there a correlation between structure and anticoagulant action of sulfated galactans and sulfated fucans. *Glycobiology* **2002**, *12*, 573–580. [CrossRef] [PubMed]

8. Pomin, V.H.; Pereira, M.S.; Valente, A.P.; Tollefsen, D.M.; Pavão, M.S.G.; Mourão, P.A.S. Selective cleavage and anticoagulant activity of a sulfated fucan: Stereospecific removal of a 2-sulfate ester from the polysaccharide by mild acid hydrolysis, preparation of oligosaccharides, and heparin cofactor II-dependent anticoagulant activity. *Glycobiology* **2005**, *15*, 369–381. [CrossRef]

9. Luo, L.; Wu, M.; Xu, L.; Lian, W.; Xiang, J.; Lu, F.; Gao, N.; Xiao, C.; Wang, S.; Zhao, J. Comparison of physicochemical characteristics and anticoagulant activities of polysaccharides from three sea cucumbers. *Mar. Drugs* **2013**, *11*, 399–417. [CrossRef] [PubMed]

10. Chen, S.; Hu, Y.; Ye, X.; Li, G.; Yu, G.; Xue, C.; Chai, W. Sequence determination and anticoagulant and antithrombotic activities of a novel sulfated fucan isolated from the sea cucumber *Isostichopus badionotus*. *Biochim. Biophys. Acta* **2012**, *1820*, 989–1000. [CrossRef] [PubMed]

11. Holtkamp, A.D.; Kelly, S.; Ulber, R.; Lang, S. Fucoidans and fucoidanases—Focus on techniques for molecular structure elucidation and modification of marine polysaccharides. *Appl. Microbiol. Biotechnol.* **2009**, *82*, 1–11. [CrossRef] [PubMed]

12. Vilela-Silva, A.C.; Castro, M.O.; Valente, A.P.; Mourão, P.A.S. Sulfated fucans from the egg jellies of the closely related sea urchins *Strongylocentrotus droebachiensis* and *S. pallidus* ensure species-specific fertilization. *J. Biol. Chem.* **2002**, *277*, 379–387. [CrossRef] [PubMed]

13. Wang, Y.; Su, W.; Zhang, C.; Xue, C.; Chang, Y.; Wu, X. Protective effect of sea cucumber (*Acaudina molpadioides*) fucoidan against ethanol-induced gastric damage. *Food Chem.* **2012**, *133*, 1414–1419. [CrossRef]

14. Kariya, Y.; Mulloy, B.; Imai, K.; Tominaga, A.; Kaneko, T.; Asari, A.; Suzuki, K.; Masuda, H.; Kyogashima, M.; Ishii, T. Isolation and partial characterization of fucan sulfates from the body wall of sea cucumber *Stichopus japonicus* and their ability to inhibit osteoclastogenesis. *Carbohydr. Res.* **2004**, *339*, 1339–1346. [CrossRef] [PubMed]

15. Mulloy, B.; Mourão, P.A.S.; Gray, E. Structure/function studies of anticoagulant sulphated polysaccharides using NMR. *J. Biotechnol.* **2000**, *77*, 123–135. [CrossRef] [PubMed]

16. Ribeiro, A.C.; Vieira, R.P.; Mourão, P.A.S.; Mulloy, B. A sulfated α-L-fucan from sea cucumber. *Carbohydr. Res.* **1994**, *255*, 225–240. [CrossRef] [PubMed]

17. Mulloy, B.; Ribeiro, A.C.; Alves, A.P.; Vieira, R.P.; Mourão, P.A.S. Sulfated fucans from echinoderms have a regular tetrasaccharide repeating unit defined by specific patterns of sulfation at the O-2 and O-4 positions. *J. Biol. Chem.* **1994**, *269*, 22113–22123. [PubMed]

18. Dietrich, C.P.; Tersariol, I.L.S.; Toma, L.; Moraes, C.T.; Porcionatto, M.A.; Oliveira, F.W.; Nader, H.B. Structure of heparan sulfate: identification of variable and constant oligosaccharide domains in eight heparan sulfates of different origins. *Cell. Mol. Biol.* **1998**, *44*, 417–429. [PubMed]

19. Wu, M.Y.; Xu, S.M.; Zhao, J.H.; Kang, H.; Ding, H. Physicochemical characteristics and anticoagulant activities of low molecular weight fractions by free radical depolymerization of a fucosylated chondroitin sulfate from sea cucumber *Thelenota ananas*. *Food Chem.* **2010**, *122*, 716–723. [CrossRef]

20. Vieira, R.P.; Mulloy, B.; Mourão, P.A.S. Structure of a fucose-branched chondroitin sulfate from sea cucumber: Evidence for the presence of 3-O-sulfo-β-D-glucuronosyl residues. *J. Biol. Chem.* **1991**, *266*, 13530–13536. [PubMed]

21. Volpi, N.; Mascellani, G.; Bianchini, P. Low molecular weight heparins (5 kDa) and oligoheparins (2 kDa) produced by gel permeation enrichment or radical process: Comparison of structures and physicochemical and biological properties. *Anal. Biochem.* **1992**, *200*, 100–107. [CrossRef] [PubMed]

22. Volpi, N. Dermatan sulfate from beef mucosa: Structure, physicochemical and biological properties of fractions prepared by chemical depolymerization and anion-exchange chromatography. *Carbohydr. Res.* **1994**, *255*, 133–144. [CrossRef] [PubMed]

23. Yoshida, K.; Minami, Y.; Nemoto, H.; Numata, K.; Yamanaka, E. Structure of DHG, a depolymerized glycosaminoglycan from sea cucumber *Stichopus japonicus*. *Tetrahedron Lett.* **1992**, *33*, 4959–4962. [CrossRef]

24. Ustyuzhanina, N.E.; Ushakova, N.A.; Zyuzina, K.A.; Bilan, M.I.; Elizarova, A.L.; Somonova, O.V.; Madzhuga, A.V.; Krylov, A.V.; Preobrazhenskaya, M.E.; Usov, A.I.; *et al.* Influence of fucoidans on hemostatic system. *Mar. Drugs* **2013**, *11*, 2444–2458. [CrossRef] [PubMed]

25. Duus, J.Ø.; Gotfredsen, C.H.; Bock, K. Carbohydrate structural determination by NMR spectroscopy: Modern methods and limitations. *Chem. Rev.* **2000**, *100*, 4589–4614. [CrossRef]

26. Karmakar, P.; Ghosh, T.; Sinha, S.; Saha, S.; Mandal, P.; Ghosal, P.K.; Ray, B. Polysaccharides from the brown seaweed *Padina tetrastromatica*: Characterization of a sulfated fucan. *Carbohydr. Polym.* **2009**, *78*, 416–421. [CrossRef]

27. Patankar, M.S.; Oehninger, S.; Barnett, T.; Williams, R.L.; Clark, G.F. A revised structure for fucoidan may explain some of its biological activities. *J. Biol. Chem.* **1993**, *268*, 21770–21776. [PubMed]

28. Wu, M.Y.; Wen, D.D.; Gao, N.; Xiao, C.; Yang, L.; Xu, L.; Lian, W.; Peng, W.L.; Jiang, J.M.; Zhao, J.H. Anticoagulant and antithrombotic evaluation of native fucosylated chondroitin sulfates and their derivatives as selective inhibitors of intrinsic factor Xase. *Eur. J. Med. Chem.* **2015**, *92*, 257–269. [CrossRef] [PubMed]

29. Sheehan, J.P.; Walke, E.N. Depolymerized holothurian glycosaminoglycan and heparin inhibit the intrinsic tenase complex by a common antithrombin-independent mechanism. *Blood* **2006**, *107*, 3876–3882. [CrossRef] [PubMed]

30. Mourão, P.A.S.; Pereira, M.S.; Pavão, M.S.G.; Mulloy, B.; Tollefsen, D.M.; Mowinckel, M.C.; Abildgaard, U. Structure and anticoagulant activity of a fucosylated chondroitin sulfate from echinoderm: Sulfated fucose branches on the polysaccharide account for its high anticoagulant action. *J. Biol. Chem.* **1996**, *271*, 23973–23984. [CrossRef] [PubMed]

31. Born, G.V.R. Aggregation of blood platelets by adenosine diphosphate and its reversal. *Nature* **1962**, *194*, 927–929. [CrossRef] [PubMed]

32. Born, G.V.R.; Cross, M.J. The aggregation of blood platelets. *J. Physiol.* **1963**, *168*, 178–195. [CrossRef] [PubMed]

33. Azevedo, T.C.G.; Bezerra, M.E.B.; Santos, M.G.L.; Souza, L.A.; Marques, C.T.; Benevides, N.M.B.; Leite, E.L. Heparinoids algal and their anticoagulant, hemorrhagic activities and platelet aggregation. *Biomed. Pharmacother.* **2009**, *63*, 477–483. [CrossRef] [PubMed]

34. Li, J.Z.; Lian, E.C. Mechanism of rabbit platelet agglutination induced by acidic mucopolysaccharide extracted from *Stichopus japonicus* Selenka. *Thromb. Haemost.* **1988**, *59*, 432–434. [PubMed]

35. Jackson, S.P. The growing complexity of platelet aggregation. *Blood* **2007**, *109*, 5087–5095. [CrossRef] [PubMed]

36. Tsukamoto, T.; Hattori, M.; Sakabe, M.; Haginaka, J. Determination of the molecular mass of new L-fucose-containing glycosaminoglycan and its distribution by high-performance gel-permeation chromatography with laser light-scattering detection. *Anal. Sci.* **2001**, *17*, 555–558. [CrossRef] [PubMed]

37. Needs, P.W.; Sevendran, R.R. Avoiding oxidative degradation during sodium hydroxide/methyl iodide-mediated carbohydrate methylation in dimethyl sulfoxide. *Carbohydr. Res.* **1993**, *245*, 1–10. [CrossRef]

38. Yang, L.; Fu, S.; Zhu, X.; Zhang, L.; Yang, Y.; Yang, X.; Liu, H. Hyperbranched acidic polysaccharide from green tea. *Biomacromolecules* **2010**, *11*, 3395–3405. [CrossRef] [PubMed]

39. Zhu, Q.; Jiang, Y.; Lin, S.; Wen, L.; Wu, D.; Zhao, M.; Chen, F.; Jia, Y.; Yang, B. Structural identification of (1→6)-α-D-glucan, a key responsible for the health benefits of longan, and evaluation of anticancer activity. *Biomacromolecules* **2013**, *14*, 1999–2003. [CrossRef] [PubMed]

40. Ciucanu, I.; Costello, C.E. Elimination of oxidative degradation during the per-O-methylation of carbohydrates. *J. Am. Chem. Soc.* **2003**, *125*, 16213–16219. [CrossRef] [PubMed]

41. Gao, N.; Wu, M.Y.; Liu, S.; Lian, W.; Li, Z.; Zhao, J.H. Preparation and characterization of O-acylated fucosylated chondroitin sulfate from sea cucumber. *Mar. Drugs* **2012**, *10*, 1647–1661. [CrossRef] [PubMed]

42. Lian, W.; Wu, M.Y.; Huang, N.; Gao, N.; Zheng, R.T.; Peng, W.L.; Zhao, J.H. Anti-HIV-1 activity and structure–activity-relationship study of a fucosylated glycosaminoglycan from an echinoderm by targeting the conserved CD4 induced epitope. *Biochim. Biophys. Acta* **2013**, *1830*, 4681–4691. [CrossRef] [PubMed]

43. Nagase, H.; Enjyoji, K.; Minamiguchi, K.; Kitazato, K.T.; Kitazato, K.; Saito, H.; Kato, H. Depolymerized holothurian glycosaminoglycan with novel anticoagulant actions: Antithrombin III-and heparin cofactor II-independent inhibition of factor X activation by factor IXa-factor VIIIa complex and heparin cofactor II-dependent inhibition of thrombin. *Blood* **1995**, *85*, 1527–1534. [PubMed]

marine drugs

MDPI

Article

Conformational Analysis of the Oligosaccharides Related to Side Chains of Holothurian Fucosylated Chondroitin Sulfates

Alexey G. Gerbst *, Andrey S. Dmitrenok, Nadezhda E. Ustyuzhanina * and Nikolay E. Nifantiev *

N.D. Zelinsky Institute of Organic Chemistry, Russian Academy of Sciences, Leninsky prospect 47, 119991 Moscow B-334, Russia; dmt@ioc.ac.ru

* Authors to whom correspondence should be addressed; alger@ioc.ac.ru (A.G.G.); ustnad@gmail.com (N.E.U.); nen@ioc.ac.ru (N.E.N.); Tel./Fax: +7-499-137-8784 (N.E.N.).

Academic Editor: Paola Laurienzo

Received: 8 December 2014; Accepted: 3 February 2015; Published: 12 February 2015

Abstract: Anionic polysaccharides fucosylated chondroitin sulfates (FCS) from holothurian species were shown to affect various biological processes, such as metastasis, angiogenesis, clot formation, thrombosis, inflammation, and some others. To understand the mechanism of FCSs action, knowledge about their spatial arrangement is required. We have started the systematic synthesis, conformational analysis, and study of biological activity of the oligosaccharides related to various fragments of these types of natural polysaccharides. In this communication, five molecules representing distinct structural fragments of chondroitin sulfate have been studied by means of molecular modeling and NMR. These are three disaccharides and two trisaccharides containing fucose and glucuronic acid residues with one sulfate group per each fucose residue or without it. Long-range C–H coupling constants were used for the verification of the theoretical models. The presence of two conformers for both linkage types was revealed. For the Fuc–GlA linkage, the dominant conformer was the same as described previously in a literature as the molecular dynamics (MD) average in a dodechasaccharide FCS fragment representing the backbone chain of the polysaccharide including GalNAc residues. This shows that the studied oligosaccharides, in addition to larger ones, may be considered as reliable models for Quantitative Structure-Activity Relationship (QSAR) studies to reveal pharmacophore fragments of FCS.

Keywords: fucosylated chondroitin sulfate; oligosaccharides; conformational analysis; glycosidic linkage; Karplus equation; long range C–H coupling constants

1. Introduction

Anionic polysaccharides of a different nature were shown to affect various biological processes, such as metastasis, angiogenesis, clot formation, thrombosis, inflammation, bacterial and viral adhesion, and some others [1–4]. In some cases, exogenous polysaccharides act *via* the mechanism of competitive inhibition of the natural ligand binding to the target protein. Thus, well known glycosaminoglycan heparin was found to reduce inflammation and metastasis in a way of inhibition of P- and L-selectins binding to their cellular ligands [4,5]. In other cases, carbohydrates could potentiate interaction of two proteins with the formation of a ternary complex, which can be illustrated by thrombin-antithrombin III-heparin complex formation playing the key role in inhibition of blood coagulation [6,7]. However, side effects of heparin treatment, such as bleeding and thrombocytopenia, force the search for and development of alternative drugs. Among the compounds under investigation are the polysaccharides fucosylated chondroitin sulfates (FCSs) isolated from different holothurian species. These biopolymers demonstrated a wide range of biological activities including antimetastatic, anti-inflammatory, anticoagulant, antithrombotic, immunostimulatory, antibacterial, antiviral, and some others activities typical of heparin but with a lack of side effects [3,8–13].

The backbone of known FCSs is built up of alternating →4)-linked β-D-glucuronic acid and →3)-linked *N*-acetyl β-D-galactosamine residues. Some of glucuronic acid units are substituted at *O*-3 by selectively *O*-sulfated fucosyl residues (Figure 1) [3,13]. The presence of side chains in FCS distinguishes them from mammalian chondroitin sulfates and was shown to be essential for biological properties of FCS [13,14]. The pattern of *O*-sulfation and the structure of branches vary accordingly to the type of holothurian species and are basically responsible for the level of biological effect of FCSs [3,13]. Fine structural characteristics of these polysaccharides also depend on the geographic range, season of harvesting, and the method of isolation [13,15]. In Figure 1 the known structures of the branched fragments of FCSs are presented.

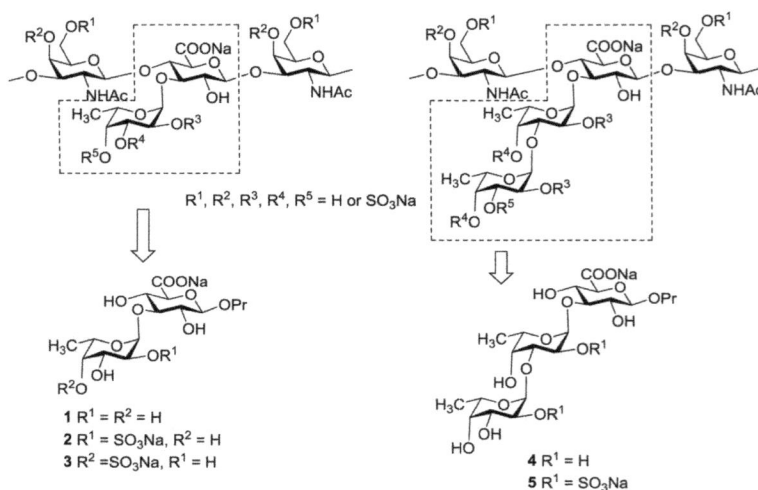

Figure 1. Branched fragments of fucosylated chondroitin sulfates and synthetic oligosaccharides related to the knots.

It is known that the presence of sulfate groups in these polysaccharides is crucial for the observed biological activity [1,3,11,13,14]. However, influence of certain sulfation patterns on the activity is poorly understood. In our laboratory, we have started a complex investigation of oligosaccharides related to various fragments of FCS, including their systematic synthesis, conformational analysis, and study of biological action, particularly the role of sulfate groups.

To understand the mechanism of FCSs action, knowledge about the spatial arrangement of these carbohydrates is required. In this communication, five molecules representing simple structural fragments of chondroitin sulfate have been studied by means of molecular modeling and NMR. These are three disaccharides (Figure 1) and two trisaccharides containing fucose and glucuronic acid residues with a single sulfate group per each fucose residue. The synthesis and detailed characterization of these compounds are described in [16].

2. Results and Discussion

The conformation of the glycosidic linkage is usually described in terms of two torsional angles, φ and ψ. These are directly connected with values of long-range ^{13}C–^{1}H spin–spin coupling constants (SSCC, $^{3}J_{C-H}$), which can be measured by various NMR techniques. The relationship between these parameters may be formalized by the Karplus type equation [17] (Figure 2), which opens the way towards the verification of the results of theoretical modeling by NMR methods.

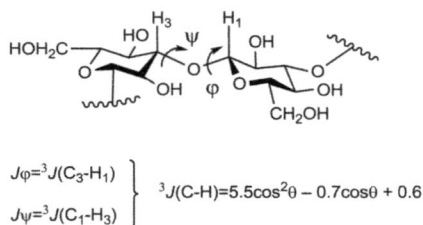

$$J\varphi = {}^3J(C_3\text{-}H_1)$$
$$J\psi = {}^3J(C_1\text{-}H_3)$$

$${}^3J(C\text{-}H) = 5.5\cos^2\theta - 0.7\cos\theta + 0.6$$

Figure 2. Torsional angles describing a glycosidic linkage and the Karplus equation [17]. Angles φ and ψ are defined as H_1–C_1–O–C_x and C_1–O–C_x–H_x correspondingly.

First, conformational maps were constructed for each oligosaccharide. A simple dihedral driving algorithm was used with a step of $10°$ for each torsion determining the conformation of the glycosidic linkage. Energies of the conformations in each step were plotted as conformational maps (Figure 3). In the modern conformational analysis of carbohydrates, such approach is considered as a very rough estimate of the potential energy surface. For each linkage in the studied compounds, only one populated area corresponding to the global minimum was revealed. Other minima had significantly higher energies and were scarcely populated. Thus, these results were not used for the modeling of the NMR parameters, and only two samples of the obtained maps are given in Figure 3.

The found minima were used as starting points for molecular dynamics (MD) calculations. These were carried out in three modes: first *in vacuo*, then with the account for solvation using the Solvent Accessible Surface Area (SASA) [18] continuum model, and with the explicit water molecules in a droplet. Such an approach was successfully used by us previously [19] during the conformational analysis of glucan fragments.

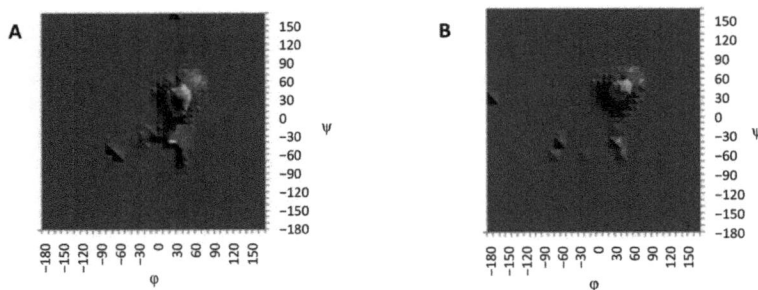

Figure 3. Sample confomational maps obtained by means of torsion scanning for Fuc–GlA (**A**) and Fuc–Fuc (**B**) linkages.

The general MD procedure included a 20,000 ps run for each starting structure with a snapshot being written at an interval of 2 ps, resulting in 10,000 snapshot structures in each trajectory. The constant temperature (298 K) protocol was used. The snapshot structures collected during MD simulations were used to calculate an ensemble of averaged SSCC values in order to compare them with the experimentally observed ones. The latter were measured employing J-HMBC technique (see the Experimental Section for details). Previously, the experimental error of thus measured ${}^3J_{C\text{-}H}$ was found by us to have the value of 0.5 Hz [20].

In vacuo calculations quite expectedly produced results inconsistent with the experiment in all the cases (Table 1). An average difference between the experimental and calculated SSCC values was about 0.8 Hz. This can be attributed to the highly charged nature of the studied molecules and, obviously, a consideration of the solvation effects was required.

Table 1. Experimental and calculated with different models spin–spin coupling constants (SSCC) values for compounds **1–5**, Hz.

Compound Number	$J\varphi$ (Fuc–GlA)	$J\psi$ (Fuc–GlA)	$J\varphi$ (Fuc–Fuc)	$J\psi$ (Fuc–Fuc)
1	3.8 (exp) 4.0 (*in vacuo*) 3.4 (SASA) 3.2 (water)	2.5 (exp) 3.8 (*in vacuo*) 2.7 (SASA) 2.5 (water)	n.a.*	n.a.*
2	2.7 (exp) 4.0 (*in vacuo*) 3.6 (SASA) 3.2 (water)	2.9 (exp) 3.7 (*in vacuo*) 2.3 (SASA) 2.4 (water)	n.a.*	n.a.*
3	3.0 (exp) 4.0 (*in vacuo*) 3.3 (SASA) 3.2 (water)	2.7 (exp) 3.8 (*in vacuo*) 2.7 (SASA) 2.6 (water)	n.a.*	n.a.*
4	3.6 (exp) 4.0 (*in vacuo*) 3.2 (SASA) 3.3 (water)	3.0 (exp) 3.8 (*in vacuo*) 2.9 (SASA) 3.4 (water)	2.8 (exp) 3.5 (*in vacuo*) 2.7 (SASA) 2.5 (water)	2.5 (exp) 3.2 (*in vacuo*) 2.6 (SASA) 2.5 (water)
5	2.7 (exp) 4.0 (*in vacuo*) 3.2 (SASA) 3.2(water)	2.5 (exp) 3.7 (*in vacuo*) 2.4 (SASA) 2.4 (water)	2.6 (exp) 3.3 (*in vacuo*) 3.2 (SASA) 2.8 (water)	2.9 (exp) 3.1 (*in vacuo*) 3.4 (SASA) 2.6 (water)

* n.a.—Not applicable.

First, the continuum solvation model SASA was tested and its use significantly improved the situation. As can be seen from Table 1, the calculated SSCC values became closer to those observed experimentally. The average deviation in this case was 0.34 Hz, which is within the experimental error. To identify the details of the conformational behavior of the studied molecules, graphs of angle change in time during MD and plots of the conformers obtained from MD simulations were constructed in coordinates φ/ψ (Figures 4 and 5).

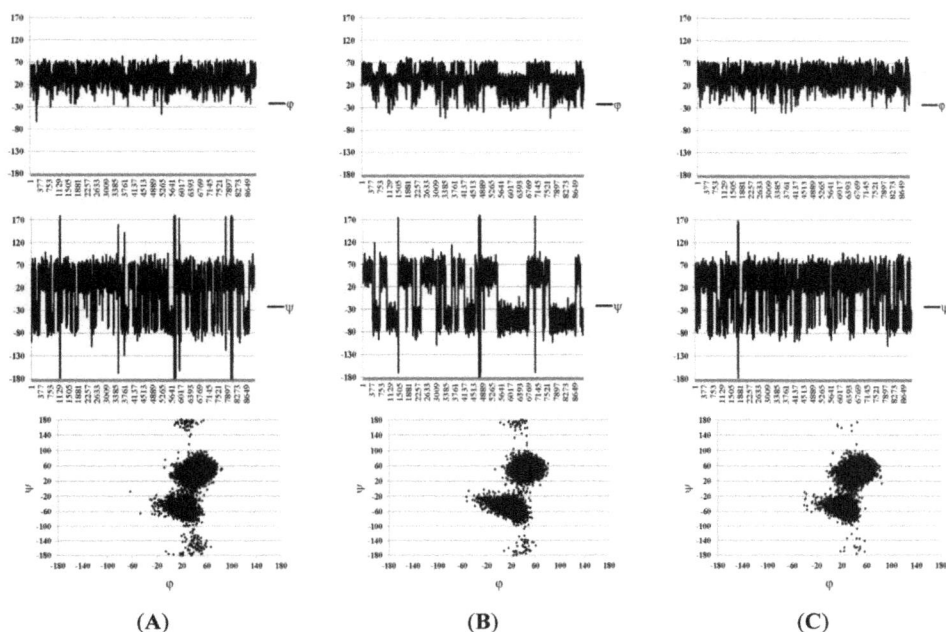

Figure 4. Molecular dynamics (MD) graphs and conformational plots obtained in the Solvent Accessible Surface Area (SASA) approximation for disaccharides **1–3** ordered from (**A**) to (**C**). Angles φ and ψ are defined as $H_1–C_1–O–C_x$ and $C_1–O–C_x–H_x$, respectively, for the sake of compatibility with the Karplus equation.

For two types of the glycosidic linkages (Fuc–Fuc and Fuc–GlcA) in these compounds, slightly different conformational distributions were observed. Both were characterized with the presence of conformers which are common among carbohydrates, with the torsional angles φ lying around $20°–40°$ (these are determined by the exo-anomeric effect), and ψ angles forming two clusters around $+40°$ (conformer **I**) and $-0°$ (conformer **II**).

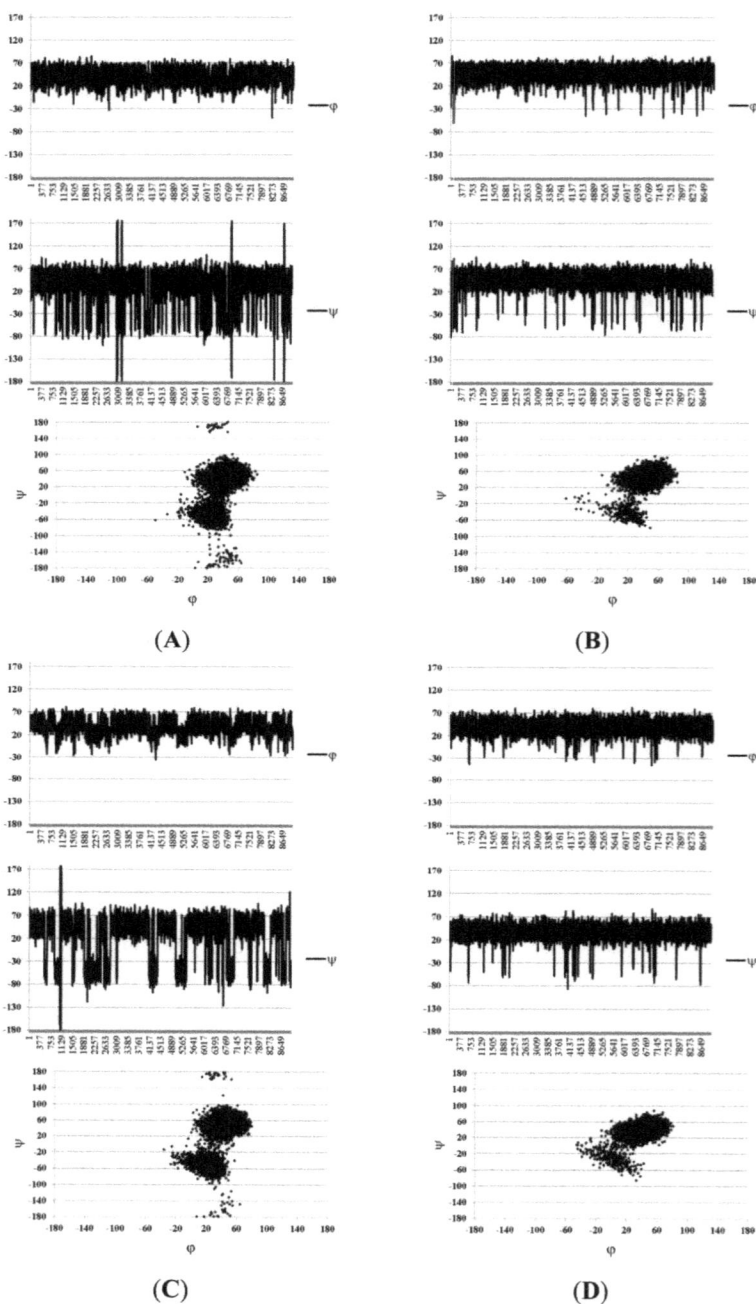

Figure 5. Conformational plots obtained in the SASA approximation for glycosidic linkages in structures **4** ((**A**), Fuc–GlA; (**B**), Fuc–Fuc) and **5** ((**C**), Fuc–GlA, (**D**), Fuc–Fuc). Angles φ and ψ are defined as H_1–C_1–O–C_x and C_1–O–C_x–H_x, respectively, for the sake of compatibility with the Karplus equation.

In the case of Fuc–GlA linkages in all compounds, trace amounts of the inverted conformers were found, which had ψ torsions in the region close to ±170–$180°$ (conformer **III**). Such conformers are sometimes encountered in the conformational distribution of carbohydrates but are usually poorly populated. In this case, their appearance may be explained by the relatively free rotation around the linkage Fuc–$(1\rightarrow3)$–GlA due to the lack of axially oriented substituents in the glucuronic residue, contrary to the case of the linkage Fuc–$(1\rightarrow3)$–Fuc. In general, these distributions resemble the conformational maps obtained by the dihedral scanning procedure (Figure 3). In this case, however, one can see that the conformers that previously appeared minor and insignificant (namely, **II** and **III**) now have become populated to a greater extent.

Graphical representations of the detected conformers are given in Figure 6. Interestingly, conformer **I**, which, according to our results, seems to be predominant for the Fuc–GlcA linkage, has the same orientation of the monosaccharide residues as determined for this fragment in [11] during the MD analysis of two dodecasaccharides representing the backbone chain of the polysaccharide including GalNAc residues.

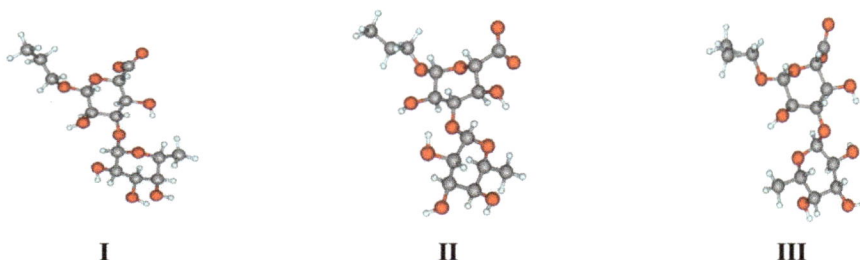

Figure 6. Principal conformers of the Fuc$(1\rightarrow3)$GlA linkage.

However, when the explicit water model was applied, the disappearance of the above mentioned inverted conformers was observed. It was accompanied also by a slight improvement of coincidence between the calculated and experimental SSCC values for all the studied compounds (Table 1).

The sample MD graphs of the glycosidic torsions and the conformational plots obtained under these conditions for the Fuc–GlcA and difucoside fragments are shown in Figure 7. They are extracted from trajectories obtained during the simulations of trisaccharide **4**. It is noteworthy that during the studies of β–$(1\rightarrow3)$–glucan related oligosaccharides [19] the same tendency for the inverted conformers was found. In fact, the dominating conformer for the Fuc–GlcA linkage was still that of type **I** (Figure 6), indicating that the conformation of the fucosyl–glucuronyl branch is essentially the same both in the presence and in the absence of the units of the backbone chain. The explanation of this fact can be achieved by closer examination of conformer **I**. The mutual orientation of fucose and glucuronic acid residue in it is such that the O-4 atom of the GlcA-unit is not hindered by the fucosyl substituent (Figure 8). Thus, it can be thought that the introduction of a galactosamine residue at this position should not change the conformation of the branch substantially.

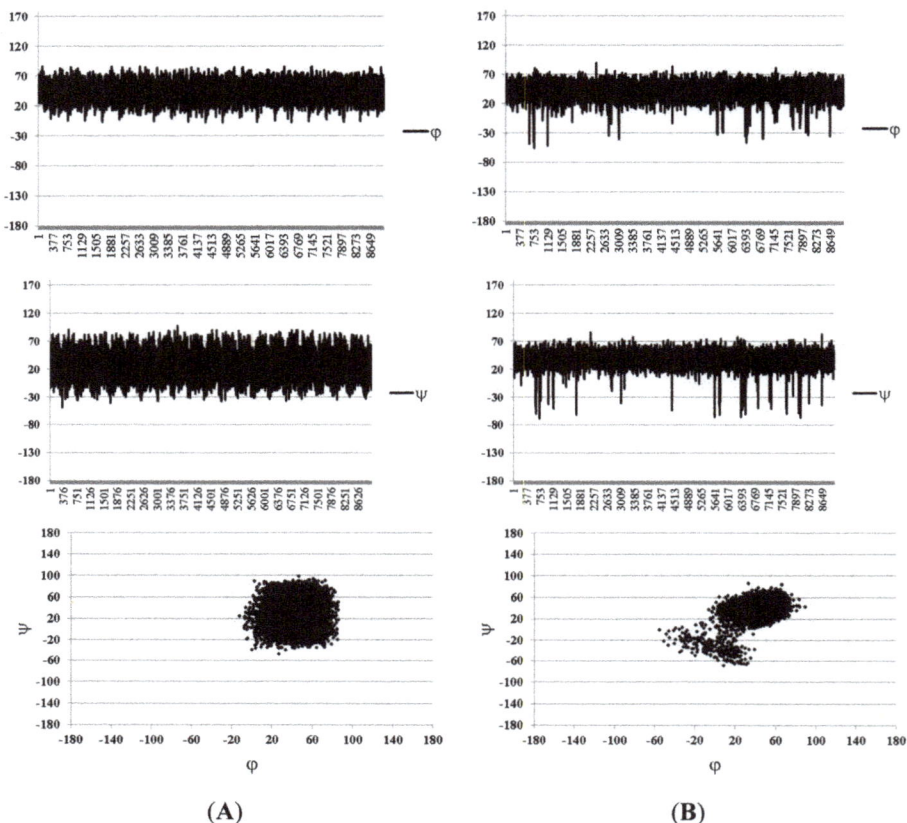

Figure 7. Torsional angles graph and conformational plots obtained using explicit water approximation for the fucosyl–glucuronide (**A**) and difucoside (**B**) fragments. Data are extracted from simulations of compound **4**. Angles φ and ψ are defined as H_1–C_1–O–C_x and C_1–O–C_x–H_x correspondingly for the sake of compatibility with the Karplus equation.

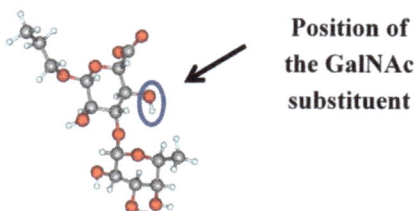

Position of the GalNAc substituent

Figure 8. The dominant conformer of the Fuc(1→3)GlA linkage with the position of GalNAc residue introduction.

Additional evidence that actually no inverted conformers were present in the conformational distribution was achieved by means of NOESY spectroscopy. The mixing time used in these experiments was 500 ms. For the studied types of molecules in diluted solutions such mixing time gave reliably measurable NOE values and still was within the linear range of the NOE build-up. This fact was confirmed by conducting a series of 1D NOESY experiments using different mixing times on compound **1** (Figure 9). Slight non-linearity at 50–100 ms could be explained by the relatively small sensitivity in this range which increased the integration error.

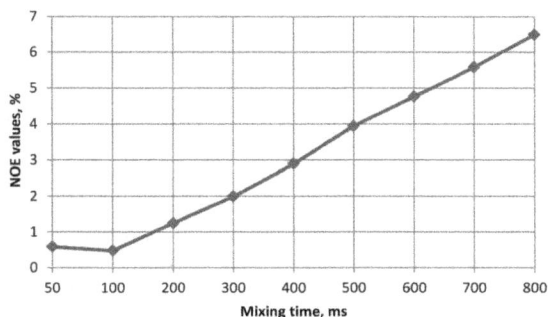

Figure 9. The NOE build-up graph at different mixing times for compound **4**. The measured NOE values are for H1(Fuc)/H3(GlA) interaction.

The relative NOE values from MD simulations were calculated as the reciprocal ratio of the averaged sixth order of the corresponding interatomic distances. The NOE were calculated with respect to H1 atoms of the glycosylating residues. The obtained results indicated that in the SASA approximation there should be a slight NOE from H-1 of the Fuc residue onto H-2 and H-3 of the GlcA unit. For non-sulfated compounds **1** and **4**, 2D-NOESY spectra were recorded and they showed no corresponding cross-peaks (Table 2).

Table 2. Relative experimental and calculated at SASA and explicit water level NOE values for structures **1** and **4**.

Compound Number	H1$_{(Fuc)}$–H3$_{(GlA)}$	H1$_{(Fuc)}$–H4$_{(GlA)}$	H1$_{(Fuc)}$–H2$_{(GlA)}$	H1$_{(Fuc')}$–H3$_{(Fuc)}$	H1$_{(Fuc')}$–H4$_{(Fuc)}$
1	1 (exp) 1 (SASA) 1 (water)	0 (exp) 0.26 (SASA) 0.05 (water)	0 (exp) 0.12 (SASA) 0.01 (water)	n.a.*	n.a.*
4	1 (exp) 1 (SASA) 1 (water)	0 (exp) 0.33 (SASA) 0 (water)	0 (exp) 0.09 (SASA) 0.01 (water)	0.62 (exp) 0.81 (SASA) 0.7 (water)	0.7 (exp) 3.0 (SASA) 1 (water)

* n.a.—Not applicable.

3. Experimental Section

3.1. Materials

The synthesis and detailed characterization of compounds **1–5** are described in [16]. For all NMR experiments, D_2O was used as a solvent (99.98% D, Merck, Darmstadt, Germany). Samples for NMR spectroscopy were dissolved in D_2O, lyophilized, dissolved in D_2O again, and then transferred to NMR tubes for analysis.

3.2. Methods

All molecular mechanics calculation were done using TINKER v. 5.1 software package [21] with the implemented MM3 force field. The detailed description of MD simulations is provided in Section 2. For explicit water simulations a structure in question was soaked into a droplet of 2500 TIP3P [22] water molecules with automatic elimination of stacking molecules. The system was then equilibrated for 1 ns and additional 20 ns simulation was run with the snapshots being written every 2 ps. All hydrogen-involving bonds were constrained using RATTLE version of SHAKE [23] algorithm and cutoff value of 5 Å was applied.

The NMR spectra of **1–5** (10–20 mg) were recorded in D_2O solutions on Bruker spectrometers AV-400 and AV-600 with 0.05% acetone as reference (^{1}H 2.225 ppm, ^{13}C 31.45 ppm) at the temperature of 303 K. Shigemi microtubes (purchased from Sigma-Aldrich, Inc., St. Louis, MO, USA) were sometimes used for sensitivity enhancement. The resonance assignment in ^{1}H and ^{13}C NMR spectra was performed by gradient enhanced 2D gCOSY, gNOESY, gHSQC, gJ-HMBC experiments as well as TOCSY and ROESY experiments.

NOE build-up was studied employing selective gradient enhanced 1D NOESY experiment with different mixing times. Other experimental NOEs were measured using a field gradient enhanced 2D gNOESY technique in D_2O solutions at 303 K, mixing time 500 ms, relaxation delay 2 s. A sinusoidal field gradient of 1 ms length and a recovery time of 1 ms were used. The processing was performed with p/2 shifted sine-square function in both dimensions.

Experimental $^3J_{C-H}$ constants were measured using J-HMBC [24] experiment. The spectral widths were about 2 ppm for ^{1}H region and 40 ppm for ^{13}C region and did not include resonances of aglycon groups. The data were collected in the echo/anti-echo mode. The length of gradients was 1 ms, and the recovery time was 100 us. The spectra were acquired with 80–120 t_1 increments and 64–256 scans per increment. 1024 points were collected during the acquisition time t_2. The HMBC preparation delay was set to 250 ms that corresponded to J_{C-H}^{min} = 2.0 Hz. The upscaling coefficient k was 25–60. The relaxation delay was 1s. The third order low-pass J-filter was introduced for the suppression of one bond constant ($^1J_{C-H}$) in the range from 125 to 180 Hz. The forward linear prediction to 1024 points was used in F1. The processing was performed with p/2 shifted sine square function in both dimensions.

4. Conclusions

Five non-sulfated and selectively sulfated oligosaccharides of definite structure representing Fuc–GlcA and Fuc–Fuc–GlcA fragments of the side chains of fucosylated chondroitin sulfates (FCS) were studied by means of theoretical molecular mechanics calculations and NMR experiments. Long-range C–H coupling constants were used for the verification of the theoretical models. The presence of two conformers for both linkages was revealed. For the Fuc–GlA linkage, the predominant conformer was the same as described by other authors [11] as the MD average in a dodechasaccharide FCS fragment representing the backbone chain of the polysaccharide including GalNAc residues. Apparently, the introduction of a GalNAc substituent slightly impacts the conformation of the Fuc–GlA branch, just reducing the population of the second conformer. This shows that the studied oligosaccharides, in addition to larger ones, may be considered as reliable models for further QSAR studies in order to reveal pharmacophore fragments of FCS and particularly to understand the role of sulfate groups.

Acknowledgments: The work was supported by the Russian Scientific Foundation (grant 14-13-01325).

Author Contributions: Nikolay E. Nifantiev and Alexey G. Gerbst conceived the study. Nadezhda E. Ustyuzhanina prepared and characterized compounds **1–5**. Alexey G. Gerbst performed molecular mechanics calculations. Andrey S. Dmitrenok recorded NMR spectra of all compounds and made signal assignment. All authors participated in interpretation of the results and preparation the final version of the manuscript.

Conflicts of Interest: The authors declare no conflict of interest.

References

1. Pomin, V.H. Fucanomics and galactanomics: Current status in drug discovery, mechanisms of action and role of the well defined structures. *Biochim. Biophys. Acta* **2012**, *1820*, 1971–1979. [CrossRef] [PubMed]
2. Jiao, G.; Yu, G.; Zhang, J.; Ewart, S. Chemical structures and bioactivities of sulfated polysaccharides from marine algae. *Mar. Drugs* **2011**, *9*, 196–223. [CrossRef] [PubMed]
3. Pomin, V.H. Holothurian fucosylated chondroitin sulfate. *Mar. Drugs* **2014**, *12*, 232–254. [CrossRef] [PubMed]

4. Stevenson, J.L.; Varki, A.; Borsig, L. Heparin attenuates metastasis mainly due to inhibition of P- and L-selectin, but non-anticoagulant heparins can have additional effects. *Thromb. Res.* **2007**, *120*, S107–S111. [CrossRef]
5. Wang, L.; Brown, J.R.; Varki, A.; Esko, J.D. Heparin's anti-inflammatory effects require glucosamine 6-O-sulfation and are mediated by blockade of L- and P-selectins. *J. Clin. Invest.* **2002**, *110*, 127–136. [CrossRef] [PubMed]
6. Petitou, M.; van Boeckel, C.A. A synthetic antithrombin III binding pentasaccharide is now a drug! What comes next? *Angew. Chem. Int. Ed. Engl.* **2004**, *43*, 3118–3133. [CrossRef] [PubMed]
7. Gray, E.; Hogwood, J.; Mulloy, B. The anticoagulant and antithrombotic mechanisms of heparin. *Handb. Exp. Pharmacol.* **2012**, *207*, 43–61. [PubMed]
8. Borsig, L.; Wang, L.; Cavalcante, M.C.; Cardilo-Reis, L.; Ferreira, P.L.; Mourão, P.A.; Esko, J.D.; Pavão, M.S. Selectin blocking activity of a fucosylated chondroitin sulfate glycosaminoglycan from sea cucumber. Effect on tumor metastasis and neutrophil recruitment. *J. Biol. Chem.* **2007**, *282*, 14984–14991. [CrossRef] [PubMed]
9. Fonseca, R.J.; Oliveira, S.N.; Pomin, V.H.; Mecawi, A.S.; Araujo, I.G.; Mourão, P.A. Effects of oversulfated and fucosylated chondroitin sulfates on coagulation. Challenges for the study of anticoagulant polysaccharides. *Thromb. Haemost.* **2010**, *103*, 994–1004. [CrossRef] [PubMed]
10. Zhao, Y.; Zhang, D.; Wang, S.; Tao, L.; Wang, A.; Chen, W.; Zhu, Z.; Zheng, S.; Gao, X.; Lu, Y. Holothurian glycosaminoglycan inhibits metastasis and thrombosis via targeting of nuclear factor-κB/tissue factor/Factor Xa pathway in melanoma B16F10 cells. *PLoS One* **2013**, *8*, e56557. [CrossRef] [PubMed]
11. Panagos, C.G.; Thomson, D.S.; Moss, C.; Hughes, A.D.; Kelly, M.S.; Liu, Y.; Chai, W.; Venkatasamy, R.; Spina, D.; Page, C.P.; *et al.* Fucosylated chondroitin sulfates from the body wall of the sea cucumber *Holothuria forskali*: Conformation, selectin binding, and biological activity. *J. Biol. Chem.* **2014**, *289*, 28284–28298. [CrossRef] [PubMed]
12. Luo, L.; Wu, M.; Xu, L.; Lian, W.; Xiang, J.; Lu, F.; Gao, N.; Xiao, C.; Wang, S.; Zhao, J. Comparison of physicochemical characteristics and anticoagulant activities of polysaccharides from three sea cucumbers. *Mar. Drugs* **2013**, *11*, 399–417. [CrossRef] [PubMed]
13. Myrona, P.; Siddiqueeb, S.; Al Azad, S. Fucosylated chondroitin sulfate diversity in sea cucumbers: A review. *Carbohydr. Polym.* **2014**, *112*, 173–178. [CrossRef] [PubMed]
14. Mourão, P.A.; Giumarães, B.; Mulloy, B.; Thomas, S.; Gray, E. Antithrombotic activity of a fucosylated chondroitin sulphate from echinoderm: Sulphated fucose branches on the polysaccharide account for its antithrombotic action. *Br. J. Haematol.* **1998**, *101*, 647–652. [CrossRef] [PubMed]
15. Chen, S.; Xue, C.; Yin, L.; Tang, Q.; Yu, G.; Chai, W. Comparison of structures and anticoagulant activities of fucosylated chondroitin sulfates from different sea cucumbers. *Carbohydr. Polym.* **2011**, *83*, 688–695. [CrossRef]
16. Ustyuzhanina, N.E.; Fomitskaya, P.A; Gerbst, A.G.; Dmitrenok, A.S.; Nifantiev, N.E. Synthesis of the oligosaccharides related to side chains of fucosylated chondroitin sulfates from sea cucumbers. *Mar. Drugs* **2015**, *13*, 770–787. [CrossRef]
17. Mulloy, B. Long-range carbon-proton coupling constants: Application to conformational studies of oligosaccharides. *Carbohydr. Res.* **1988**, *184*, 39–46. [CrossRef] [PubMed]
18. Ooi, T.; Oobatake, M.; Némethy, G.; Scheraga, H.A. Accessible surface areas as a measure of the thermodynamic parameters of hydration of peptides. *Proc. Natl. Acad. Sci. USA* **1987**, *84*, 3086–3090. [CrossRef] [PubMed]
19. Gerbst, A.G.; Grachev, A.A.; Yashunsky, D.V.; Tsvetkov, Y.E.; Shashkov, A.S.; Nifantiev, N.E. Theoretical and experimental conformational studies of oligoglucosides structurally related to fragments of fungal cell wall β-(1→3)-D-Glucan. *J. Carbohydr. Chem.* **2013**, *32*, 205–221. [CrossRef]
20. Grachev, A.A.; Gerbst, A.G.; Ustuzhanina, N.E.; Shashkov, A.S.; Usov, A.I.; Nifantiev, N.E. NMR Investigation of the influence of sulfate groups at C-2 and C-4 on the conformational behavior of fucoidan fragments with homo-(1→3)-linked backbone. *J. Carbohydr. Chem.* **2006**, *25*, 315–330. [CrossRef]
21. TINKER Molecular Modeling. Available online: http://dasher.wustl.edu/tinker (accessed on 6 February 2015).
22. Jorgensen, W.L.; Chandrasekhar, J.; Madura, J.D.; Impey, R.W.; Klein, M.L. Comparison of simple potential functions for simulating liquid water. *J. Chem. Phys.* **1983**, *79*, 926–935. [CrossRef]

Mar. Drugs **2015**, *13*, 936–947

23. Leimkuhler, B.; Skeel, R. Symplectic numerical integrators in constrained Hamiltonian systems. *J. Comp. Phys.* **1994**, *112*, 117–125. [CrossRef]

24. Meissner, A.; Sørensen, O.W. Measurement of $J_{(H,H)}$ and long-range $J_{(X,H)}$ coupling constants in small molecules. Broadband XLOC and J-HMBC. *Magn. Reson. Chem.* **2001**, *39*, 49–52.

marine drugs

MDPI

Article

Isolation and Structural Characterization of a Novel Antioxidant Mannoglucan from a Marine Bubble Snail, *Bullacta exarata* (Philippi)

Donghong Liu [1,2,†], Ningbo Liao [1,†], Xingqian Ye [1,2], Yaqin Hu [1,2,*], Dan Wu [1], Xin Guo [1], Jianjun Zhong [1], Jianyong Wu [3] and Shiguo Chen [1,3]

[1]   College of Biosystem Engineering and Food Science, Zhejiang University, 866 Yuhangtang Road, Xihu District, Hangzhou 310058, China; dhliu@zju.edu.cn (D.L.); liaoningbo2010@sina.com (N.L.); psu@zju.edu.cn (X.Y.); processing@zju.edu.cn (D.W.); crystalgx0509@gmail.com (X.G.); hendle527@yahoo.com (J.Z.); chenshiguo210@zju.edu.cn (S.C.)

[2]   Fuli Institute of Food Science, Zhejiang University, Hangzhou 310058, China

[3]   Department of Applied Biology & Chemical Technology, The Hong Kong Polytechnic University, Hung Hom 990077, Kowloon, Hong Kong; jian-yong.wu@polyu.edu.hk

*   Author to whom correspondence should be addressed; yqhu@zju.edu.cn; Tel./Fax: +86-571-8898-2154.

†   These authors contributed equally to this work.

Received: 11 September 2013; in revised form: 12 October 2013; Accepted: 12 October 2013; Published: 11 November 2013

Abstract: *Bullacta exarata* is one of the most economically important aquatic species in China, noted for not only its delicious taste and nutritional value, but also for its pharmacological activities. In order to explore its potential in medical applications, a mannoglucan designated as BEPS-IB was isolated and purified from the foot muscle of *B. exarata* after papain digestion. Chemical composition analysis indicated BEPS-IB contained mainly D-glucose and D-mannose in a molar ratio of 1:0.52, with an average molecular weight of about 94 kDa. The linkage information was determined by methylation analysis, and the anomeric configuration and chain linkage were confirmed by IR and 2D NMR. The results indicated BEPS-IB was composed of Glcp_6Man$p$ heptasaccharide repeating unit in the backbone, with occasional branch chains of mannose residues (14%) occurring in the backbone mannose. Further antioxidant assay indicated BEPS-IB exhibited positive antioxidant activity in scavenging superoxide radicals and reducing power. This is the first report on the structure and bioactivity of the mannoglucan from the *B. exarata*.

Keywords: *Bullacta exarata*; structure; polysaccharide; mannoglucan

1. Introduction

There has been an increasing interest in recent years in the structures and functions of natural polysaccharides from various sources because of their nutraceutical and pharmaceutical potential. Glucans, a kind of neutral polysaccharides mainly consisting of glucose monomers, have been widely isolated from medical herb or fungus. They have been reported to stimulate the immune system and decrease infectious complications in humans [1–3] and experimental animals [4]. These activities were related to their structural features. For example, β-(1→6)-linked side chains of glucose residues increases antitumor activity [5], and a β-(1→3)-linked backbone seems essential for the antioxidant effect [6,7]. However, most of the reports on the glucans were focused on medical herb or fungi sources, with little knowledge on the structure and function of the glucans in the animals, especially the marine invertebrates.

Polysaccharides from mollusks have attracted increasing attention in recent years because of their potential pharmaceutical values, such as anti-tumor [8], immunity-enhancement [9], anti-inflammation [10] and anti-aging effects [11]. Several bioactive glucans from the mollusks have also been reported, e.g., Zhang, Ye and Wang [12] reported a water soluble α-glucan from the soft body of *Bellamya purificata* with a main chain of (1→4)-linked D-glucopyranosyl and branching points at O-6 of (1→6)-linked D-glucopyranosyl residues, which showed significant anti-inflammatory activity. A glucan from *Cyclina sinensis* also showed antioxidant and hepatoprotective activities [13].

Bullacta exarata (Philippi), generally called Tutie or Niluo in China, belongs to the Mollusca phylum, Gastropoda class, and Haminoeidae family, which is a species of bubble snail with a bullate, spirally striate shell found in the coastlines of the South and East China Seas. As an important economic resource in eastern China, *B. exarata* is noted not only for its delicious taste and nutritional value, but also for its pharmacological activities. It is a highly acclaimed species in traditional Chinese medicine (TCM) with a broad spectrum of health promoting effects on the kidney, lung, liver and immune functions [14–16]. The foot muscle is the main edible part of *B. exarata*, and its abundance in proteins and carbohydrates makes it attractive for exploiting. However, the structure and bioactivities of the polysaccharides from this species are rarely researched and little is known about them.

In the present study, a mannoglucan was isolated from *B. exarata* and purified by anion-exchange and gel-filtration chromatography. The sequence of the purified polysaccharide was determined by a combination of composition analysis, methylation analysis, IR and NMR, and its antioxidant activity was investigated by the reducing power assay and the superoxide radical scavenging assay.

2. Results and Discussion

2.1. Isolation and Purification of Polysaccharides

The extraction of *B. exarata* foot muscle by papain digestion retained an extract yield of 7.3% (w/w) by dry weight, which was named as CBEPS. It was further fractionated by ion exchange chromatography on a DEAE-52 column (Figure 1a). The major peak labeled as BEPS-I (3.62%, w/w) was collected and further purified on a Sephacryl S-300 HR gel-permeation chromatography, two factions designated as BEPS-IA (1.03%, w/w) and BEPS-IB (2.12%, w/w) was collected (Figure 1b). The main fraction BEPS-IB showed a single peak on the GPC (Figure 1c), corresponding to an average molecular weight around 94 kDa.

Chemical composition analysis indicated the BEPS-IB showed a total carbohydrate of 98.7% by the phenol-sulfuric acid method, indicating the removing of the protein part after papain digestion. Further monosaccharide composition analysis by the HPLC-PMP method indicated BEPS-IB had a simpler composition than the crude polysaccharides after extensive purification, which was mainly composed of Glc and Man, with a proportion of 1:0.52. A small proportion of Gal and Fuc was also detected (Table 1). The protein content of BEPS-IB decreased from 10.7 to 2.1 after purification, whereas the sulfated was around 0.47%–1.5% (w/w) in all the tested fractions.

Table 1. Yields, protein contents, sugar contents, sulfate contents and Mw of *B. exarata* polysaccharides.

Composition	Samples			
	[a] CBEPS	BEPS-I	BEPS-IA	BEPS-IB
Yield (%)	7.3	3.6	1.03	2.2
Neutral sugar (%)	80.12	89.31	76.29	98.76
Protein (%)	10.7	2.4	2.7	2.1
Sulfate (%)	1.2	1.5	0.47	1.23
Mw (kDa)	_ [b]	_ [b]	127	94
Molar ratio of monosaccharides				
Mannose	0.74	1.69	0.32	0.52
Glucose	0.43	1.26	0.57	1
Galactose	0.32	0.03	0.04	0.03
Fucose	0.18	0.12	ND	0.01
Rhamnose	0.56	0.22	0.31	ND
Arabinose	0.17	0.07	0.03	ND

[a] Crude *B. exarata* polysaccharide; [b] The polysaccharide was a mixture; ND, not detected.

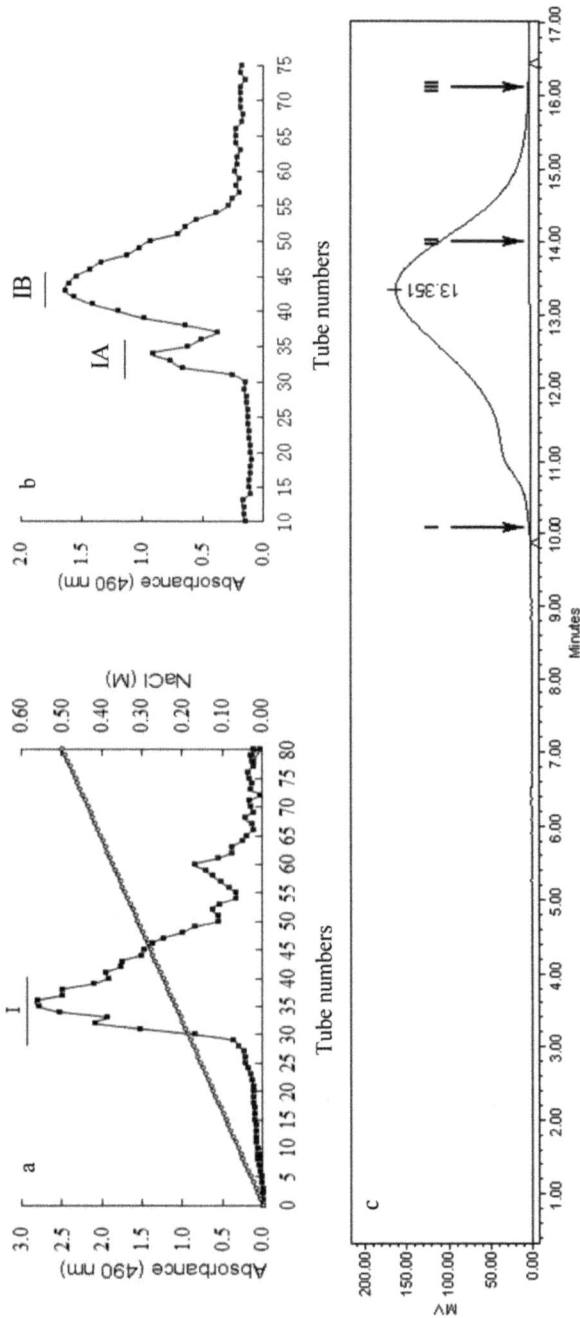

Figure 1. Isolation of the polysaccharides present in the aqueous extract of *B. exarata*. The crude extract was fractionated by ion-exchange chromatography on a DEAE ion-exchange column (**a**) and the collected fraction was further purified by gel filtration chromatography on a Sephacryl S-300 HR column (**b**). Solid bars indicate the fractions collected. The molecular weight of polysaccharide fraction BEPS-IB (Mw = 94 kDa) was determined by HPLC on a TSK-Gel G4000 PWXL column, eluted with 0.2 mol/L NaCl at 0.5 mL/min (**c**). Range of molecular weight in kDa: I = 500; II = 66.9; III = 40.

2.1.1. IR Spectrum and Elucidation of BEPS-IB

Figure 2 presents the IR spectrum of BEPS-IB. The broad and intense stretching at 3400 cm^{-1} is characteristic of hydroxyl groups, and the weak stretching at 2930 cm^{-1} is attributed to the C–H bond [17]. The band at 1647 cm^{-1} can be attributed to water bound to the polysaccharide molecule, and the bands between 950 and 1200 cm^{-1} are mostly attributed to C–O–C and C–O–H linkages [18]. Absorptions at 916 cm^{-1} are typical for D-Glc in the pyranose form. The fraction also exhibited an obvious characteristic absorption at 920 and 809 cm^{-1} corresponding to the existence of mannose [19]. Moreover, the characteristic absorptions at 845 cm^{-1} in the IR spectra indicated the presence of α-glycosidic linkages.

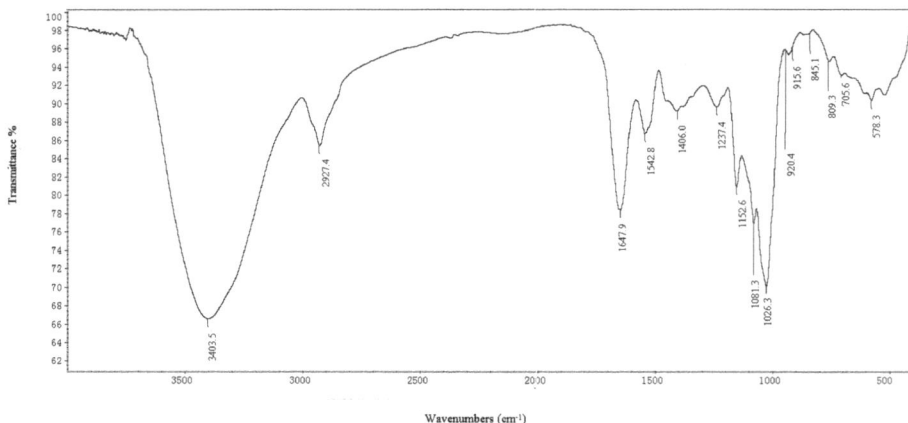

Figure 2. Infrared spectra of polysaccharide (BEPS-IB) from *B. exarata*.

2.1.2. Methylation Analysis

Methylation analysis GC-MS suggested BEPS-IB contained terminal Glc, 6-linked Glc, 3-linked Man and 3,6-linked Man (assigned as X, V, U and W) at a molar ratio of 1.32:6.43:3.16:0.97 (Table 2). The presence of the 3,6-linked Man linkages in the polysaccharide chain was indicative of the substitution of the Man residues.

2.1.3. Structure Characteristics of BEPS-IB from NMR

Based on the methylation analysis, the structure characteristics of BEPS-IB, especially the chain linkage and conformation of the sugar units, were completed and confirmed by 1D and 2D NMR.

The ^{1}H NMR spectrum (600 MHz) (Figure 3a) of BEPS-IB was recorded at 60 °C. It showed four anomeric proton signals at δ 5.338, 5.192, 4.943 and 4.603 ppm in a molar ratio of about 6:1:3:1 (Figure 3a), which were assigned to the four types of sugar units obtained in the methylation analysis (U, V, W and X). The signals around 3.5–4.5 ppm were assigned to cross ring protons (Figure 3a). Similarly, the ^{13}C spectrum (Figure 3b) showed four signals δ 102.2, 100.8, 98.2 and 94.6 ppm in a molar ratio of nearly 6:1:1:3, which were assigned to the anomeric carbon of sugar units U, V, W and V, respectively. The other signals around 60–85 ppm were also assigned to the carbon signals for C2–C5 (Figure 3b).

Table 2. GC-MS data for alditol acetate derivatives from methylated polysaccharide (BEPS-IB) isolated from *B. exarata*.

Methylated sugar	Retention time (min)	Molar ratio	Mass fragment (*m/z*)	Type of linkage
2,3,4,6-tetra-*O*-Me-Glc [a]	14.80	1.32	43, 45, 71, 87, 101, 117, 129, 145, 161, 205	Glc-(1→
2,4,6-tri-*O*-Me-Man	16.95	3.16	43, 45, 87, 101, 117, 129, 161, 233	→3)-Man-(1→
2,3,4-tri-*O*-Me-Glc	17.31	6.43	43, 45, 71, 87, 101, 117, 129, 161, 173, 189, 233	→6)-Glc-(1→
2,4-tri-*O*-Me-Man	19.16	0.97	43, 87, 101, 117, 129, 189	→3,6)-Man-(1→

[a] 2,3,4,6-tetra-*O*-Me-Glc = 1,5-di-*O*-acetyl-2,3,4,6-tetra-*O*-methyl-glucose.

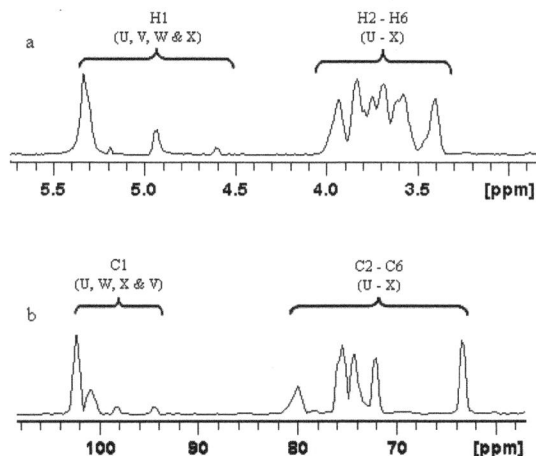

Figure 3. ^{1}H NMR (**a**) and ^{13}C NMR (**b**) spectrum (600 MHz, D_2O, 60 °C) of BEPS-IB isolated from *B. exarata*.

The complete assignment of the chemicals-shifts of the sugar units were obtained from 2D NMR (Table 3), including ^{1}H-^{1}H COSY (Supplementary Figure S1), TOCSY (Supplementary Figure S2a), NOESY (Supplementary Figure S2b), ^{1}H-^{13}C HMBC (data not shown) and HMQC (Figure 4), according to the published methodology [20]. The down-shifts in carbon signals of the sugar units compared to the native glucan may suggest possible linkage information, e.g., the down-shift in the C-3 position of the unit W indicated it may be 3-linked Man. Similarly, unit U, V and X were deduced as 6-linked Glc, terminal Glc and 3,6-linked Man, respectively.

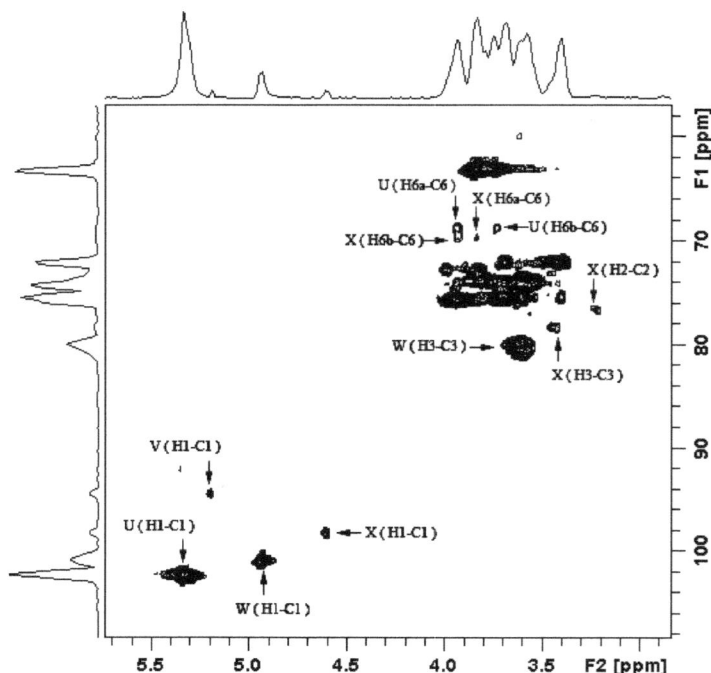

Figure 4. HMQC spectrum of BEPS-IB isolated from *B. exarata*.

Table 3. The ^{1}H and ^{13}C NMR chemical shifts for the polysaccharide BEPS-IB isolated from *B. exarata*.

Glycosidic linkage		Chemical shifts (ppm)					
		H-1 (C-1)	H-2 (C-2)	H-3 (C-3)	H-4 (C-4)	H-5 (C-5)	H-6 (C-6)
U	→6)-α-Glc*p*-(1→	5.338 (102.2)	3.587 (72.1)	3.742 (76.8)	3.632 (73.2)	3.964 (76.8)	3.937/3.736 (68.9)
V	α-Glc*p*-(1→	5.192 (94.6)	3.517 (73.8)	3.811 (72.4)	3.342 (73.2)	3.623 (75.3)	3.831/3.792 (63.2)
W	→3)-α-Man-(1→	4.943 (100.8)	3.542 (75.4)	3.605 (80.1)	3.614 (74.8)	3.963 (76.3)	3.972/3.752 (69.2)
X	→3,6)-α-Man-(1→	4.603 (98.2)	3.211 (76.5)	3.729 (78.5)	3.412 (73.9)	3.574 (76.7)	3.833/3.925 (69.6)

The linkage information of the sugar units were further confirmed by NOESY and HMBC experiments. In the HMBC spectrum, the intra- and inter-residual connectivities of both anomeric protons and carbons of each of the glycosyl residues were summarized in Table 4. Cross peaks indicated the correlation signals were found between C-1 of residue U (δ 102.2) with H-6 of residue X (U C-1, X H-6), H-1 of residue U (δ 5.338) with C-6 of residue X (U H-1, X C-6), indicating the U was linked to the 6-position of the X. Similarly, the correlation signals W H-1–X C-3 indicated the unit W linked to the 3-position of X. Thus, both 3 and 6 positions of X were substituted. The correlation signals V H-1–W C-3 and V C-1–W H-3 indicated unit V linked to the 3-position of the unit W; an intraresidual coupling between H-1 of residue U with its own C-6 (U H-1, U C-6) indicated that the 6-linkage was among different U repeats. In the NOESY spectrum (Supplementary Figure S2b), similar correlations of the protons were observed: U1–X6, V1–W3, W1–X3 and X1–U6, which confirmed the above results from HMBC.

Table 4. The connectivities observed in an HMBC spectrum for the anomeric protons/carbons of the sugar residues of BEPS-IB from *B. exarata*.

Residue	Sugar linkage	Anomeric atom (δ_H/δ_C)	Observed connectivities		
			δ_H/δ_C	Residue	Atom
U	→6)-α-Glc*p*-(1→	5.338	69.6	U: H-1	X: C-6
			73.2	U: H-1	U: C-4
		102.3	3.83	U: C-1	X: H-6a
			3.63	U: C-1	U: H-4
V	α-Glc*p*-(1→	5.192	80.1	V: H-1	W: C-3
			73.8	V: H-1	V: C-2
			72.4	V: H-1	V: C-3
		94.6	3.61	V: C-1	W: H-3
			3.52	V: C-1	V: H-2
			3.81	V: C-1	V: H-3
W	→3)-α-Man-(1→	4.943	78.5	W: H-1	X: C-3
			76.3	W: H-1	W: C-5
		100.8	3.96	W: C-1	W: H-5
X	→3,6)-α-Man-(1→	4.603	68.9	X: H-1	U: C-6
			3.94	X: C-1	U: H-6a
		98.2	3.21	X: C-2	X: H-2

Based on all these results from methylation analysis and 2D NMR, the main repeating unit structure of BEPS-IB was deduced and is shown in Figure 5.

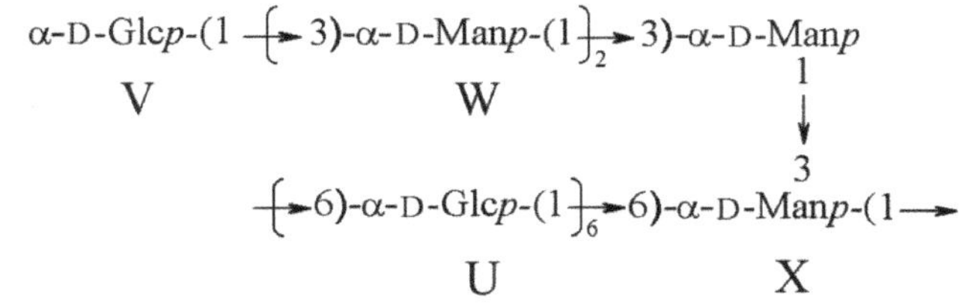

$$\alpha\text{-D-Glc}p\text{-(1}\ \xrightarrow{}\ \text{3)-}\alpha\text{-D-Man}p\text{-(1}\big]_2\xrightarrow{}\text{3)-}\alpha\text{-D-Man}p$$

V   W   1

$$\big[\xrightarrow{}\text{6)-}\alpha\text{-D-Glc}p\text{-(1}\big]_6\xrightarrow{}\text{6)-}\alpha\text{-D-Man}p\text{-(1}\xrightarrow{}$$

U   X

Figure 5. Proposed structural features of the BEPS-IB isolated from *B. exarata*.

2.2. Antioxidant Activity

Figure 6 showed the antioxidant activity of BEPS-IB determined by scavenge superoxide radical assay and reducing power assay. As shown in Figure 6a, BEPS-IB can scavenge superoxide radicals at concentrations between 2 and 12 mg/mL in a dose dependent way. The IC_{50} was 6.23 mg/mL, similar to the glucans isolated from other mollusks, which were usually around 4–10 mg/mL [13]. However, the IC_{50} was much lower than those from medical fungus, which were usually bound with pigments and caused an increase in the antioxidant activity.

In the reducing power assay (Figure 6b), reducing capacity was expressed as a percentage of the activity shown by vitamin C. The reducing capacity was positively correlated with sample concentration. At a concentration of 12 mg/mL, the reducing capacity of BEPS-IB was 75%. The reducing properties are generally associated with the presence of reductones, which have been shown to exert antioxidant action by breaking the free radical chain by donating a hydrogen atom. Our data of the reduction potential suggested that there might be a direct correlation between antioxidant activity and reducing capacity in BEPS-IB.

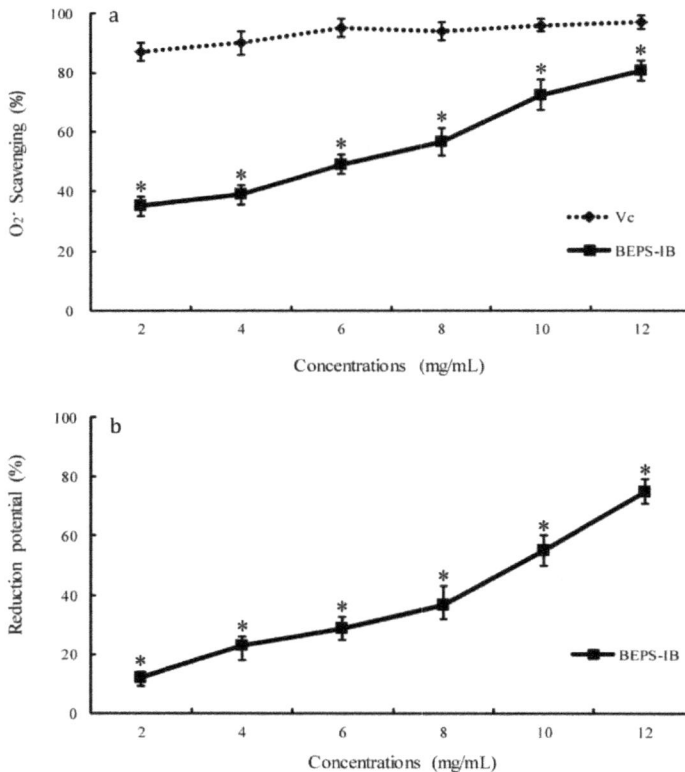

Figure 6. Antioxidant activity of the BEPS-IB. (**a**) Scavenging effects of BEPS-IB on superoxide radical ($O_2^\bullet$); (**b**) Reducing power. Values are means $\pm$ SD ($n = 3$). Significant differences from the control were evaluated using Student's t-test: * $p < 0.05$. Reducing power was expressed as a percentage of the activity shown by vitamin C.

Reactive oxygen species (ROS), capable of causing damage to DNA, have been associated with carcinogenesis, coronary heart disease, and many other health problems related to advancing age [21,22]. The antioxidant activities of polysaccharides were not a function of a single factor but a combination of several factors, such as content of sulfuric radicals, molecular weight, protein content and type of sugar. The relatively lower potential of the BEPS-IB to other glucans may attribute to the high content of mannose. However, it has been reported that addition of mannose in glucan can also improve the antitumor action of the polysaccharides, as a polysaccharide receptor has been found on human macrophages, which demonstrated high specificity for mannose [23]. Thus, the antitumor potential of polysaccharides needs to be further investigated.

3. Materials and Methods

3.1. Materials and Reagents

Bullacta exarata snails were supplied by Huzhou Lurong Seafood Co., Ltd., China and stored at $-20\ ^\circ$C before use. TSK G4000PWXL columns were sourced from TOSOH BIOSEP (Tokyo, Japan), and Sephacryl S-300 HR from Amersham Biosciences (Uppsala, Sweden). Diethyaminoethyl ion-exchange gel was from Whatman (Brentford, UK). Monosaccharides standards and disaccharide lactose were purchased from Sigma (St. Louis, MO, USA). Papain and cystein (Cys) were purchased from Fluka (Seelze, Germany). The derivatization reagent 1-phenyl-3-methyl-5-pyrazolone (PMP) was from Sinopharm Chemical Reagent (Shanghai, China). All other reagents used were analytical grade.

3.2. Isolation and Purification of Polysaccharides

The procedure used for the isolation of polysaccharides was similar to previously described [24]. *B. exarata* snails (30 kg) were shelled and the foot muscle was homogenized, and treated with acetone to remove fats (1:1). After centrifugation (6000 rpm, 20 min) and overnight drying, the resulting pellets were kept in distilled water at 60 °C for 8 h with constant stirring. The process was repeated three times. The supernatant was concentrated and precipitated in 4 volumes of ethanol. The precipitate was collected by centrifugation (6000 rpm, 20 min) and dissolved in distilled water and protein was removed by the Sevag method [25]. Then, the crude polysaccharide fraction was obtained by precipitation in 4 volumes of ethanol and washed with acetone and ethyl ether several times. The crude polysaccharide preparation was separated using a DEAE ion-exchange column (2.6 cm × 30 cm), followed with a Sephacryl S-300 gel filtration column (1.6 cm × 100 cm). Carbohydrate content was determined by the phenol/sulfuric acid assay. The isolation procedures were repeated three times and showed no significant influenceon the composition of the polysaccharides.

3.3. Chemical Analysis of Polysaccharide Fractions

Estimation of average molecular weights was performed on HPLC using TSK-G4000 and -G3000 PWXL columns, at a sample injection volume of 20 μL (1 mg/mL) and flow rate of 0.5 mL/min on a Waters 2870 system (Milford, MA, USA), with a 2414 refractive index detector. The mobile phase consisted of 0.2 M NaCl. The column was maintained at 40 °C. Gel permeation chromatography (GPC) were recorded on a computer with liquid chromatography (LC) solution version 1.25 software, preliminary calibration of the column was performed using dextrin in a range of molecular weights measured in kDa (I = 500; II = 66.9; III = 48; IV = 20; V = 5; Showa-Denko, Tokyo, Japan). The Breeze™ 2 software was utilized for data acquisition and analysis.

The monosaccharide composition was determined using a PMP-HPLC method [26]. HPLC analyses were performed on an Agilent ZORBAX Eclipse XDB-C18 column (5 μm, 4.6 mm × 150 mm) at 25 °C and UV detection at 250 nm. The mobile phase was 0.05 M KH_2PO_4 (pH 6.9) with 15% (solvent A) and 40% (solvent B) acetonitrile in water. A gradient of B from 8% to 19% in 25 min was used. Protein concentration was determined by the Lowry method [27].

3.4. Methylation Analysis

BEPS-IB (10 mg) was methylated with the method reported by Needs and Selvendran [28]. The partially methylated sample hydrolyzed by 4 M trifluoroacetic acid at 100 °C for 4 h. The resultant aldoses were reduced to their corresponding alditols by sodium borodeuteride ($NaBD_4$). The partially methylated alditols were then acetylated with a pyridine:acetic anhydride (1:1) solution at 100 °C for 1 h. The alditol acetates were analyzed by GC-MS, and the methylated sugar linkages were identified by the retention time and fragmentation pattern [28,29].

3.5. Measurement of IR and NMR Spectra

Purified BEPS-IB was deuterium-exchanged by freeze-drying three times and then dissolved in D_2O to a final concentration of 60 mg/mL. ^1H NMR spectra of BEPS-IB were measured at 600 MHz, in D_2O on a Bruker AVANCE III 600 spectrometer at 60 °C, the ^{13}C NMR was recorded at room temperature. Signals at δ_H 2.22 and δ_C 31.1 for acetone were used as external standards. The ^1H-^1H and ^1H-^{13}C connectivities were established by two-dimensional NMR (COSY, HMQC, NOESY, HMBC, and TOCSY).

An infrared spectrum of the polysaccharide (2 mg) was recorded on a Perkin-Elmer instrument in KBr pellets at room temperature.

3.6. Antioxidant Activity Assays

The antioxidant activity of BEPS-IB was tested with two *in vitro* assays: The reducing power assay and the superoxide radical scavenging assay. The reducing power assay was performed as described by Li, Zhou and Li [30] with modifications. Briefly, 1 mL samples of different concentrations (2–12 mg/mL) in phosphate buffer (0.2 M, pH 6.6) were mixed with 1 mL potassium ferricyanide (1%, w/v), and incubated at 50 °C for 20 min. The reaction was terminated by the addition of 1 mL trichloroacetic acid (10%, w/v) to the mixture and the solution was mixed with 0.2 mL ferric chloride (0.1%, w/v) and the absorbance was measured at 700 nm. Reducing power was expressed as a percentage of the activity shown by a 1 mM solution of vitamin C.

The superoxide radical scavenging activity assay was performed using the method of photoreduction of NBT (nitroblue tetrazolium) [31], with some modifications. Superoxide radicals were generated in 3 mL phosphate buffer (0.1 M, pH 7.4) containing 156 μM nicotinamide adenine dinucleotide (NADH) (reduced form), 52 μM nitrotetrazolium blue chloride (NBT), 20 μM phenazin methosulfate, and varying concentrations of polysaccharides (2–12 mg/mL). The color reaction of superoxide radicals and NBT was detected by monitoring the absorbance at 560 nm. Vitamin C was used as reference material. In the essential control, NADH was substituted with phosphate buffer. The inhibition percentage was calculated using the following formula:

$$\text{Scavenging effect (\%)} = (1 - A_{\text{Sample 560}}/A_{\text{Control 560}}) \times 100$$

3.7. Statistical Analysis

The data were reported as mean $\pm$ standard deviation (SD) ($n = 3$) and evaluated by one-way analysis of variance (ANOVA) followed by the Student's *t*-test. Differences were considered to be statistically significant if $p < 0.05$. All statistical analyses were carried out using Statistical Product and Service Solutions (SPSS) for Windows, Version 16.0 (SPSS Inc., Chicago, IL, USA).

4. Conclusions

A novel mannoglucan designated as BEPS-IB has been isolated and purified from the foot muscle of *B. exarata*. Chemical composition analysis indicated that BEPS-IB mainly consisted of glucose, mannose and minor contents of galactose and fucose, with a molar ratio of 1:0.52:0.03:0.01. The complete molecular structure was established through several experiments, including methylation analysis, NMR and IR spectra. The results indicated that BEPS-IB was a heptasaccharide backbone ([−(1→6Glcα)$_6$1→6Manα−]$_n$), with a tetrasaccharides branch ([−(1→3Manα)$_3$1→3Glcα−]$_n$) occurring at the O-3 position of mannose residues in the backbone. The purified polysaccharide also showed significant antioxidant activity in scavenging superoxide radicals ($O_2^{\bullet}$) and reducing power. The present study has demonstrated the potential of *B. exarata* as a rich and promising source of novel bioactive polysaccharides.

Acknowledgments: This program was part of the Hong Kong Scholar Program jointly funded by the Chinese Government (XJ201115) and National Science Foundation of China (31301417) and National Project (2012BAD38B09).

Abbreviations

BEPS	crude polysaccharide extracted from *Bullacta exarata*
BEPS-IB	purified *B. exarata* polysaccharide of molecular weight 94 kDa
Fuc	fucose
Gal	galactose
Man	mannose
Glc	glucose
Arb	arabinose
Rha	rhamnose
Xyl	xylose

GalN	galactosamine
GlcN	glucosamine
GlcA	glucuronic acid
GalA	galacturonic acid

Conflicts of Interest: The authors declare no conflict of interest.

References

1. Babineau, T.J.; Hackford, A.; Kenler, A.; Bistrian, B.; Forse, R.A.; Fairchild, P.G.; Heard, S.; Keroack, M.; Caushaj, P.; Benotti, P. A phase II multicenter, double-blind, randomized, placebo-controlled study of three dosages of an immunomodulator (PGG-glucan) in high-risk surgical patients. *Arch. Surg.* **1994**, *129*, 1204–1210. [CrossRef]

2. Babineau, T.J.; Marcello, P.; Swails, W.; Kenler, A.; Bistrian, B.; Forse, R.A. Randomized phase I/II trial of a macrophage-specific immunomodulator (PGG-glucan) in high-risk surgical patients. *Ann. Surg.* **1994**, *220*, 601–609. [CrossRef]

3. Browder, W.; Williams, D.; Sherwood, E.; McNamee, R.; Jones, E.; DiLuzio, N. Synergistic effect of nonspecific immunostimulation and antibiotics in experimental peritonitis. *Surgery* **1987**, *102*, 206–214.

4. Williams, D.L.; Mueller, A.; Browder, W. Glucan-based macrophage stimulators. *Clin. Immunother.* **1996**, *5*, 392–399. [CrossRef]

5. Misaki, A.; Kakuta, M.; Sasaki, T.; Tanaka, M.; Miyaji, H. Studies on interrelation of structure and antitumor effects of polysaccharides: Antitumor action of periodate-modified, branched (1→3)-β-D-glucan of *Auricularia auricula-judae*, and other polysaccharides containing (1→3)-glycosidic linkages. *Carbohydr. Res.* **1981**, *92*, 115–129. [CrossRef]

6. Kogan, G.; Staško, A.; Bauerová, K.; Polovka, M.; Šoltés, L.; Brezová, V.; Navarová, J.; Mihalová, D. Antioxidant properties of yeast (1→3)-β-D-glucan studied by electron paramagnetic resonance spectroscopy and its activity in the adjuvant arthritis. *Carbohydr. Polym.* **2005**, *61*, 18–28. [CrossRef]

7. Tsiapali, E.; Whaley, S.; Kalbfleisch, J.; Ensley, H.E.; Browder, I.W.; Williams, D.L. Glucans exhibit weak antioxidant activity, but stimulate macrophage free radical activity. *Free Radic. Biol. Med.* **2001**, *30*, 393–402. [CrossRef]

8. Zhang, L.; Liu, W.; Han, B.; Sun, J.; Wang, D. Isolation and characterization of antitumor polysaccharides from the marine mollusk *Ruditapes philippinarum*. *Eur. Food Res. Technol.* **2008**, *227*, 103–110. [CrossRef]

9. Vetvicka, V.; Yvin, J.-C. Effects of marine β-1,3-glucan on immune reactions. *Int. Immunopharmacol.* **2004**, *4*, 721–730. [CrossRef]

10. Mikheiskaya, L.; Molchanova, V.; Ovodova, R.; Santana, V.F. Branched α-1,4-glucan from the gastropod mollusk *Strombus gigas*. *Chem. Nat. Compd.* **1988**, *24*, 29–32. [CrossRef]

11. Vetvicka, V.; Sima, P. β-Glucan in invertebrates. *ISJ* **2004**, *1*, 60–65.

12. Zhang, H.; Ye, L.; Wang, K. Structural characterization and anti-inflammatory activity of two water-soluble polysaccharides from *Bellamya purificata*. *Carbohydr. Polym.* **2010**, *81*, 953–960. [CrossRef]

13. Jiang, C.; Wang, M.; Liu, J.; Gan, D.; Zeng, X. Extraction, preliminary characterization, antioxidant and anticancer activities *in vitro* of polysaccharides from *Cyclina sinensis*. *Carbohydr. Polym.* **2011**, *84*, 851–857. [CrossRef]

14. Zhu, J.; Xu, J.; Jin, X.; Xie, J. Effect of extract from bullacta exarata on murine melanoma cell B16. *Biotechnol. Bull.* **2009**, *2*, 124–128.

15. Zhang, D.; Wu, H.; Xia, Z.; Wang, C.; Cai, J.; Huang, Z.; Du, L.; Sun, P.; Xie, J. Partial characterization, antioxidant and antitumor activities of three sulfated polysaccharides purified from *Bullacta exarata*. *J. Funct. Foods* **2012**, *4*, 784–792. [CrossRef]

16. Zhang, D.; Wang, C.; Wu, H.; Xie, J.; Du, L.; Xia, Z.; Cai, J.; Huang, Z.; Wei, D. Three sulfated polysaccharides isolated from the mucilage of mud snail, *Bullacta exarata* Philippi: Characterization and antitumor activity. *Food Chem.* **2013**, *1*, 306–314.

17. Santhiya, D.; Subramanian, S.; Natarajan, K. Surface chemical studies on sphalerite and galena using extracellular polysaccharides isolated from *Bacillus polymyxa*. *J. Colloid Interface Sci.* **2002**, *256*, 237–248. [CrossRef]

18. Kacurakova, M.; Capek, P.; Sasinkova, V.; Wellner, N.; Ebringerova, A. FT-IR study of plant cell wall model compounds: Pectic polysaccharides and hemicelluloses. *Carbohydr. Polym.* **2000**, *43*, 195–203. [CrossRef]
19. Mathlouthi, M.; Koenig, J.L. Vibrational spectra of carbohydrates. *Adv. Carbohydr. Chem. Biochem.* **1987**, *44*, 7–89. [CrossRef]
20. Duus, J.Ø.; Gotfredsen, C.H.; Bock, K. Carbohydrate structural determination by NMR spectroscopy: Modern methods and limitations. *Chem. Rev.* **2000**, *100*, 4589–4614. [CrossRef]
21. Cadenas, E.; Davies, K.J. Mitochondrial free radical generation, oxidative stress, and aging. *Free Radic. Biol. Med.* **2000**, *29*, 222–230. [CrossRef]
22. Uchida, K. Role of reactive aldehyde in cardiovascular diseases. *Free Radic. Biol. Med.* **2000**, *28*, 1685–1696. [CrossRef]
23. Stahl, P.D.; Ezekowitz, R.A.B. The mannose receptor is a pattern recognition receptor involved in host defense. *Curr. Opin. Immunol.* **1998**, *10*, 50–55. [CrossRef]
24. Chen, S.; Xue, C.; Yin, L.A.; Tang, Q.; Yu, G.; Chai, W. Comparison of structures and anticoagulant activities of fucosylated chondroitin sulfates from different sea cucumbers. *Carbohydr. Polym.* **2011**, *83*, 688–696. [CrossRef]
25. Staub, A. Removal of protein-Sevag method. *Methods Carbohydr. Chem.* **1965**, *5*, 5–6.
26. Wu, N.; Ye, X.; Guo, X.; Liao, N.; Yin, X.; Hu, Y.; Sun, Y.; Liu, D.; Chen, S. Depolymerization of fucosylated chondroitin sulfate from sea cucumber, *Pearsonothuria graeffei*, via ^{60}Co irradiation. *Carbohydr. Polym.* **2013**, *2*, 604–614.
27. Lowry, O.H.; Rosebrough, N.J.; Farr, A.L.; Randall, R.J. Protein measurement with the Folin phenol reagent. *J. Biol. Chem.* **1951**, *193*, 265–275.
28. Needs, P.; Selvendran, R. An improved methylation procedure for the analysis of complex polysaccharides including resistant starch and a critique of the factors which lead to undermethylation. *Phytochem. Anal.* **1993**, *4*, 210–216. [CrossRef]
29. Björndal, H.; Lindberg, B.; Svensson, S. Mass spectrometry of partially methylated alditol acetates. *Carbohydr. Res.* **1967**, *5*, 433–440. [CrossRef]
30. Li, X.; Zhou, A.; Li, X. Inhibition of *Lycium barbarum* polysaccharides and *Ganoderma lucidum* polysaccharides against oxidative injury induced by γ-irradiation in rat liver mitochondria. *Carbohydr. Polym.* **2007**, *69*, 172–178. [CrossRef]
31. Li, X.; Zhou, A.; Han, Y. Anti-oxidation and anti-microorganism activities of purification polysaccharide from *Lygodium japonicum in vitro*. *Carbohydr. Polym.* **2006**, *66*, 34–42. [CrossRef]

marine drugs

MDPI

Article

Structural and Immunological Activity Characterization of a Polysaccharide Isolated from *Meretrix meretrix* Linnaeus

Li Li [†], Heng Li [†], Jianying Qian, Yongfeng He, Jialin Zheng, Zhenming Lu, Zhenghong Xu and Jinsong Shi *

School of Pharmaceutical Science, Jiangnan University, Wuxi 214122, China; li792051095@163.com (L.L.); eternal83@163.com (H.L.); jackieqian@163.com (J.Q.); 15061888901@163.com (Y.H.); 18262280354@163.com (J.Z.); zhenming_lu@163.com (Z.L.); zhenghxu@jiangnan.edu.cn (Z.X.)
* Correspondence: shijs@163.com; Tel.: +86-510-8532-8177; Fax: +86-510-8591-8206
† These authors contributed equally to this work.

Academic Editor: Paola Laurienzo
Received: 2 December 2015; Accepted: 21 December 2015; Published: 29 December 2015

Abstract: Polysaccharides from marine clams perform various biological activities, whereas information on structure is scarce. Here, a water-soluble polysaccharide MMPX-B2 was isolated from *Meretrix meretrix* Linnaeus. The proposed structure was deduced through characterization and its immunological activity was investigated. MMPX-B2 consisted of D-glucose and D-galctose residues at a molar ratio of 3.51:1.00. The average molecular weight of MMPX-B2 was 510 kDa. This polysaccharide possessed a main chain of $(1\rightarrow4)$-linked-α-D-glucopyranosyl residues, partially substituted at the C-6 position by a few terminal β-D-galactose residues or branched chains consisting of $(1\rightarrow3)$-linked β-D-galactose residues. Preliminary immunological tests *in vitro* showed that MMPX-B2 could stimulate the murine macrophages to release various cytokines, and the structure-activity relationship was then established. The present study demonstrated the potential immunological activity of MMPX-B2, and provided references for studying the active ingredients in *M. meretrix*.

Keywords: *Meretrix meretrix* Linnaeus; polysaccharide; structure; immunological activity

1. Introduction

Meretrix meretrix Linnaeus (*M. meretrix*, Meretrix, Veneridae) is a historically marine food and a valuable source of traditional Chinese medicine (TCM), which is widely distributed in coastal areas of South and Southeast Asia [1]. Documented in the ancient Chinese pharmacopeia *Compendium of material* (the 16th century, by Li Shizhen), *M. meretrix* could diminish inflammation, treat typhoid fever, hangover and relieve pain. Modern research has verified the effects of *M. meretrix* on immuno-regulating [2], antioxidant [3], antineoplastic [4], antihypertensive and hypolipidemic activities [5]. The noteworthy biological activities relied on the functional ingredients it contains, including peptides, polysaccharides, amino acids, and enzyme inhibitors. Among these compounds, polysaccharide is one of the major components which accounted for 4.1%–8.3% of *M. meretrix* [2]. Various biological functions of polysaccharides including immunomodulation, anti-inflammation, anti-coagulation, and anti-tumor have been explored and evaluated [6]. A series of immunological indicators, including the phagocytic power, the number of leukocytes, and the level of hemolysin antibody, were all ameliorated in the rats with damaged immune systems induced by cyclophosphamide through the oral administration of a polysaccharide isolated from *M. meretrix* [2]. Zhang *et al.* also reported that the ethanol extract of *M. meretrix* could enhance the expression of T-

and B- lymphocytes by 18% and 43%, respectively [7]. The underlying mechanism through which polysaccharides exerted their physiological activities has experienced a long exploring process, but a rapid advancement has been achieved recently, mainly because of the discovery of roles of gut microbiota. Polysaccharides which could not be absorbed into the small intestine could be further hydrolyzed and metabolized in the colon, which might exert some influences on the physiological features and community structure of gut microbiota [8,9]. Chang *et al.* showed comprehensively that the high molecular weight polysaccharides (4300 kDa) isolated from *Ganoderma lucidum* (a medicinal mushroom) mycelium reduced body weight, inflammation and insulin resistance in mice fed a high-fat diet (HFD) by reversing HFD-induced gut dysbiosis, maintaining the intestinal barrier integrity, and reducing metabolic endotoxemia [10].

The progress in the mechanism elucidation further forced the interests in the research of novel polysaccharides in clams and also in their structure-activity relationship. A polysaccharide extracted from *Cyclina sinensis* performed high inhibitory activity *in vitro* against human gastric cancer cells. It was composed of glucose linked by α-(1→4) glycosidic bonds, with branches attached to the backbone chain by (1→6) glycosidic bonds [11]. Liao *et al.* also isolated a polysaccharide from the clam of *Corbicula fluminea* with significant inhibitory effects on growth of human gastric cancer cells and human ovarian carcinoma cells [12]. In addition, Vidhyanandhini *et al.* obtained purified glycosaminoglycans from *Meretrix casta*, whose structural characterization was carried out by Fourier transform-infrared (FT-IR) and ^{1}H-NMR spectroscopy [13]. These glycosaminoglycans showed comparable anticoagulant activity with heparin [14]. However, the current research on polysaccharides of *M. meretrix* is still confined to the extraction method improvement and primary bio-activity evaluation. Detailed characterization of structure and elucidation of structure-activity relationship are still scarce.

In the present study, a polysaccharide was isolated from *M. meretrix* and purified. The structure was deduced and proposed based on detailed chemical characterization, and the immuno-regulatory activity was evaluated. An attempt to find a preliminary structure-activity relationship was carried out as well.

2. Results and Discussion

2.1. Extraction and Purification of MMPX-B2

The crude polysaccharide (MMPX) was extracted from *M. meretrix* following an enzymatic extraction method with addition of 2% trypsin. Under the optimized conditions, the yield of MMPX reached 12.0%. As preliminary evaluation showed, this crude polysaccharide could increase NO production in RAW264.7 cell; however, a higher purity and homogeneity of the polysaccharide was needed in order to analyze its structure and bio-activity analysis. Therefore, MMPX was then isolated by DEAE-52 cellulose anion-exchange chromatography. The chromatogram (Figure 1a) showed three peaks, in which the major part eluted by 0.1 M NaCl was collected and denoted as MMPX-B. MMPX-B was further purified by Superdex 200 dextran gel permeation chromatography, and two fractions were isolated. As shown in Figure 1b, two peaks arose, and the constituent which performed a symmetrical and sharper peak was collected for subsequent studies and denoted as MMPX-B2. The purity and homogeneity of MMPX-B2 were further investigated. There was no absorption at 260 nm and 280 nm (data not shown) in UV spectrum, indicating no impurities of protein and nucleic acids left. The high performance gel permeation chromatography (HP-GPC) profile in Figure 1c showed only one symmetrical peak, indicating that MMPX-B2 was a homogenous polysaccharide, with an apparent molecular weight of 510 kDa. Furthermore, the polydispersity index (PDI) was determined as 1.11 corresponding to a narrow molecular weight distribution [15], suggesting a relative simple composition of MMPX-B2.

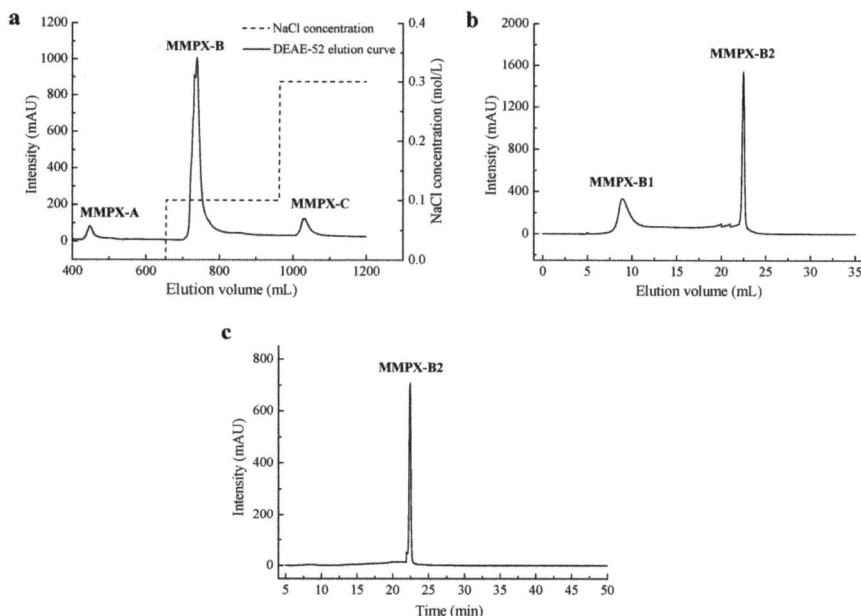

Figure 1. Purification of MMPX-B2. (**a**) Elution profile of crude polysaccharides by DEAE-52 cellulose; (**b**) Purification profile of MMPX-B by Superdex 200; (**c**) HP-GPC profile of MMPX-B2.

2.2. Chemical Composition of MMPX-B2

A gas chromatography (GC) analysis (Figure 2) showed that MMPX-B2 was composed of D-Glucose and D-Galactose with the molar ratio of 7.13:1.00, indicating that this polysaccharide was neutral.

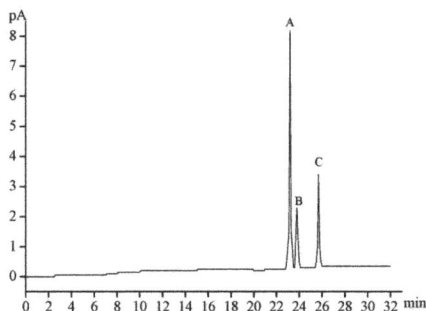

Figure 2. GC profile of MMPX-B2 with acid hydrolysis and acetylation. (**A**) D-glucose; (**B**) D-galactose; (**C**) internal standard inositol.

The structure was further analyzed by FT-IR. As shown in Figure 3, a broad and intense characteristic peak at 3390 cm^{-1} was attributed to the stretching vibration of O–H. The bands at 2930 cm^{-1} were assigned to C–H$_2$ and C–H stretching vibrations [16]. The relatively strong absorption at 1640 cm^{-1} was due to associated water [17]. Additionally, 1410 cm^{-1} and 1078 cm^{-1} corresponded to exocyclic and endocyclic C–O stretching bands, respectively [18]. The peaks around 1400–1200 cm^{-1} were also the characteristic absorptions of C–H bonds. The absorptions at 1200–950 cm^{-1} were due to the vibrations of C–O–H side groups and C–O–C glycosidic bonds [16]. The band around 900 cm^{-1} was the characteristic absorption of β-linkage of pyranose [19,20]. The relatively intensive bonds at 700–900 cm^{-1} were assigned to skeletal modes of a pyranose ring, among which the absorption at 844 cm^{-1} suggested the presence of α-type glycosidic bonds [21].

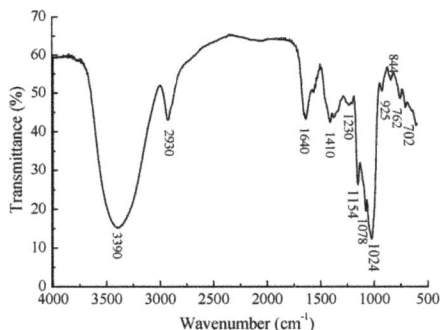

Figure 3. FT-IR spectrum of MMPX-B2.

2.3. Linkage Analysis and Structure Speculation

Preliminary analysis of the glycosidic linkage locations was carried out by following periodate oxidation and Smith degradation methods.

The results of periodate oxidation showed that, per mole sugar, 0.960 mol of periodate was consumed and 0.112 mol formic acid was produced. The formation of formic acid suggested the presence of pyranohexose in 1→ or 1→6 linked forms in 11.2%. As the amount of periodate consumption was more than two fold of the amount of formic acid produced, linkages which only consumed periodate without formic acid production were therefore deduced to exist as 1→2, 1→2,6, 1→4, and 1→4,6 forms, which occupied 73.6% of the total glycosyl linkages. The ratio of other linkages as 1→3-linked forms were 15.2% which did not consume periodate.

Additionally, large amounts of glycerol and erythritol, examined by gas chromatography-mass spectrometry (GC-MS) after the periodate-oxidized products of MMPX-B2, were reduced and hydrolyzed indicating that most of the linkages were in 1→, 1→6, 1→2, 1→2,6, 1→4, and 1→4,6 forms. Combined with the results of periodate oxidation analysis, it could be inferred that linkages of 1→2, 1→2,6, 1→4, 1→4,6 were the major types. Nevertheless, small amounts of 1→3 glycosyl linkages also exist.

Methylation analysis was employed for further derivation of glycosyl linkages in MMPX-B2. The results (Table 1) showed that the glucose residues were linked together as (1→4)-Glc*p* and (1→4,6)-Glc*p* forms, while the galactose residues were present as terminal and (1→3)-linked Gal*p* forms. Further, the ratio of four types of glycosyl linkages were determined by integrating peak areas. The number of (1→4)-linked and (1→4,6)-linked Glc*p* residues accounted for 77.22% of the total methylated sugar residues, suggesting that MMPX-B2 was probably consisted of a backbone of 1,4-linked and 1,4,6-linked Glc*p*, with terminal Gal*p* residues (majority) and branches of 1,3-linked Gal*p* (minority) attached to the C-6 of some 1,4,6-linked Glc*p* residues. There is also some possibility of a few 1,3-linked Gal*p* residues inside the Glc*p* main chain. Further structural data are needed for structure speculation.

Table 1. GC-MS analysis of the methylated products of MMPX-B2.

Methylated Sugar Residue	Molar Ratio	Ratio (%)	Type of Linkage
2,3,4,6-Me$_4$-Gal	1.03	17.64	Gal*p*-(1→
2,4,6-Me$_3$-Gal	0.30	5.14	→3)-Gal*p*-(1→
2,3,6-Me$_3$-Glc	3.51	60.10	→4)-Glc*p*-(1→
2,3-Me$_2$-Glc	1.00	17.12	→4,6)-Glc*p*-(1→

As the MMPX-B2 structure deduced was not complicated, 1D NMR spectroscopy was chosen to provide a more exact structural information (Figure 4). The major chemical shifts are listed in Table 2, and the assignments were mainly based on literature values. As MMPX-B2 was composed almost exclusively by glucose, the signals of glucose and those of galactose could be easily differentiated. In the ^{13}C NMR spectrum, the resonances in the anomeric region (δ = 95–110 ppm) allowed for quick assignments of the configurations of the sugar residues, as α-configuration at δ = 95–102 ppm and β-configuration at δ = 103–110 ppm [22]. The signals at δ 98.70 and 99.92 ppm were attributed to the anomeric carbon atoms of (1→4,6)-linked α-D-Glc*p* and (1→4)-linked α-D-Glc*p* [23,24], while the signals at δ 103.78 and 102.39 ppm were ascribed to the terminal β-D-Gal*p* and (1→3)-linked β-D-Gal*p* [25]. The signal in the lower magnetic field at δ 60.50 ppm was attributed to C-6 resonance of →4)-α-D-Glc*p*-(1→. In addition, ^{1}H-NMR spectrum showed intensive and broad peaks in the range of 3.0–4.5 ppm, which gave additional information for structure derivation. The signals at δ 4.49, 4.75, 5.25, and 5.30 ppm were assigned to the anomeric protons of the terminal β-D-Gal*p*, (1→3)-linked β-D-Gal*p*, (1→4)-linked α-D-Glc*p*, and (1→4,6)-linked α-D-Glc*p*, respectively. The chemical shifts at δ 3.4 to 4.2 ppm were assigned to the proton signals of carbons C-2 to C-6 of the sugar rings. NMR information was consistent with the FT-IR results, and the possibility of the structure of a few 1,3-linked Gal*p* residues inside the Glc*p* main chain could be excluded.

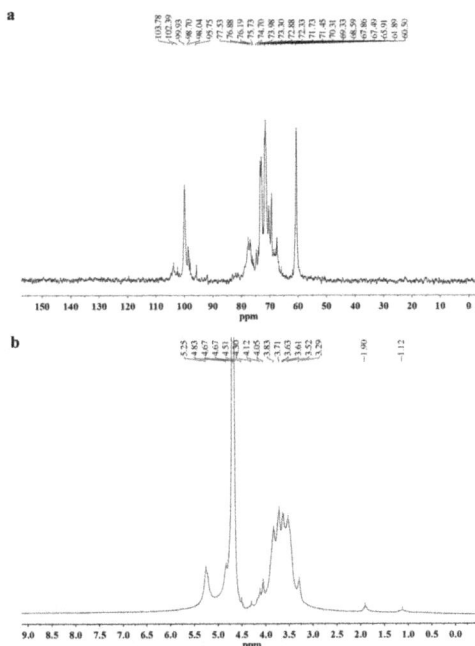

Figure 4. NMR spectra of MMPX-B2. (a) ^{13}C-NMR; (b) ^{1}H-NMR.

Table 2. ^{13}C-NMR and ^{1}H-NMR chemical shifts for resonances of glycosyl residues.

Residues	Chemical Shift (δ, ppm)					
	C1/H1	C2/H2	C3/H3	C4/H4	C5/H5	C6/H6
→4)-α-Glc*p*-(1→	99.92/5.25	71.45/3.63	73.30/3.97	76.88/3.63	71.25/3.83	60.50/3.71
→4,6)-α-Glc*p*-(1→	98.70/5.30	72.33/3.52	74.70/3.71	78.45/3.61	71.28/3.63	61.20/3.45
β-Gal*p*-(1→	103.78/4.49	72.30/3.54	74.10/3.63	69.94/4.12	76.70/3.71	61.90/3.73
→3)-β-Gal*p*-(1→	102.39/4.75	71.73/3.83	83.10/3.90	69.90/4.17	75.73/3.73	61.90/3.71

Based on the above chemical composition and structural characterizations, the structure of MMPX-B2 might then be inferred and demonstrated as shown in Figure 5. This polysaccharide consisted of (1→4)-linked α-D-Glc*p* residues, partially substituted at C-6 by branches consisting of (1→3)-linked β-D-Gal*p* with terminal β-D-Gal*p* residues.

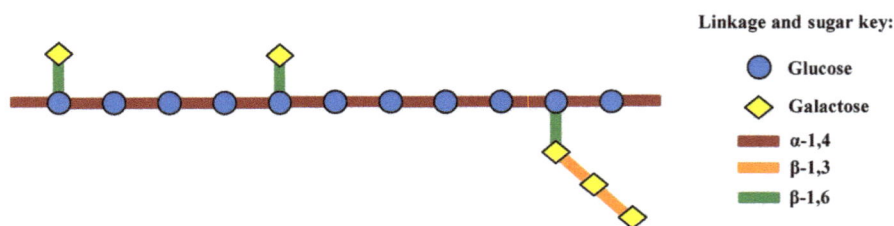

Figure 5. Proposed structural model of MMPX-B2.

As the main chain of MMPX-B2 seemed to be composed of (1→4)-linked α-D-Glc*p* residues, which was similar to amylose, a verification experiment was conducted using α-amylase which could cut off the α-(1→4) glycosyl linkage between Glc*p* residues randomly. After being treated with amylase at 40 °C for 5 h, the molecular weight of the MMPX-B2 hydrolysate reduced sharply from 510 kDa to 4.8 kDa. The results verified the deduction that the backbone of MMPX-B2 was composed of (1→4)-linked α-D-Glc*p* residues, which coincided well with the structure derived from the stoichiometric method.

Polysaccharides with β-(1→3) and β-(1→6) glycosyl linkages from clams have rarely been reported. However, a galactan sulfate with a β-(1→3)-glycosidic linkages that was isolated from the marine clam species *Meretrix petechialis* showed anti-HIV activity by inhibiting the syncytia formation [26]. In another experiment, Dai *et al.* obtained a polysaccharide from the clam of *Hyriopsis cumingii* Lea, whose main chain showed to be composed by (1→4)-linked β-D-glucopyranosyl residues, performed immuno-stimulatory activity [27]. Nevertheless, polysaccharides with β-(1→3) and β-(1→6) glycosyl linkages from mushrooms have relatively been deeply studied. Besides, it is known that polysaccharides with β-(1→3) linkages in the main chain and additional β-(1→6) branch points seemed to be necessary for the activity of immunomodulation [28,29]. Thus, considering its structural similarity, MMPX-B2 might possess a similar immuno-regulating function.

2.4. Cell-Mediated Immunological Activity

Cell-mediated immunological tests were then conducted in order to verify the immunological properties of MMPX-B2. The macrophages-mediated immunity plays an important role in the innate immune system. The process of macrophage activation is associated with the production of various inflammatory mediators and cytokines including interleukin (IL), tumor necrosis factor (TNF), and NO [30]. The effects of the immunological activity of MMPX-B2 are indicated in Figure 6. NO, as a messenger or effector, plays important roles in cardiovascular, neural, and immune systems [31]. TNF-α is especially produced by activated macrophages and is involved in systemic inflammation.

It is often induced together with IL-6, which plays a major regulatory role in acute local and systemic inflammatory responses [28]. IL-1β is also produced by activated macrophages as a pro-protein and involved in a variety of cellular activities, including cell proliferation, differentiation, and apoptosis [32]. The results showed in Figure 6 indicate that MMPX-B2 possesses immuno-stimulating properties. The levels of NO, IL-6, and IL-1β induced by 125 μg/mL MMPX-B2 matched well with lipopolusaccharide (LPS) and performed in a dose-dependent manner. TNF-α secretion seemed not to be significantly affected by the concentration of the polysaccharide: 25 μg/mL MMPX-B2 already showed comparable effects with 1 μg/mL LPS. Therefore, it seems that there is a correlation between β-(1→3; 1→6) linkages and the immune-regulating effect of MMPX-B2. However, further studies on the immuno-regulation mechanisms are being carried out in order to confirm the initial hypothesis.

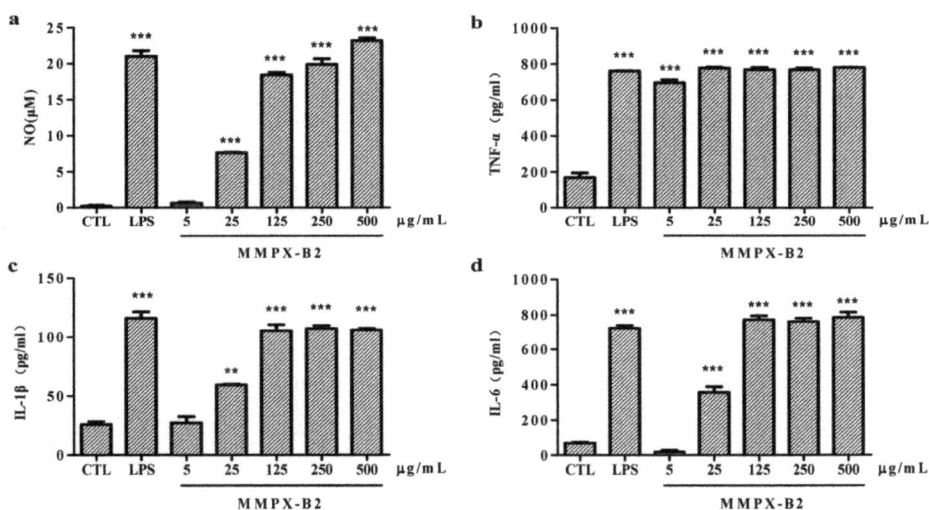

Figure 6. Effects of MMPX-B2 on macrophages-mediated immunity *in vitro*. (**a**) Nitrate accumulation; (**b**) TNF-α; (**c**) IL-1β; (**d**) IL-6. ** is representative of $p < 0.01$ and *** is representative of $p < 0.001$, when compared to the control group.

3. Experimental Section

3.1. Materials

The fresh tissue of *M. meretrix* was purchased from Nan Tong Changhua Aquatic Food Co., Ltd (Nantong, Jiangsu, China). Murine monocyte macrophage RAW264.7 was purchased from Cobioer Biosciences Co., Ltd (Nanjing, Jiangsu, China). DEAE-52 cellulose, phosphate buffered saline (PBS), 3-(4,5-dimethyl-2-thiazolyl)-2,5-diphenyl-2H-tetrazolium bromide (MTT), LPS and Griess reagent were purchased from Sigma Chemical Company (St. Louis, MO, USA). The enzymes of trypsin and α-amylase were purchased from Sinopharm Chemical Reagent Co., Ltd (Shanghai, China). Superdex-200 gel prepacked column was purchased from GE Healthcare (Uppsala, Sweden). Enzyme-linked immunosorbent assays (ELISA) kits for measurement of TNF-α, IL-1β, and IL-6, were purchased from Ebioscience Biotechnology Co., Ltd (SanDiego, CA, USA). Penicillin-streptomycin solution (PS) was from Gibco Company (Auckland, Newzealand). Dulbecco's modified eagle medium (DMEM) was picked from Corning Biotechnology Co., Ltd (Tewksbury, MA, USA). Fetal calf serum (FBS) was from PAA Laboratories (Pasching, Austria). The other chemical reagents were of analytical grade and obtained from Sinopharm Chemical Reagent Co., Ltd (Shanghai, China).

3.2. Extraction, Isolation and Purification of MMPX-B2

The crude polysaccharide of *M. meretrix* was extracted by modified water extraction method coupled with enzyme hydrolysis. The fresh tissue of *M. meretrix* was firstly homogenized and then extracted with distilled water which was four times the volume of the tissue with additional 2% trypsin (calculated by the wet weight basis of tissue) at 50 °C for 4 h three times [33]. The obtained aqueous extracts were centrifugated at $5000 \times g$ at room temperature for 10 min. The collected supernatants were concentrated under vacuum at 55 °C. Then, the concentrate was precipitated by 75% ethanol at 4 °C overnight, followed by centrifugation at $5000 \times g$ for 10 min. The precipitate was collected, dissolved with water and protein impurities were removed with Sevage reagent (chloroform and butanol in the ratio of 3:1) [34]. The mixture was centrifugated and the precipitate was washed twice with absolute ethanol. The crude polysaccharide MMPX was obtained after lyophilization.

MMPX was dissolved in distilled water in order to obtain a solution with the concentration of 5.0 mg/mL. After being loaded onto a DEAE-52 cellulose column (2.6 × 20 cm), MMPX was eluted stepwise with 0, 0.1, and 0.3 M NaCl at a flow rate of 2.0 mL/min. The elutes were collected and lyophilized. Then, the major fraction was further purified by a Superdex 200 dextran column (1.0 × 24 cm) with distilled water as the eluent at a flow rate of 0.3 mL/min. The elutes were collected and lyophilized. The purified polysaccharide MMPX-B2 was then obtained.

3.3. Homogeneity and Molecular Weight Determination

The homogeneity and molecular weight of MMPX-B2 were determined by HP-GPC method on a Waters HPLC system (Allances 2695, Waters, Milliford, MA, USA) equipped with a TSK-GEL G3000SWxl column (7.5 × 300 mm) and a refractive index detector (RID) [35]. The PDI was calculated by the ratio of weight-average molecular weight and number-average molecular weight (Mw/Mn). The purified polysaccharide was dissolved in distilled water and eluted with 0.1 M NaCl solution at a flow rate of 0.6 mL/min. The column was calibrated with the Dextran T-series standard samples with different molecular weights (Dextran T 2000, T 500, T 70, T 40, T 10, and T 5).

3.4. Analysis of Monosaccharide Composition

GC analysis was employed to detect the monosaccharide composition. MMPX-B2 was hydrolyzed with 2 mol/L trifluoroacetic acid (TFA) (2 mL) at 100 °C for 6 h. The hydrolysate was repeatedly concentrated with methanol until drying, and was then acetylated with 10 mg hydroxylamine hydrochloride and 1 mL pyridine at 90 °C for 30 min. After the addition of acetic anhydride (1 mL), the aldononitrile acetate derivatives were obtained and analyzed by Agilent 6820 GC system (Agilent, Santa Clara, CA, USA) equipped with an OV-17 capillary column (Agilent, USA). The oven was kept at 200 °C for 2 min, then raised to 220 °C at a rate of 15 °C·min^{-1} where it remained for 10 min, and finally increased to 240 °C at a rate of 10 °C·min^{-1} and held for 10 min [36].

3.5. FT-IR Analysis

An aliquot of MMPX-B2 was vacuum-dried overnight before being submitted to infrared spectroscopy analysis. FT-IR spectra were recorded on a Nicolet 6700 FT-IR spectrometer (Thermo Scientific, Madison, WI, USA) within the frequency range of 4000–400 cm^{-1} with potassium bromide (KBr) disc method [37].

3.6. Periodate Oxidation and Smith Degradation

For the experiment, 20 mg MMPX-B2 was oxidized with 15 mmol/L NaIO$_4$ and kept in the dark for nearly 48 h [38]. The absorbance was read at 223 nm every 6 h until the consumption of NaIO$_4$ reached a constant value. The reaction was stopped by the addition of 2 mL ethylene glycol. The amount of formic acid generated was titrated with 5 mmol/L NaOH. The solution was dialyzed (Mw cut off: 3000 Da) against distilled water for 48 h and was further reduced by NaBH$_4$ for 24 h at 25 °C. The pH value was adjusted to 5.5 with 0.1 mol/L acetic acid. After being hydrolyzed with TFA, the derivatives of alditol acetates were analyzed by GC. The GC method was the same as that of monosaccharide composition analysis.

3.7. Methylation Analysis

According to the method of Ciucanu and Kerek, MMPX-B2 was methylated three times using powdered NaOH in DMSO-MeI [39]. The absence of the peak (3700–3200 cm^{-1}) corresponding to the hydroxyl group on FT-IR spectra indicated complete methylation. The methylated products were hydrolyzed with 2 mol/L TFA (2 mL) at 100 °C for 6 h and reduced with NaBH$_4$. After correcting the pH to 5.5 with 0.1 M acetic acid, the solution was acetylated with equivalent amounts of acetic anhydride pyridine. The resulting products were then examined by GC-MS with a HP-1 capillary column. The following temperature program was employed: 150 °C to 180 °C at a rate of 10 °C/min, then 180 °C to 260 °C at a rate of 15 °C/min and held for 5 min.

3.8. NMR Spectroscopy

^{13}C-NMR and ^{1}H-NMR spectra were obtained on a Bruker DRX-500 NMR spectrometer (Bruker Daltonics, Rheinstetten, Germany). MMPX-B2 was dissolved in D$_2$O, and the spectra were recorded at 50 °C. Acetone (not deuterated) was used as the internal standard for the ^{13}C analysis.

3.9. Hydrolysis of MMPX-B2 with α-Amylase

The polysaccharide of MMPX-B2 was hydrolyzed using α-amylase to verify the backbone structure speculated from chemical characterization. MMPX-B2 was dissolved in distilled water to obtain a solution with the concentration of 1 mg/mL. After the addition of 1% α-amylase (calculated by the weight of MMPX-B2), the enzymatic hydrolysis reaction was carried out at 40 °C for 5 h. The α-amylase in the product was removed using Sevage reagent. The molecular weight of MMPX-B2 hydrolysate was determined with the same method as that of MMPX-B2. The standard samples were Dextran T 40, T 10, T 5, and T 1.

3.10. Assay of Cell-Mediated Immunity

3.10.1. Cell Culture

Murine monocyte macrophages RAW264.7 were grown in DMEM medium supplemented with 10% FBS and 1% penicillin-streptomycin (PS) mixed solution under a humidified atmosphere of 5% CO$_2$ at 37 °C.

3.10.2. Quantitative Determination of NO and Cytokines

A density of 1.0×10^5 of cells/well were seeded into a 96-well plate and incubated for 16 h at 37 °C. Thereafter, 100 μL of LPS (1 μg/mL, the positive control) or polysaccharide samples (5, 25, 125, 250, 500 μg/mL) were added in, while the equivalent DHEM medium was used as the control. Each group was set with six replicates. The 96-well plate was incubated for 24 h and centrifuged at 1500× *g* for 5 min to collect the supernatant.

To estimate NO level in RAW264.7 cells, nitrite was measured by Griess assay which is a stable reaction product of NO with molecular oxygen [30]. Each cultured supernatant (100 μL) was collected and mixed with 100 μL of Griess reagent. The optical density was measured at 540 nm (SpectraMax M2, Danaher, CA, USA) after 15 min of incubation. The nitrate accumulation was then calculated according to a calibration curve made with a series of known concentrations (2–60 μmol/L) of NaNO$_2$ standard solution. The level of TNF-α, IL-1β, and IL-6 were investigated by ELISA kits.

4. Conclusions

A water-soluble polysaccharide MMPX-B2 was isolated and purified from the clam of *M. meretrix*. MMPX-B2 showed to be a homogenous polysaccharide, with an apparent molecular weight of 510 kDa. This polysaccharide was composed of residues of D-glucose and D-galactose at a molar ratio of 3.51:1.00. After characterizing the chemical composition and glycosidic linkages, it was hypothesized that MMPX-B2 possessed a main chain of (1→4)-linked α-D-glucopyranosyl residues, partially substituted at the C-6 position with branches consisting of (1→3)-linked β-D-Gal*p* and terminal β-D-Gal*p* residues. The *in vitro* immunological tests with murine macrophage RAW264.7 cells showed that the levels of NO, TNF-α, IL-6, and IL-1β released increased with the addition of MMPX-B2 in a dose-dependent manner, indicating that MMPX-B2 possessed immuno-stimulating properties. The structure-activity relationship was also established.

Acknowledgments: This work was supported by grants from the "twelfth five-year" National Science and Technology Support Program (No. 2012BAD33B06), Public Science and Technology Research Funds Projects of the Ocean (No. 201305007), National Natural Science Foundation of China (No. 31300026) and the Ministry of Education of the People's Republic of China (JUSRP51516).

Author Contributions: Jinsong Shi, Zhenghong Xu, and Jianying Qian conceived and designed the experiments; Li Li performed the experiments; Li Li, Heng Li, and Jinsong Shi analyzed the data; Jianying Qian, Yongfeng He and Jialin Zheng contributed reagents/materials/analysis tools; Li Li, Heng Li, Zhenming Lu and Jinsong Shi wrote the paper.

Conflicts of Interest: The authors declare no conflict of interest.

References

1. Supatra, K.; Soottawat, B.; Hideki, K.; Yung-Hsiang, T. Chemical compositions and nutritional value of Asian hard clam (*Meretrix lusoria*) from the coast of Andaman Sea. *Food Chem.* **2013**, *141*, 4138–4145.
2. Xie, W.Y.; Chen, C.; Liu, X.S.; Wang, B.; Sun, Y.; Ma, Y.; Zhang, X. *Meretrix meretrix*: Active components and their bioactivities. *Life Sci. J.* **2012**, *9*, 756–762.
3. Jiang, C.X.; Xiong, Q.P.; Gan, D.; Jiao, Y.; Liu, J.; Ma, L.; Zeng, X. Antioxidant activity and potential hepatoprotective effect of polysaccharides from *Cyclina sinensis*. *Carbohydr. Polym.* **2013**, *91*, 262–268. [CrossRef] [PubMed]
4. Jiang, C.X.; Wang, M.; Liu, J.; Gan, D.; Zeng, X. Extraction, preliminary characterization, antioxidant and anticancer activities *in vitro* of polysaccharides from *Cyclina sinensis*. *Carbohydr. Polym.* **2011**, *84*, 851–857. [CrossRef]
5. Xu, X.L.; Li, T.M.; Zhang, C.R. Experimental study on the stability of thrombin activity. *Chin. J. Biochem. Pharm.* **1999**, *20*, 298–299.
6. Xie, H.; Zhong, Z.W.; Zhu, W.C.; Cao, K.G. Process of research on the medicinal value of *Meretrix meretrix* Linnaeus. *J. Chengde Pet. Coll.* **2005**, *7*, 9–12.
7. Zhang, L.X.; Fan, X.; Niu, R.L. Immunomodulatory activity determination of ethanol extracts from some marine invertebrates. *Chin. J. Immunol.* **2003**, *19*, 739–743.
8. Flint, H.J.; Bayer, E.A.; Rincon, M.T.; Lamed, R.; White, B.A. Polysaccharide utilization by gut bacteria: Potential for new insights from genomic analysis. *Nat. Rev. Microbiol.* **2008**, *6*, 121–131. [CrossRef] [PubMed]
9. Barlow, G.M.; Yu, A.; Mathur, R. Role of the gut microbiome in obesity and diabetes mellitus. *Nutr. Clin. Pract.* **2015**, *30*, 787–797. [CrossRef] [PubMed]
10. Chang, C.J.; Lin, C.S.; Lu, C.C.; Martel, J.; Ko, Y.F.; Ojcius, D.M.; Tseng, S.F.; Wu, T.R.; Chen, Y.Y.M.; Young, J.D.; *et al*. Ganoderma lucidum reduces obesity in mice by modulating the composition of the gut microbiota. *Nat. Commun.* **2015**, *6*, 7489–7506. [CrossRef] [PubMed]
11. Jiang, C.X.; Xiong, Q.P.; Li, S.L.; Zhao, X.R.; Zeng, X.X. Structural characterization, sulfation and antitumor activity of a polysaccharide fraction from *Cyclina sinensis*. *Carbohydr. Polym.* **2015**, *115*, 200–206. [CrossRef] [PubMed]
12. Liao, N.B.; Chen, S.G.; Ye, X.Q.; Zhong, J.J.; Wu, N.; Dong, S.L.; Yang, B.; Liu, D.H. Antioxidant and antitumor activity of a polysaccharide from freshwater clam *Corbicula flumineat*. *Food Funct.* **2013**, *4*, 539–548. [CrossRef] [PubMed]

13. Vidhyanandhini, R.; Vairamani, S. The anticoauulant activity and structural characterization of fractionated and purified glycosaminoglycans from venerid clam *Meretrix casta* (Chemnitz). *J. Liq. Chromatogr. Relat. Technol.* **2014**, *37*, 917–929. [CrossRef]

14. Saravanan, R.; Shanmugam, A. Isolation and characterization of heparin sulfate from marine scallop *Amusium pleuronectes* (Linne) an alternative source of heparin. *Carbohydr. Polym.* **2011**, *86*, 1082–1084. [CrossRef]

15. Pan, T.T.; Peng, S.H.; Xu, Z.L.; Xiong, B.; Wen, C.R.; Yao, M.; Pang, J. Synergetic degradation of konjacglucomannan by γ-ray irradiation and hydrogen peroxide. *Carbohydr. Polym.* **2013**, *93*, 761–767. [CrossRef] [PubMed]

16. Hu, H.B.; Liang, H.P.; Wu, Y. Isolation, purification and structural characterization of polysaccharide from *Acanthopanax brachypus*. *Carbohydr. Polym.* **2015**, *127*, 94–100. [CrossRef] [PubMed]

17. Cai, W.R.; Xie, L.L.; Chen, Y.; Zhang, H. Purification, characterization and anticoagulant activity of the polysaccharides from green tea. *Carbohydr. Polym.* **2013**, *92*, 1086–1090. [CrossRef] [PubMed]

18. Zheng, C.P.; Dong, Q.; Chen, H.J.; Cong, Q.F.; Ding, K. Structural characterization of a polysaccharide from *Chrysanthemum morifolium* flowers and its antioxidant activity. *Carbohydr. Polym.* **2015**, *130*, 113–121. [CrossRef] [PubMed]

19. Huang, Y.T.; Li, N.S.; Wan, J.B.; Zhang, D.Z.; Yan, C.Y. Structural characterization and antioxidant activity of a novel heteropolysaccharide from the submerged fermentation mycelia of *Ganoderma capense*. *Carbohydr. Polym.* **2015**, *134*, 752–760. [CrossRef] [PubMed]

20. Liao, W.Z.; Luo, Z.; Liu, D.; Ning, Z.X.; Yang, J.G.; Ren, J.Y. Structure characterization of a novel polysaccharide from *Dictyophora indusiata* and its macrophage immunomodulatory activities. *J. Agric. Food Chem.* **2015**, *63*, 535–544. [CrossRef] [PubMed]

21. Wang, Z.Y.; Zhao, Y.Y.; Su, T.T.; Zhang, J.; Wang, F. Characterization and antioxidant activity *in vitro* and *in vivo* of polysaccharide purified from *Rana chensinensis* skin. *Carbohydr. Polym.* **2015**, *126*, 17–22. [CrossRef] [PubMed]

22. Li, C.; You, L.J.; Fu, X.; Huang, Q.; Yu, S.J.; Liu, R.H. Structural characterization and immunomodulatory activity of a new hetero polysaccharide from *Prunella vulgaris*. *Food Funct.* **2015**, *6*, 1557–1567. [CrossRef] [PubMed]

23. Wei, Y.N.; Niu, Y.G.; Lv, J.L.; Xie, Z.H.; Jin, L.; Yao, W.B.; Gao, X.D.; Yu, L.L. Characterization of a hetero polysaccharide isolated from diploid *Gynostemma pentaphyllum* Makino. *Carbohydr. Polym.* **2013**, *92*, 2111–2117.

24. Dobruchowska, J.M.; Meng, X.F.; Leemhuis, H.; Gerwig, G.J.; Dijkhuizen, L.; Kamerling, J.P. Glucooligomers initially formed by the reuteransucrase enzyme of *Lactobacillus reuteri* 121 incubated with sucrose and malto-oligosaccharides. *Glycobiology* **2013**, *23*, 1084–1096. [CrossRef] [PubMed]

25. Bushneva, O.A.; Ovodova, R.G.; Shashkov, A.S.; Ovodov, Y.S. Structural studies on hairy region of pectic polysaccharide from campion *Silene vulgaris* (Obernabehen). *Carbohydr. Polym.* **2002**, *49*, 471–478. [CrossRef]

26. Amornrut, C.; Toida, T.; Imanari, T.; Woo, E.R.; Park, H.; Linhardt, R.; Wu, S.J.; Kim, Y.S. A new sulfated β-galactan from clams with anti-HIV activity. *Carbohydr. Res.* **1999**, *321*, 121–127. [CrossRef]

27. Dai, Z.Y.; Zhang, H.; Zhang, Y.P.; Wang, H.H. Chemical properties and immunostimulatory activity of a water-soluble polysaccharide from the clam of *Hyriopsis cumingii* Lea. *Carbohydr. Polym.* **2009**, *77*, 365–369. [CrossRef]

28. Papawee, S.; Kazuhiro, N.; Toshihiko, T.; Leo, J.L.D.; Van, G. Structural characterization and immunomodulatory effects of polysaccharides from *Phellinus linteus* and *Phellinus igniarius* on the IL-6/IL-10 cytokine balance of the mouse macrophage cell lines (RAW 264.7). *Food Funct.* **2015**, *6*, 2834–2844.

29. Wasser, S. Medicinal mushrooms as a source of antitumor and immunomodulating polysaccharides. *J. Microbiol. Biotechnol.* **2002**, *60*, 258–274.

30. Xu, X.; Wu, X.T.; Wang, Q.Q.; Cai, N.; Zhang, H.X.; Jiang, Z.D.; Wan, M.; Oda, T. Immunomodulatory effects of alginate oligosaccharides on murine macrophage RAW264.7 cells and their structure-activity relationships. *Food Chem.* **2014**, *62*, 3168–3176. [CrossRef] [PubMed]

31. Mao, K.R.; Chen, S.Z.; Chen, M.K.; Ma, Y.L.; Wang, Y.; Huang, B.; He, Z.Y.; Zeng, Y.; Hu, Y.; Sun, S.H.; *et al.* Nitric oxide suppresses NLRP3 inflammasome activation and protects against LPS-induced septic shock. *Cell Res.* **2013**, *23*, 201–212. [CrossRef] [PubMed]

32. Yakut, E.; Jakobs, C.; Peric, A.; Michel, G.; Baal, N.; Bein, G.; Brüne, B.; Hornung, V.; Hackstein, H. Extracorporeal photopheresis promotes IL-1β production. *J. Immunol.* **2015**, *194*, 2569–2577. [CrossRef] [PubMed]

33. Jia, D.D.; Zhang, J.; Lan, R.; Yang, H.L.; Sun, Y.Y. A simple preparative method for isolation and purification of polysaccharides from mulberry (*Morusalba*, L.) leaves. *Int. J. Food Sci. Technol.* **2013**, *48*, 1275–1281. [CrossRef]

34. Zhao, T.; Zhou, Y.; Mao, G.H.; Zou, Y.; Zhao, J.L.; Bai, S.Q.; Yang, L.Q.; Wu, X.Y. Extraction, purification and characterisation of chondroitin sulfate in Chinese sturgeon cartilage. *J. Sci. Food Agric.* **2013**, *93*, 1633–1640. [CrossRef] [PubMed]

35. Tong, H.B.; Liu, X.M.; Tian, D.; Sun, X. Purification, chemical characterization and radical scavenging activities of alkali-extracted polysaccharide fractions isolated from the fruit bodies of *Tricholoma matsutake*. *World J. Microbiol. Biotechnol.* **2013**, *29*, 775–780. [CrossRef] [PubMed]

36. Li, N.S.; Yan, C.Y.; Hua, D.S.; Zhang, D.Z. Isolation, purification, and structural characterization of a novel polysaccharide from *Ganoderma capense*. *Int. J. Biol. Macromol.* **2013**, *57*, 285–290. [CrossRef] [PubMed]

37. Wang, M.C.; Jiang, C.X.; Ma, L.P.; Zhang, Z.J.; Cao, L. Preparation, preliminary characterization and immunostimulatory activity of polysaccharide fractions from the peduncles of *Hovenia dulcis*. *Food Chem.* **2013**, *138*, 41–47. [CrossRef] [PubMed]

38. Dixon, J.; Lipkin, D. Spectrophotometric determination of vicinal glycols. *Anal. Chem.* **1954**, *26*, 1092–1093. [CrossRef]

39. Ciucanu, I.; Kerek, F. A simple and rapid method for permethylation of carbohydrates. *Carbohydr. Res.* **1984**, *131*, 209–217. [CrossRef]

marine drugs

MDPI

Review

Bacterial Exopolysaccharides from Extreme Marine Habitats: Production, Characterization and Biological Activities

Annarita Poli [1], Gianluca Anzelmo [1,2] and Barbara Nicolaus [1,*]

[1] Institute of Biomolecular Chemistry, C.N.R., Via Campi Flegrei 34, 80078 Pozzuoli, Naples, Italy; annarita.poli@icb.cnr.it(A.P.)

[2] Department of Environmental Sciences, Parthenope University of Naples, Centro Direzionale, Isola C4, (80143) Naples, Italy; gianluca.anzelmo@uniparthenope.it(G.A.)

* *Author to whom correspondence should be addressed; barbara.nicolaus@icb.cnr.it;
Tel.: +39-081-8675245; Fax: +39-081-8041770.

Received: 8 May 2010; in revised form: 25 May 2010; Accepted: 2 June 2010; Published: 3 June 2010

Abstract: Many marine bacteria produce exopolysaccharides (EPS) as a strategy for growth, adhering to solid surfaces, and to survive adverse conditions. There is growing interest in isolating new EPS producing bacteria from marine environments, particularly from extreme marine environments such as deep-sea hydrothermal vents characterized by high pressure and temperature and heavy metal presence. Marine EPS-producing microorganisms have been also isolated from several extreme niches such as the cold marine environments typically of Arctic and Antarctic sea ice, characterized by low temperature and low nutrient concentration, and the hypersaline marine environment found in a wide variety of aquatic and terrestrial ecosystems such as salt lakes and salterns. Most of their EPSs are heteropolysaccharides containing three or four different monosaccharides arranged in groups of 10 or less to form the repeating units. These polymers are often linear with an average molecular weight ranging from 1×10^5 to 3×10^5 Da. Some EPS are neutral macromolecules, but the majority of them are polyanionic for the presence of uronic acids or ketal-linked pyruvate or inorganic residues such as phosphate or sulfate. EPSs, forming a layer surrounding the cell, provide an effective protection against high or low temperature and salinity, or against possible predators. By examining their structure and chemical-physical characteristics it is possible to gain insight into their commercial application, and they are employed in several industries. Indeed EPSs produced by microorganisms from extreme habitats show biotechnological promise ranging from pharmaceutical industries, for their immunomodulatory and antiviral effects, bone regeneration and cicatrizing capacity, to food-processing industries for their peculiar gelling and thickening properties. Moreover, some EPSs are employed as biosurfactants and in detoxification mechanisms of petrochemical oil-polluted areas. The aim of this paper is to give an overview of current knowledge on EPSs produced by marine bacteria including symbiotic marine EPS-producing bacteria isolated from some marine annelid worms that live in extreme niches.

Keywords: chemical composition; exopolysaccharides; extremophiles; marine bacteria; biological activity

1. Introduction

Exopolysaccharides (EPSs) are high molecular weight carbohydrate polymers that make up a substantial component of the extracellular polymers surrounding most microbial cells in the marine environment. They constitute a large fraction of the reduced carbon reservoir in the ocean and enhance the survival of marine bacteria influencing the physicochemical environment in proximity of the bacterial cell. Moreover, they assist the microbial communities to endure extremes of temperature, salinity, and nutrient availability. In recent years the increased demand for natural polymers for pharmaceutical, food and other industrial applications has led to a remarkable interest in polysaccharides produced by microorganisms. Indeed, a substantial interest has aroused with regard to the isolation and identification of new microbial polysaccharides that might have innovative applications as gelling, emulsifier and stabilizer agents [1]. These biopolymers have emerged as new polymeric materials with novel and unique physical characteristics and therefore they have found extensive applications. Many microorganisms (many species of Gram-positive and Gram-negative Bacteria, Archaea, Fungi and some Alga) are known to produce extracellular polysaccharides or exopolysaccharides (EPS). This term was used for the first time by Sutherland in 1972 [2] in order to describe carbohydrate polymers produced by marine bacteria with high molecular weight. The advantages of microbial EPSs with regard to plants or marine macroalgal polymers are related to their chemical and physical properties and they embody a stable supply as well [3]. As far as the microbial biodiversity is concerned, bacterial EPSs exhibit a wide range of chemical structures: several EPSs display a high molecular weight as well as a heteropolymeric composition. Bacterial EPS usually occur in two forms: as capsular polysaccharides (CPS) where the polymers are covalently bound to the cell surface, and as slime polysaccharides which either remain attached (loosely bound) to the cell surface or are found in the extracellular medium as amorphous matrix [4,5].

In their natural environment, most bacteria occur in microbial aggregates whose structural and functional integrity is based on the presence of a matrix of extracellular polymeric substances and the EPS production seems to be essential for their survival [6]. In particular, marine polysaccharides, together with other macromolecules such as proteins, lipids and nucleic acids, comprise the organic matrix present in the intracellular space of microbial biofilms, which represents one of the largest reservoirs of reduced carbon on earth [6,7]. Moreover, the recent focus on extreme marine habitat has raised increasing attention on bacteria thriving in these conditions: extremophiles. These organisms, adopting special metabolic pathways and protective mechanisms to survive, represent a model to study the stability and the possibleroles of their biomolecules.

2. Roles of Microbial EPS in the Marine Environment

The physiological role of EPS depends on the ecological niches and the natural environment in which microorganisms have been isolated as well. Indeed, the EPS production is a process that requires a noticeable energy cost of up to 70%, representing a significant carbon investment for microorganisms. However, the benefits related to EPSs are significantly higher than costs considering the increasing growth and survival of microorganisms in their presence [8]. Indubitably, they possess a protective nature: the EPSs, forming a layer surrounding a cell, provide an effective protection against high or low temperature and salinity or against possible predators. They are essential in the aggregate formation, in the mechanism of adhesion to surfaces and to other organisms, in the formation of biofilm and in the uptake of nutrients [1,9,12]. In particular, studies of sea ice microbial communities have also found bacteria strongly associated to particles and have pointed out that microbial EPS played an important role in cryoprotection [13,14]. Moreover, the rate of synthesis and the amount of exopolysaccharide accumulated as capsular form in pathogenic bacteria influence their pathogenicity, in that capsular polysaccharides do not stimulate the immune system for their ability to mimic the cell surface of host cells.

EPSs display an important role in biofilm matrix in regard to the biochemical interactions between bacteria and surrounding cells [5,15]. The hydrated biofilms offer a stable micro-environment in which extracellular enzymes can find storage and in the same time facilitating cellular uptake of small molecules [11]. In addition, in a natural aquatic environment, the nutrients can interact with exopolymers in order to increase the rate of substance uptake and concentrate dissolved organic compounds, making them available to support microbial growth [5,15]. Finally, it has been proved that strains isolated from deep-sea hydrothermal vents show resistance to heavy metals and their purified EPSs presented the capacity to bind metals and toxic substances [16,18]. Indeed, these exopolymers exhibit a polyanionic state in marine environment displaying a high binding affinity for cations as well as trace metals. Since they generally contain uronic acids with a percentage between 20–50% of the total polysaccharide content, the acidic carboxyl groups attached are ionizable at seawater pH. Furthermore, EPS produced by some Antarctic bacterial isolates contain uronic acids and sulfate groups and may act as ligands for cations present as trace metals in the Southern Ocean environment, enhancing the primary production of microbial communities usually limited by poor availability of trace metals such as iron (Fe^{+3}) [19].

3. Structure and Production of EPS by Marine Bacteria

Most EPSs produced by marine bacteria are heteropolysaccharides containing three or four different monosaccharides arranged in groups of 10 or less to form repeating units [5]. Components most commonly found in marine EPS are monosaccharide such as pentoses (as D-arabinose, D-Ribose, D-Xylose), hexoses (D-Glucose, D-Galactose, D-Mannose, D-Allose, L-Rhamnose, L-Fucose), amino sugars (D-Glucosamine and D-Galactosamine) or uronic acids (D-Glucuronic acids, D-Galacturonic acids). Organic or inorganic substituents such as sulfate, phosphate, acetic acid, succinic acid and pyruvic acid may also be present [20]. Most polymers are linear, with an average molecular weight ranging from 1×10^5 to 3×10^5 Da [21]. Some EPS are neutral macromolecules, but the majority of them are polyanionic for the presence of uronic acids or ketal-linked pyruvate or inorganic residues such as phosphate or sulfate as well [22]. Moreover, the linkages between monosaccharides that have been most commonly found are 1,4-β- or 1,3-β- linkages in the backbones characterized by strong rigidity and 1,2-α- or 1,6-α-linkages in the more flexible ones. The physical properties of polysaccharides are deeply influenced by the way the monosaccharides are arranged together and the assemblage of the single polymer chains [23]. It has been widely accepted that many EPSs undergo transition from an ordered state, especially at low temperature or in the presence of some ions, to a less ordered state at high temperature or in absence of ions, enabling the conversion of extracellular matrices from a gel to a solid state. For instance, solutions of the EPS produced by *Alteromonas* strain 1644, originated from a polychaete annelid living in the proximity of deep-sea hydrothermal vents, displayed very low viscosity values at low ionic concentration (below 0.03 M) [24]. As a result, it formed a gel at higher ionic concentrations or exhibited an unusual high temperature-dependent viscosity in solution at low polymer concentration. This behavior could also depend on the nature of the ions. In addition, *Alteromonas* strain 1644 is able to produce two kind of EPS, and this a very peculiar microbial characteristic, as only a few reports are known about simultaneous production of two different EPSs by the same microorganism [25]. This may be attributed to the difficulty in separating possible mixtures of exopolysaccharides but, in this case, the separation of the two polysaccharides produced by strain 1644 has been greatly facilitated by their different gelling properties.

Since the analysis of the extracellular matrix polysaccharides in the natural marine environment is difficult due to the low abundance of each polymer, the possibility to grow a single isolated strain under controlled laboratory conditions represents a suitable approach to investigate microbial EPS production [19]. However, there is no single set of culture conditions that guarantees high EPS yields since microorganisms differ in the critical factors for maximum EPS production: carbon and nitrogen source utilization, mineral requirements, temperature and optimal pH. Moreover, it is possible to modulate the molecular mass, the number of residues and the degree of branching of EPS by using a physiological control. Indeed, the nutritional and environmental conditions (culture conditions) can affect the yield and quality of microbial EPS [26]. Sutherland [4] showed that the EPS production increased if marine bacteria were grown in laboratory conditions on limited nutrients (such as nitrogen, phosphorus, sulfur and potassium). In general, suboptimal temperature of growth, osmotic stress or other physical factors that restrict the growth may enhance the EPS production. Moreover, the choice of selected carbon source (sugar or non-sugar sources) in the growth medium represents the first step for the optimization of EPS production. Actually, *Hahella chejuensis*, a microorganism isolated from marine sediment collected from Marado, Cheju Island, Republic of Korea, produced the highest EPS yield

Mar. Drugs **2008**, *8*, 1779–1802

in a growth media emended with sucrose [27,28] while acetate was the most efficient carbon source for EPS production in *Halomonas alkaliantarctica*, strain CRSS, an haloalkalophilic bacteria isolated from saline lake in Antarctica [29,30]. In particular, strain CRSS was able to synthesize EPSs with different chemical composition on various substrates utilized as nutrients: a mannan or a xylo-mannan was produced in two different complex media and a fructo-glucan when the strain was grown on minimal media containing acetate as sole carbon source [29]. This is a common occurrence described for other strains, since culture conditions can modify the yield and the structure of polymers [31]. In regard to the nitrogen source, in general for EPS production, ammonium sulfate, peptone, sodium nitrate, urea and yeast extract are the components most currently used. In general, the presence of an organic nitrogen source promotes both the growth rate and the EPS production [32], even if there is some evidence showing that EPS production was higher at lower nitrogen concentration as suggested by Gorret *et al.* [33]. Samain *et al.* [24] used the fed-batch method in order to improve the EPS yield by an *Alteromonas* strain 1644: they controlled the growth by substituting the complex medium with a mineral defined medium containing ammonium chloride as a sole nitrogen source (nitrogen limitation). The initial ammonium chloride concentration was 0.4 g/L, and after the initial exponential phase, an ammonium chloride solution was continuously fed into the medium in order to maintain a small level of protein synthesis. During the entire feeding phase, the ammonium did not accumulate indicating that the growth was nitrogen-limited. In these conditions, EPS production started as soon as the ammonium had been depleted and continued almost linearly until the end of the fermentation, reaching a yield of 50% higher than obtained previously with the complex medium.

Sutherland [4] stated that there is a competition between EPS and cell-wall polymer (lipopolysaccharides, peptidoglycan, *etc.*) biosynthesis since the isoprenoid glycosyl lipid carriers and, consequently, EPS production, is not growth-associated. Recently, it has been found that most bacteria release the highest quantity of EPS in the stationary growth phase: *Alteromonas* strain 1644 [24], strain 4004, a thermophilic bacterium belonging to *Geobacillus* genus isolated from sea sand in Ischia Island (Italy) [34] and strain B3-15, isolated from marine hot spring at Vulcano Island, classified as a new strain of *Bacillus licheniformis* [35], are examples of microorganisms that produce high amounts of EPS in this phase. However, there are microorganisms that produce the maximum EPS during the exponential growth phase as reported for *Pseudoalteromonas antarctica* strain NF3 isolated from a glacial marine sludge at the South Shetland Island in Antarctica [36]. In *Alteromonas macleodii* subsp. *fijiensis*, isolated from a deep sea vent, the production of EPS began at the end of the exponential phase and continued throughout the stationary phase, reaching 6 g/L at 60 h of incubation [37].

Another shifting parameter during the EPS production is the rheology of the growth media: the broth develops non-Newtonian characteristics acting as a pseudoplastic fluid where the viscosity decreases with increasing shear rate. This change can be caused by the presence of EPS and their metabolic products, as well as the lack of homogeneity in terms of mixing, mass and oxygen. In these conditions, the EPS producer bacteria, exposed to such variable gradients, can produce heterogeneous EPS in terms of molecular weight, branching capacity, *etc.* The rheological shifts during the fermentation process could be used as a parameter to monitor the constancy of EPS quality and production [38]. Indeed, the aeration rate could be a parameter that affects EPS production as well. Lee *et al.* [28] reported that high aeration rates generally enhanced EPS production and increased the viscosity of the culture broth regarding *Hahella chejuensis*. Moreover, the use of some detergents such as Tween 40 (polyoxyethylene sorbital monopalmitate), Tween 80 (polyoxyethylene sorbital monooleate), CHAPS (3-[(3-cholamidopropyl) dimethyl ammonio]-1-hydroxypropane-sulfonate) and Triton X 100 (nonaethylene glycol octylphenol ether) may ameliorate oxygen concentration in the growth media, increasing the production of EPS [38].

4. Marine EPS-Producing Microorganisms Isolated from

4.1. Deep-Sea Hydrothermal Vents, Volcanic and Hydrothermal Marine Areas, Shallow Submarine Thermal Springs

To date, mainly mesophilic heterotrophic bacteria have been investigated rather than thermophilic microorganisms, even though the latter microorganisms possess thermostable enzymes with interesting biotechnological application for large scale industrial production. Actually, although considerable information related to the chemical composition, the rheological properties and metal binding capability have been determined, only a few polymers have been fully characterized so far.

Deep-sea hydrothermal vents, volcanic and hydrothermal marine areas, and shallow submarine thermal springs offer a new source of EPS producer bacteria (Table 1). *Pseudoalteromonas* strain 721 produced an EPS containing an octasaccharide repeating unit with two side chains (Figure 1), [39,40]. This EPS exhibited gel formation and viscoelastic behavior at increasing temperature.

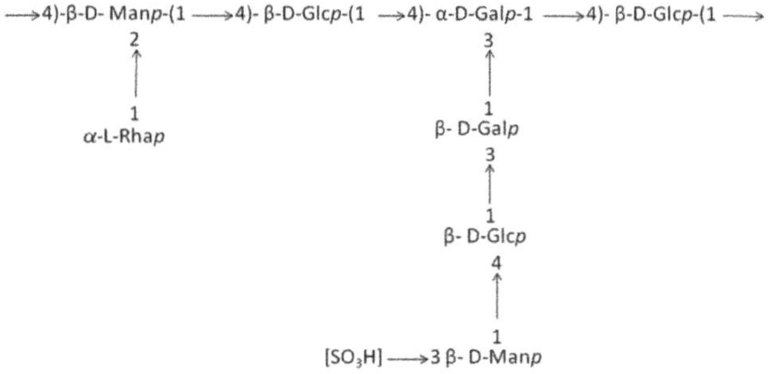

Figure 1. The repeating unit of EPS secreted by *Pseudoalteromonas* strain 721 [39].

Alteromonas macleodii subsp. *fijiensis* is an aerobic, mesophilic bacterium isolated from a hydrothermal vent at a depth of 2600 m in a rift system of the North Fiji basin [37,41]. This strain produced an EPS with an high metal-binding maximum capacity of up to 316 mg Pb(II)/g polymer [16,17]. The repeating unit consists of an hexasaccharide containing three uronosyl residues with a branch point at a galacturonosyl residue and a side chain terminated by a 4,6-*O*-(1-carboxyethylidene)-β-D-Man*p* [41]. Proposed uses for this polymer include water treatment and removal of heavy metal pollutants. Moreover, the xanthan-like EPS produced by *Alteromonas macleodii* subsp. *fijiensis* could find application as a food-thickening agent [41]. In addition, experiments conducted with this EPS encouraged its use in the bone healing application as it showed ability to promote the adhesion of rat calvaria osteoblastic cells *in vivo* [42]. Additionally, it could also be used in cardiovascular diseases as suggested by Colliec Jouault *et al.* [43].

Indeed, a valuable source of EPS is embodied by bacteria isolated from volcanic and hydrothermal marine areas and, in particular, some thermophiles, including *Methanosarcina*, *Haloferax*, *Haloarcula*, *Sulfolobus* and *Bacillus* species, and more recently *Thermotoga marittima* and *Thermococcus litoralis* have been studied so far as producers of unusual extra-cellular polysaccharides [35,44]. In particular, *Thermococcus litoralis* has been showed to produce EPS in a sulfur-free, defined growth medium with growth rates and cell yields comparable to those obtained on complex media. This microorganism is a heterotrophic facultative sulfur-dependent hyperthermophilic Archaeon, isolated from a shallow submarine thermal spring with an optimal growth temperature of 88 °C. Although extracellular polysaccharides have been identified in several marine extreme environments, the production of polysaccharides by extremophiles has been investigated only to a minor extent. The EPS produced by *T. litoralis* contains mannose as the only monosaccharidic constituent and this is a very peculiar feature for a prokaryote since the production of mannan-like constituents are typically produced by eukaryotes such as plants or yeasts [45]. According to the initial dry weight of the material, 1 to 2% sulfate and 1.5 to 4.5% phosphorus were also found in *T. litoralis* EPS. Sulfated EPSs are common in all domains of life and EPSs such as sulphoevernan, chondroitin sulfate, dextran sulfate, heparin, and mannan sulfate may enable protection of eukaryotic cells from viruses, including human immunodeficiency virus type 1, by inhibiting virus particle adsorption to host cells [46]. In addition, *T. litoralis* displayed biofilm formation on hydrophilic surfaces under a variety of conditions and, particularly, cell adhesion reached high levels in culture media emended with maltose and/or yeast extract. Indeed, polysaccharide levels in biofilms formed under different growth conditions were similar. The adhesion of marine bacteria to surfaces - a process in which EPSs are likely to play a leading role - is not unusual. However, this occurrence has not been investigated to any extent in hyperthermophilic archaea, though an increasing number of organisms are isolated from hydrothermal vent sites. Further study regarding the EPS productions could display new insights on the natural growth of these organisms providing information on essential ecological interactions in microbial world [47].

Additionally, thermophilic bacteria belonging to *Geobacillus* species have been isolated from shallow, marine hydrothermal vents of flegrean area in Italy and characterized as EPS producers as well [48]. Accordingly, a thermophilic strain from the genus *Geobacillus* was isolated from sediment samples in a marine hot spring near the seashore of Maronti (Ischia Island, Italy) and the EPS produced was characterized (Figure 2).

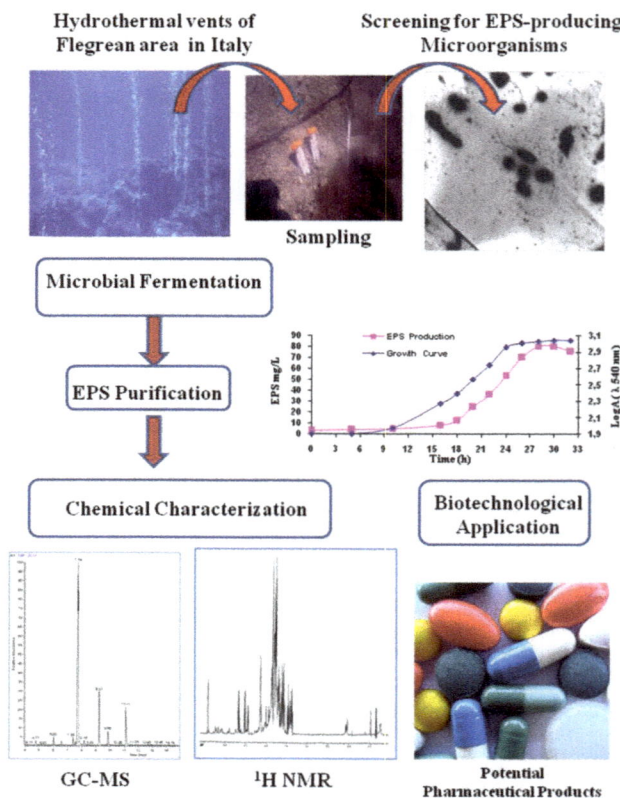

Figure 2. Schematic steps involved in the studies of *Geobacillus* strain 4004 EPS [48].

Among different strains isolated, the strain *Geobacillus* sp. 4004 was able to produce an EPS in yields of 90 mg/L in Bacto Marine Broth 2216 (Difco) at the optimal temperature of 60 °C at pH 7.0. Additionally, it was established that, by modulating the conditions of growth as well as emending the culture medium with different carbon sources, the yield of the biopolymer increased and the biomass production was directly proportional to EPS production. Therefore, the EPS produced by strain 4004 in the sucrose medium was isolated via ethanol precipitation of the cell-free medium. The polysaccharide fraction, tested for carbohydrate (85%), protein (10%), and nucleic acid (2%) content, was purified by gel filtration chromatography. The average molecular mass of the strain 4004 EPS was determined to be about 1×10^6 Da and the ^{1}H and ^{13}C-NMR spectra showed that there are five different residues within the repeating saccharidic unit; two of them with a *gluco/galacto* configuration and three with a *manno* configuration. One of the residues is an acetamido-sugar and at least one uronic acid is present. Sugar analysis was performed on hydrolysed polysaccharide fraction and the EPS was reported to be composed of Gal/Man/GlcN/Ara in a relative ratio of 1.0:0.8:0.4:0.2, respectively. The successful application of EPS largely depends on their physicochemical properties rather than on yield alone. Therefore, in order to gain insight about the viscometric properties, specific viscosity (η) was studied with an Ubbelohde type viscometer under different water concentrations of EPS, in particular at 1% $CaCl_2$ and 1% NaCl. It was observed that at increasing concentration of EPS there was an increase in viscosity in all the solutions and the maximum viscosity was found to be 1.08 dL/g in 1% $CaCl_2$.

Furthermore, marine shallow hydrothermal vents around volcanic Eolian islands, close to Sicily coasts (Italy), represent accessible fields for the isolation of thermophilic bacteria. Previous studies described diversity and distribution of bacterial communities within deep and shallow hydrothermal systems at Porto di Levante, Vulcano, revealing the presence of chemosynthetic, thermophilic, archaeal and bacterial strains [49]. In addition, a thermophilic aerobic microorganism, able to produce two exocellular polysaccharides (EPS1 and EPS2), was isolated from sea water of a shallow hydrothermal vent at Vulcano island (Eolian Islands, Italy). This new eolian thermophilic isolate was identified as *Bacillus thermodenitrificans* strain B3-72. This strain displayed the highest EPS production in a Bacto Marine Broth 2216 (Difco) at an optimal temperature of 65 °C and pH 7.0 in aerobic conditions. The production started at the end of the exponential phase of growth and continued during the stationary phase. The highest concentrations of polysaccharide harvested after three days of culture was approximately of 70 mg/L. In particular, EPS2, the polymeric fraction obtained after DEAE Gel filtration (80% carbohydrate and 3% proteins content), displayed a molecular weight of approximately 4.0×10^5 Da. Moreover, the IR spectrum of EPS2 suggested the absence of uronic acid and sulfate residues. Hydrolysed EPS2 from *B. thermodenitrificans* yielded mannose and glucose as principal constituents in a relative ratio of 1:0.2. The ^{13}C and ^{1}H-NMR spectra of this polymer were performed and it was possible to conclude that it possesses a trisaccharide repeating unit essentially constituted of sugars having a *manno-pyranosidic* configuration. In subsequent experiments, the immunomodulatory and antiviral effects of the *B. thermodenitrificans* EPS2 were evaluated, since polysaccharides with high molecular weight have exhibited immunogenic activity so far [50]. *In vitro* studies have demonstrated that sulfated polysaccharides, such as dextran sulfated, have antiviral effects against enveloped viruses such as herpes simplex virus [51]. The sulfated-EPS are known to interfere with the absorption and penetration of viruses into host cell and to inhibit various retroviral reverse transcriptases [52]. It was established that the *B. thermodenitrificans* EPS2 obstructs HSV -2 replication in human peripheral blood mononuclear cells (PBMC). Actually, high levels of IFN-α, IL-12, IFN-γ, TNF-α, IL-18 (IFN, interferons; TNF, tumor necrosis factor; IL, interleukine) were detected in supernatants of EPS-2 treated PBMC and, additionally, this effect was dose-dependent. Those results highlight the potential role of *B. thermodenitrificans* EPS2 toward equilibrating the immune response during viral infection and that the immunological disorders determined by HSV-2 could be partially restored by treatment with EPS [53].

Table 1. EPSs produced by microorganisms isolated from marine hot springs and hydrothermal vents.

Microorganisms	Source Environment	Description of EPS and Chemical Composition	Suggested Ecological Role and Biotechnological Application	References
Pseudoalteromonas strain 721	Deep-sea hydrothermal vent	Octasaccharide repeating unit with two side chains, (Figure 1)	Gelling properties	[39,40]
Alteromonas macleodii subsp. *fijiensis*	Deep-sea hydrothermal vent, North Fijian Basin	Sulfated heteropolysaccharide, high uronic acids with pyruvate. The repeating unit is a branched hexasaccharide containing Glc, Man, Gal, GlcA, GalA, pyruvated mannose	Thickening agent in food- processing industry, biotoxification and waste-water treatment, bone healing, treatment of cardiovascular diseases	[16,17,37,41,43]
Thermococcus litoralis	Shallow submarine thermal spring	Man is the only monosaccharide	Biofilm formation	[47]
Geobacillus sp. strain 4004	Sediment in marine hot spring near the seashore of Maronti, Ischia Island, Italy	A pentasaccharide repeating unit (two of them with a gluco-galacto configuration and three with a *manno* configuration. Gal:Man:GlcN:Arab (1.0:0.8:0.4:02)	Pharmaceutical application	[48]
Bacillus thermodenitrificans strain B3-72	Water of a shallow hydrothermal vent, Vulcano Island, Italy	Trisaccharide repeating unit and a *mannopyranosidic* configuration. Man:Glc (1.0:2)	Immunomodulatory and antiviral activities	[49,53]
Bacillus licheniformis strain B3-15	Water of a shallow marine hot spring, Vulcano Island, Italy	Man is the main monosaccharide. Tetrasaccharide repeating unit and a *mannopyranosidic* configuration	Antiviral activity	[35,54]

Besides, a thermotolerant *Bacillus licheniformis* strain (B3-15), isolated from water of a shallow, marine hot spring at Vulcano Island (Eolian Islands, Italy), produced an EPS with immuno-modulating properties. This strain exhibited a production of 165 mg/L exocellular polysaccharide when grown in liquid mineral medium with the addition of glucose. Furthermore, *Bacillus licheniformis* strain B3-15 showed a high growth rate in media containing kerosene as sole carbon source. The purified exopolymers fraction displayed a single fraction with high carbohydrate content. According to the chemical structure analyses, the EPS was a tetrasaccharide repeating unit essentially constituted by sugars having a *manno-pyranosidic* configuration [35]. Solutions of the *Bacillus licheniformis* EPS added to *in vitro* cultures of human peripheral blood mononuclear cells (PBMC) exhibited a marked, dose-dependent decrease in HSV-2 (Herpes simplex virus type 2) replication [54]. In order to assess whether the antiviral activity induced by EPS in PBMC could be related to an immune-modulatory mechanism, the production of different cytokines involved in the immune response toward virus infection, such as IFN-a, IL-12, IFN-g, TNF-a, IL-18, was evaluated. As a result, high levels of all these cytokines were detected in supernatants from PBMC treated with EPS. On the other hand, IL-4, a strong hallmark of Th2 responses, was not detected in any of the supernatants tested. According to the data collected, the effect of EPS was dose dependent when PBMC were treated with EPS and simultaneously infected with HSV-2 and cytokine production was down-regulated. Consequently, those data suggest that EPS may contribute to improve immune surveillance of PBMC toward virus infection eliciting a therapeutic Th1-like response in clinical settings of viral diseases as well as in immune-compromised host. Actually, a large number of immune-modulatory compounds have been developed in recent years in order to enhance the response of the host to invading infectious agents. However, the accomplishment of such therapeutic efforts relies upon the complete knowledge of cell types and factors involved in the development of a protective antiviral response [54].

4.2. Cold Marine Environments: Deep-sea, Arctic and Antarctic Sea Ice

Psychrophilic ("cold loving") bacteria prefer a growth temperature of less than 15 °C. Bacteria that can grow at such cold temperatures, but which prefer a high growth temperature, are known as psychrotrophs or psychrotolerants.

It has become widely accepted that bacteria isolated from deep-sea vents will represent a valuable resource of unusual molecules for biotechnological purpose and at the same time provide insight into the role of the peculiar molecules produced (Table 2). Most deep-sea environments are influenced by high pressure, low temperature and low nutrient concentration. However, many kinds of deep-sea psychrotolerant bacteria live in the abyssal ecological community and they have been largely investigated as they may reveal important ecological characteristics enabling the morphological, physiological and metabolic adaptation of bacteria to the continuously changing deep-sea ecosystem [55].

Pseudoalteromonas sp. SM9913 is a gamma-proteobacterium isolated from 1855 m deep-sea sediment in the Bohai Gulf, the innermost gulf of the Yellow Sea on the coast of north eastern China. Qin *et al.* [56] studied the EPS producing capacity in well established laboratory conditions and reported that the yield of the EPS increased at decreasing culture temperatures in the range of 30–10 °C, and it reached a yield 5.25 g/L (dry weight) under optimal growth conditions (15 °C, 52 h). This is a very high yield compared with the data reported for the EPSs produced by other psychrotolerant microorganisms [57]. Its structure is a linear arrangement of ;-(1→6) linkage of glucose with a high degree of acetylation and with a molecular mass of 4×10^4 Da. It was established that glucose is the main sugar component (61.8%) with minor monosaccharide units including terminal arabinofuranosyl (t-Ara f, 11.0%) and terminal glucopyranosyl (t-Glc, 11.2%) as well as a small amount of t-Gal (3.1%), 4-Xyl f (3.9%), 4-Glc (5.0%) and 3,6-Glc (4.0%). Furthermore, this EPS has been investigated for its flocculation behavior and bio-sorption capacity, providing insight into its ecological rule [58]. Particularly, this polyanionic polymers with high levels of acetyl groups has been shown to

Mar. Drugs **2008**, *8*, 1779–1802

bind a wide range of metal cations, such as Fe^{2+}, Zn^{2+}, Cu^{2+} and Co^{2+}, indicating an helpful role in concentrating metal ions to the surface of the strain in the microenvironment around the cell.

Besides, EPSs have a leading role for the survival of psychrophile microorganisms in permanently cold environments. In line with their capacity to protect cells from freezing, high concentrations of EPSs have been found in Antarctic marine bacteria [19] and in Arctic winter sea ice [14]. It has been demonstrated that EPSs produced by sea-ice isolates possess a molecular weight 5–50 times larger than the average of the other isolated marine EPSs and it ranges between 1–3×10^5 Da [5]. The structure and physicochemical properties of the EPSs are influenced by the length of the polymer chain. As the length of the polymer increases, the opportunity for complex entanglement of the chains and intra-molecular associations increases, and these contribute to the tertiary structure and physical behavior of the polymer [22].

It is very interesting to note that particulate aggregates are ubiquitous and abundant in the world's oceans. Marine bacteria benefit from living in aggregates since their proximity to other cells and surfaces provides opportunities for interaction and nutrient uptake. EPSs excreted by bacteria are polymeric substances that offer a network to hold these structures together [59]. Bacteria are found in abundance in the bottom layers of the ice or in brine channels and are often attached to detrital particles or living microalgal cells. Delille and Rosier [60] also suggested that the high numbers of particle associated bacteria found in sea ice may explain observations of underlying water enriched in bacterial biomass relative to the open ocean. More recently, studies of Arctic sea ice in winter showed that active bacteria could be found in brine channels at temperatures as low as $-20\,°C$ and that they were particle associated [13]. EPSs may provide a cryoprotectant role in these environments of high salinities and low temperature [14]. According to this kind of investigation, Mancuso Nichols *et al.* [61] studied the EPS production of two different strains isolated from particulate material and melted sea ice collected in the Southern Ocean. In particular, the bacterial strain CAM025 was isolated from particles collected in melted Antarctic sea ice and CAM036 was isolated from particles captured by a plankton net towed through the Ocean. Both the psychrotolerant strains (growth at $4\,°C$ and $25\,°C$) displayed an enhanced mucoid morphology on marine agar medium supplemented with glucose. According to the results of 16S rDNA gene sequencing and whole cell fatty acid analyses, these two isolates were shown to be closely related to the genus *Pseudoalteromonas*. Both strains have been cultivated in a Marine Broth Medium enriched with glucose in temperatures ranging from 4–$20\,°C$ and the EPS yield data suggest that there is a decreased production of EPS at higher temperature ($20\,°C$) for the Antarctic sea ice strain tested; particularly, strain CAM025 was observed to produce 30-fold more EPS at $-2\,°C$ and $10\,°C$ (100 mg/g dry cell weight) than at $20\,°C$. This result supports the proposed assumption that EPS production by psychrotolerant bacteria may play an important role in the sea ice microbial communities since bacterial EPS production in brine channels and, maybe, other deep cold ecosystems, may offer a fence against the environmental extremes experienced by bacterial cells attempting to modify water properties near the cells. Preliminary characterizations showed that the structure of the EPS from CAM025 and CAM036 includes sulfate as well as high levels of uronic acids, such as galacturonic acid, along with acetyl groups. In addition, the EPS from CAM036 was shown by NMR data to include a succinyl group. Monosaccharide composition was estimated to be Glc/GalA/Rha/Gal (1:0.5:0.1:0.08) and GalA/Glc/Man/GalNAc/Ara (1:0.8:0.84:0.36:0.13) for strain CAM025 and CAM036, respectively. These features suggest an overall polyanionic or 'sticky' capacity of the EPSs in the marine environment, since at the pH of seawater (pH 8.0) many of the acidic groups present on these polymers are ionized. This 'stickiness' is important since these EPS can bind cations, such as Fe^{+3}, whose availability in the Southern Ocean is known to limit primary metabolite production as well as for the uptake of other dissolved metals [62]. According to relative monosaccharide molar ratio, those two EPS displayed some similarities in the composition including the presence of the sole acidic sugar, galacturonic acid, in significant proportion, and glucose as one of the major monosaccharide components as well. Nevertheless, further investigations on the ecological role of the Antarctic EPS could provide insight into possible commercial uses of these novel polymers.

Table 2. EPSs produced by microorganisms isolated from cold marine environments.

Microorganisms	Source environment	Description of EPS and Chemical composition	Suggested Ecological Role and Biotechnological Application	References
Pseudoalteromonas strain SM9913	Deep-sea sediment in the Bohai Gulf, gulf of the Yellow Sea, China	Linear arrangement of α-(1→6) linkage of glucose with a high degree of acetylation	Flocculation behavior and bio-sorption capacity	[56,58]
Pseudoalteromonas strain CAM025	Isolated from particles collected in melted Antarctic sea	Sulfated heteropolysaccharide, high levels of uronic acids with acetyl groups Glc:GalA:Rha:Gal (1:0.5:0.1:0.08)	Cryoprotection	[61]
Pseudoalteromonas strain CAM036	Isolated from particles captured by a plankton net towed through the Southern Ocean	Sulfated heteropolysaccharide, high levels of uronic acids with acetyl and succinyl groups GalA:Glc:Man:GalNAc:Ara (1:0.8:0.84:0.36:0.13)	Trace metal binding	[61]
Colwellia psychrerythraea strain 34H	Arctic marine sediments	n.r.	Cryoprotection	[63]

n.r. not reported.

Some EPS producing bacteria have been found to colonize either the Antarctic or Artic ocean sea ice. Therefore, the study of the EPS production and the optimal growth condition could provide insight into the ecological role of the EPSs. Recently, Marx *et al.* [63] examined the relative effects of temperature, pressure, and salinity on EPS production by a marine and psychrophilic gamma-proteobacterium, *Colwellia psychrerythraea* strain 34H. This flagella-containing microorganism can be found in the persistently cold marine environments including Arctic and Antarctic sea ice. Strain 34H, in particular, was isolated from Arctic marine sediments. It has been exposed that at growth temperatures ranging from −8 to −14 °C, the EPS production rose dramatically. Similarly, at higher pressures of 400 and 600 atm, the EPS production increased noticeably as well as in salinity tests at 10–100 parts per million (at temperatures of both −1 and −5 °C). Extreme environmental conditions hence appeared to stimulate EPS production by this strain. Furthermore, strain 34H recovered best from deep-freezing to −80 °C if first supplemented with a preparation of its own EPS, rather than other cryoprotectants like glycerol, suggesting that the EPS production is both a survival strategy and a source of compounds with potentially novel properties for biotechnological applications.

4.3. Hypersaline Marine Environment: Salt Lakes and Marine Salterns

Hypersaline environments are found in a wide variety of aquatic and terrestrial ecosystems. They are inhabited by halotolerant microorganisms as well as halophilic microorganisms (0.5 M and 2.5 M NaCl) and extreme halophiles (up above 2.5 M NaCl). Moderate and extreme halophiles have been isolated not only from hypersaline ecosystems (salt lakes, marine salterns and saline soils) but also from alkaline ecosystems (alkaline lakes). Halophilic microorganisms have developed various biochemical strategies in order to survive in high saline conditions, including compatible solute synthesis to maintain cell structure and function. Their products such as ectoine, bacteriorhodopsins, exopolysaccharides, hydrolases, biosurfactants are noticeably of industrial interest.

Indeed, the genus *Halomonas* has received increasing interest as several species are able to produce significant quantities of EPS with high surface activity and/or rheological properties [64,65]. Intriguing properties of the EPS derived from *Halomonas* species, such as emulsification activity, appear to be worthwhile for an ample range of products and application. Actually, large quantities of surface-active agents, indispensable components in the food, cosmetic and pharmaceutical sector, are synthesized from hydrocarbons and this represents a problem because they are derived from a non renewable resource arousing concerns about their environmental impact and potential health risks. Conversely, biosurfactants and bioemulsifiers of biological origin have received increasing interest since they are renewable and less toxic compounds. Microbial production of surface-active agents offers a sustainable and potentially cost-effective alternative to chemical synthesis [66]. In addition, according to recent developments in industrial processing technologies, there is an increasing demand for new types of biopolymers with novel or enhanced functionalities and new microbial isolates, particularly of a marine origin, offer a relatively underexploited resource [67].

However, the first EPS to be described was produced by the extremely halophilic archaea *Haloferax mediterranei* (Table 3), isolated from the Mediterranean Sea [68]. In laboratory conditions, the cells were detected to be surrounded by an amorphous matrix in unshaken cultures. In addition, it was recognized, by means of electron microscopy using thin sections and negative staining, a thick layer of EPS surrounds the cells. The EPS production was scarcely influenced by the growth conditions studied and, although the amount of polymer produced with yeast extract or glucose was not strikingly different; the aggregation of the cells on top in unshaken cultures appeared to be dependent on the presence of glucose. This EPS was obtained from the supernatant of shaken liquid cultures by cold ethanol precipitation, and yielded 3 mg/mL. Three neutral sugars were detected such as glucose, galactose, and mannose as major component. The IR spectrum of the EPS indicated the presence of sulfate groups as well as carboxyl group of sterified organic acids such as uronic acids, indicating that the EPS produced is an acidic heteropolysaccaride. The study of rheological properties displayed a pseudo plastic behavior as the EPS solutions indicated a high apparent viscosity, increasing noticeably with the concentration, but the viscosity remained quite constant over wide ranges of pH, temperature and different salinities. Indeed, the extreme salt tolerance of this polysaccharide and the producer organism showed that this microorganism is a valuable candidate for the recovery of oil, especially in oil deposits with high salinity concentrations. In subsequent experiments, Parolis *et al.* [69] elucidated completely the structure of the repeating unit of the EPS produced by *Haloferax mediterranei* by using a combination of chemical and spectroscopic approaches. The EPS was hydrolysed with 4M trifluoroacetic acid and the hydrolysate was converted into alditol acetates to be analysed by GLC. GLC analysis showed only mannitol hexacetate, indicating that mannose was the only neutral sugar present in the polymer. A further portion of the EPS was methanolysed, then the methyl esters were reduced and the products were hydrolysed and converted into alditol acetates as before. GLC analysis showed mannitol hexaacetate and 2-acetamido-2-deoxyglucitol pentaacetate in the molar ratio 1.0:1.1, respectively, indicating that the acidic sugar present in the EPS is 2-amino-2-deoxyglucuronic acid. 1D and 2D NMR spectroscopic analysis of the native and periodate-oxidized/reduced polysaccharide provided the complete structure of repeating unit of EPS as: →4)-β-D-GlcpNAcA-(1→6)-α-D-Manp-(1→4)-β-D-GlcpNAcA-3-*O*-SO$_3^-$-(1→.

Table 3. EPSs produced by microorganisms isolated from hypersaline marine environments.

Microorganisms	Source environment	Description of EPS and Chemical composition	Suggested ecological role and Potential Biotechnological Application	References
Haloferax mediterranei	Mediterranean Sea	→4)-β-D-GlcpNAcA-(1→6)-α-D-Manp-(1→4)-β-D-GlcpNAcA-3-*O*-SO$_3$-(1→	Candidate in oil recovery, especially in oil deposits with high salinity concentrations	[68,69]
Hahella chejuensis	Marine sediment sample collected from Marado, Cheju Island, Republic of Korea	EPS named EPS-R Glc:Gal (0.68:1.0)	Biosurfactant and detoxification of polluted areas from petrochemical oils	[28]
Halomonas alkaliantarctica strain CRSS	Salt lake in Cape Russell in Antarctica	Glc:Fru:GlcN:GalN (1.0:0.7:0.3:trace)	High viscosity	[29,30]

A novel EPS, called EPS-R, produced by a slightly halophilic marine bacterium *Hahella chejuensis*, was isolated from a marine sediment sample collected from Marado, Cheju Island, Republic of Korea [28] and displayed a specific emulsifying capacity higher than that observed in commercial polysaccharides such as xanthan gum, gellan gum or sodium alginate. EPS-R contains glucose and galactose as the main sugars components in a molar ratio of 0.68:1.0, respectively; traces of minor sugar components such as xylose and ribose have also been found. The average molecular mass, as determined by gel filtration chromatography, is 2.2×10^3 KDa. The rheological behavior of EPS-R indicated that this EPS formed a structure intermediate between a random-coil polysaccharide and a weak gel. Furthermore, EPS-R was stable to pH and salts, demonstrating proper capacity for future applications as biosurfactant which may enhance hydrophobic substrate utilization and detoxification of polluted areas from petrochemical oils. In fact, biosurfactants from microorganisms are worthwhile more than their chemical counterparts as they are biodegradable and can be synthesized under user-friendly conditions (e.g., low temperatures and pressures) and are effective over a wide range of temperature, pH and salinity conditions [70].

A new halo-alkalophilic *Halomonas* strain has been shown to produce EPS [29,30] The new strain, *Halomonas alkaliantarctica* strain CRSS, was isolated from salt sediments near the salt lake in Cape Russell in Antarctica. The strain CRSS grew aerobically in a complex medium containing 100 g/L NaCl and, among different carbon sources tested, acetate was established to be the most efficient one displaying a yield of 2.9 g/g dry cells at 30 °C after 48 h of growth. The cell-free supernatant was precipated with ethanol and the soluble fraction was further investigated. The hydrolysed EPS produced by strain CRSS presented a monosaccharide composition of Glc/Fru/GlcN/GalN in relative proportion of 1.0:0.7:0.3:tr, respectively. The ^{1}H and ^{13}C-NMR spectra showed the presence of six different residues in the repetitive saccharidic unit, five of them with α configuration and one with a *manno* configuration indicating a complex primary structure of the biopolymer. Furthermore, in accordance with the measurements of viscosity (η) performed on aqueous solutions of the polysaccharide (1% w/v), an increase of solution viscosity at 2.0–3.0 pH values and at 2.5% (w/v) NaCl was detected according to the size and number of macromolecules in solution [29].

4.4. Polychaete Annelid in Symbiotic Relationships

Mutually beneficial symbiotic relationships are widespread in marine ecosystems and complex positive interactions have been shown to be common especially in extreme marine conditions such as hydrothermal and deep sea vents. Recent insights have noticeably expanded the ecological niche biodiversity and increase the number of microorganisms involved in a variety of systems. Accordingly, some marine annelid worms have completely lost the digestive tract of their relatives (the common earthworm) and it has been established that some species get their sustenance from a large population of at least five different species of bacteria living underneath their outer skin. Particularly a polychaete annelid, *Alvinella pompejana*, commonly known as the Pompeii worm, resides in tubes near hydrothermal vents along the seafloor. While in the tube, the worm's tail end might be immersed in temperatures as hot as 81 °C, while its head rests in cooler water, as moderate as 22 °C. Living in a symbiotic relationship, the worms secrete mucous from tiny glands on their backs to feed the bacteria, and in return they are protected by some degree of insulation. Besides the temperatures, the Pompeii worm has to deal with lethal chemicals too, like sulfides and heavy metals such as lead, cadmium, zinc, and copper. It has been supposed that symbiotic bacteria making the worm's back their home may detoxify the water within the worm's tube. Therefore, those bacteria could prove useful in cleaning up toxic waste sites. An EPS-producing strain, designated as HYD657 (Table 4), isolated from epidermis of a polychaete annelid, *Alvinella pompejana*, was collected near an active hydrothermal vent of the East Pacific Rise [71]. Phylogenetic analyses showed that strain HYD657 belonged to the gamma-subdivision of the phylum *Proteobacteria* and that it was closely related to *Alteromonas macleodii*. According to the measurement of DNA-DNA homology, the name of *A. macleodii* subsp. *fijiensis* biovar deepsane was assigned to the bacterium strain HYD657. The monosaccharide composition of the polysaccharide excreted was determined by means of GC, analysing monosaccharide derivates as trimethylsilyl glycosides following methanolysis, and it presented Gal/Glc/Rha/Fuc/Man/GlcA/GalA/3-0-(1 carboxyethyl)-D-GlcA, an unusual diacidic hexose found only in another EPS excreted by a bacterial strain originating from deep-sea vents [72] in a molar ratio of 1:0.42:0.85:0.5:0.42:0.5:0.5:0.5, respectively. Preliminary structural studies on the sugar derivatives by Nuclear Magnetic Resonance and GC-Mass Spectrometry revealed the repeating unit of this polysaccharide was an undesaccharide with three side-chains. Although the definitive structure determination of the polymer requires additional studies, this EPS has been shown to possess

Mar. Drugs **2008**, *8*, 1779–1802

interesting biological activities and applications have already been found in cosmetics (patentPCT 94907582-4).

Furthermore, a new facultative anaerobic, heterotrophic and mesophilic bacterium was also isolated from Pompei worm tube collected from a deep-sea hydrothermal field of the East Pacific Rise and named *Vibrio diabolicus* [75]. During the stationary phase of growth in batch cultures on a medium enriched with glucose, *V. diabolicus* strain HE800 produced an EPS characterized by equal amounts of uronic acid and hexosamines (N-acetyl glucosamine and N-acetyl galactosamine) in a molar ratio of approximately 1:0.5:0.5, respectively. Structural studies have recently been conducted on this polymer demonstrating that it consists of a linear tetrasaccharide repeating unit [76]. This novel bacterial polysaccharide has been recently investigated as a bone regeneration and cicatrizing material and a patent has been obtained (patent US 7015206B2). As a result, the EPS excreted by *V. diabolicus* make it possible to fill critical size bone defects in rat calvaria. It can form an attractive extracellular matrix for the direct adhesion of osteoblasts, osteoprogenitor cells and pericytes and may protect the growth and hormonal factors associated with the healing process. The high binding capacity of calcium by this polysaccharide can also be another important parameter in its efficiency to induce fast bone healing. Moreover, it was established that this EPS did not cause any inflammatory reaction and it is completely absorbed by the surrounding tissues. Therefore, this EPS constitutes a material which potentiates bone repair and its particular activity is probably due to its original physicochemical characteristics [76].

Alteromonas infernus, strain GY785 was isolated from a sample of fluid collected among a dense population of *Riftia pachyptila* in the proximity of an active hydrothermal vent of the Southern depression of the Guaymas basin (Gulf of California). This strain, in a glucose enriched medium, secreted a water-soluble acidic heteropolysaccharide consisting of glucose, galactose, glucuronic and galacturonic acids (1:1:0.7:0.4), respectively. This high-molecular-weight polysaccharide (1×10^6 Da) differed in monosaccharide content and/or ratio and sulfate content (10%) from other EPS isolated from deep-sea hydrothermal bacteria. Different fractions of the purified EPS have been chemically modified by sulfation and acidic depolymerization leading to an elevated increase of the attached sulfated groups (33–40%) producing low-molecular-weight, low-viscous EPSs in yields ranging from 25–50% [43]. In addition, the overall monosaccharide composition of all the fractions is not chemical modified by those reactions and the high sulfate content up to 40% of the low-molecular-weight fractions along with the nature of the sugar units should give rise to original bioactive compounds. However, regioselectivity of the reaction is an important issue as the biological activity can be related to the position of the sulfate groups in the polymers as it is reported for other highly sulfated polysaccharides, such as chondroitin sulfate, mannans, dermatan sulfates, fucoidan [77,78] k-carrageenan and heparin [79], which have been shown to possess important physiological functions and they have been applied as therapeutic anticoagulant and antithrombotic agents. Moreover, the original backbone structure of *A. infernus* EPS is made up primarily of glucose, galactose and uronic acid residues bound by→13 or/and 1→4 linkages, in the manner of most other anticoagulant 1 polysaccharides. Original sulfation patterns were obtained due to this unique backbone structure consisting of many different monosaccharide building units, glycosidic linkage types and unit branches. Consequently, this kind of marine EPS can be used as a library of compounds useful for the elucidation of the structure-activity relationship and would represent a valid alternative to mammalian sources of anticoagulant avoiding the risk of contamination with pathogenic agents potentially present in mammalian tissues [43].

Mar. Drugs **2008**, *8*, 1779–1802

Table 4. EPSs produced by marine microorganisms involved in symbiotic relationships.

Microorganisms	Source environment	Description of EPS and Chemical composition	Suggested ecological role and Potential Biotechnological Application	References
Alteromonas macleodii subsp. *fijiensis* biovar deepsane strain HYD657	Isolated from epidermis of a polychaete annelid, *Alvinella pompejana*, hydrothermal vent of the East Pacific Rise	The repeating unit is an undesaccharide with three side-chains. Gal:Glc:Rha:Fuc:Man:GlcA:GalA:3-0-(1 carboxyethyl)-D-GlcA (1:0.42:0.85:0.5:0.42:0.5:0.5:0.5)	Cosmetics (patent PCT 94907582-4)	[71]
Alteromonas strain 1644	Isolated from *Alvinellidae* collected near hydrothermal vent of the East Pacific Rise	Main chain of five sugars with a side chain of three sugars including a dicarboxylic acid. Glc:Gal:GlcA:3Lac-GlcA:GalA	Heavy metal binding	[73,74]
Vibrio diabolicus strain HE800	Isolated from a Pompei worm tube collected from a deep-sea hydrothermal field of the East Pacific Rise	A linear tetrasaccharide repeating unit. Uronic acid :GlcN:GalN 1:0.5:0.5 →3)-β-D-Glc$p$NAc-(1→4)-β-D-Glc$p$A-(1→4)-β-D-Glc$p$A-(1→4)-α-D-Gal$p$NAc-(1→	Bone regeneration and cicatrizing material (patent US 7015206B2)	[75,76,80,81]
Alteromonas infernus strain 785	Isolated from a fluid sample collected among a dense population of *Riftia pachyptila* in the proximity of an active hydrothermal vent, Guaymas basin (Gulf of California)	Glc:Gal:GlcA:GalA (1:1:0.7:0.4)	Anticoagulant activity	[43]

Mar. Drugs **2008**, *8*, 1779–1802

5. Conclusion

Microbial EPSs are ubiquitous in the extreme marine environment where they are essential for microbial survival. Most of the functions ascribed to EPS are of a protective nature and their precise roles are dependent on the ecological niches in which the microorganisms live. They could assist the microbial communities to endure extremes of temperature, salinity and nutrient availability, creating a boundary between the bacterial cell and its immediate environment. Several EPSs produced by microorganisms from extreme habitats show biotechnological promise. By examining their structure and chemical-physical characteristics, it is possible to gain insight into their commercial application and they are employed in several industries ranging from pharmaceutical to food-processing fields, through to the detoxification capability of polluted areas from petrochemical oils.

Considering that the microbial biodiversity of marine eco-systems is relatively unexplored, it is reasonable to hypothesize that the isolation and identification of new microorganisms will provide wide opportunities for new industrial fields.

Acknowledgments: This work was performed with the contribution of the Ministry of Foreign Affairs (MAE-Italy).

References

1. Sutherland, IW. Microbial polysaccharides from Gram-negative bacteria. *Int. Dairy J* **2001**, *11*, 663–674.
2. Sutherland, IW. Bacterial exopolysaccharides. *Adv. Microb. Phys* **1972**, *8*, 143–213.
3. MacCormick, CA; Harris, JE; Jay, AJ; Ridout, MJ; Colquhoun, IJ; Morris, VJ. Isolation and characterization of new extracellularpolysaccharide from an *Acetobacter* species. *J. Appl. Bacteriol* **1996**, *81*, 419–426.
4. Sutherland, IW. Biosynthesis of microbial exopolysaccharides. *Adv. Microb. Phys* **1982**, *23*, 79–150.
5. Decho, AW. Barnes, M, Ed.; Microbial exopolymer secretions in ocean environments: their role(s) in food webs and marine processes. In *Oceanography and Marine Biology: an Annual Review*; Aberdeen Univ Press: Aberdeen, UK, 1990; pp. 73–153.
6. Wingender, J; Neu, TR; Flemming, H-C. Wingender, J, Neu, TR, Flemming, H-C, Eds.; What are bacterial extracellular polymer substances? In *Microbial Extracellular Polymer Substance*; Springer: Berlin, Germany, 1999; pp. 1–19.
7. McCarthy, M; Hedges, J; Benner, R. Major biochemical composition of dissolved high molecular weight organic matter in sea water. *Mar.Chem* **1996**, *55*, 281–297.
8. Wolfaardt, GM; Lawrence, JR; Korbe, DR. Wingender, J, Neu, TR, Flemming, H-C, Eds.; Function of EPS. In *Microbial Extracellular Polymeric Substances: Characterization, Structure and Function*; Springer-Verlag: New York, NY, USA, 1999; pp. 171–200.
9. Alldredge, A. Interstitial dissolved organic carbon (DOC) concentrations within sinking marine aggregates and their potential contribution to carbon flux. *Limnol. Oceanogr* **2000**, *45*, 1245–1253.
10. Holmstrom, C; Kjelleberg, S. Marine *Pseudoalteromonas* species are associated with higher organisms and produce biologically active extracellular agents. *FEMS Microbiol. Ecol* **1999**, *30*, 285–293.
11. Decho, AW; Herndl, GJ. Microbial activities and the transformation of organic matter within mucilaginous material. *Sci. Total Environ* **1995**, *165*, 33–42.
12. Mancuso Nichols, CA; Guezennec, J; Bowman, JP. Bacterial exopolysaccharides from extreme marine environments with special consideration of the southern ocean, sea ice, and deep-sea Hydrothermal Vents: a review. *Mar. Biotechnol* **2005a**, *7*, 253–271.
13. Junge, K; Eicken, H; Deming, JW. Bacterial activity at −2 to −20 °C in Arctic Wintertime sea ice. *Appl. Environ. Microbiol* **2004**, *70*, 550–557.
14. Krembs, C; Eicken, H; Junge, K; Deming, JW. High concentrations of exopolymeric substances in Arctic winter sea ice: implication for the polar ocean carbon cycle and cryoprotection of diatoms. *Deep-Sea Res. Part I* **2002**, *49*, 2163–2181.
15. Logan, BE; Hunt, JR. Advantages to microbes of growth in permeable aggregates in marine systems. *Limnol. Oceanogr* **1987**, *32*, 1034–1048.
16. Loaec, M; Olier, R; Guezennec, J. Uptake of lead, cadmium and zinc by a novel bacterial exopolysaccharide. *Water Res* **1997**, *31*, 1171–1179.

17. Loaec, M; Olier, R; Guezennec, J. Chelating properties of bacterial exopolysaccharides from deep-sea hydrothermal vents. *Carbohydr. Polym* **1998**, *35*, 65–70.

18. Wuertz, S; Muller, E; Spaeth, R; Pfleiderer, P; Flemming, H-C. Detection of heavy metals in bacterial biofilms and microbial floes with the fluorescent complexing agent Newport Green. *J. Ind. Microbiol. Biotechnol* **2000**, *24*, 116–123.

19. Mancuso Nichols, C; Garon lardière, S; Bowman, JP; Nichols, PD; Gibson, JAE; Guézennec, J. Chemical characterization of exopolysaccharides from Antarctic marine bacteria. *Microb. Ecol* **2005b**, *49*, 578–589.

20. Kenne, L; Lindberg, B. Aspinall, GO, Ed.; Bacterial polysaccharides. In *The Polysaccharides*; Academic Press: New York, NY, USA, 1983; Volume 2, pp. 287–363.

21. Sutherland, IW. Sanford, PA, Laskin, A, Eds.; Microbial exopolysaccharide synthesis. In *Extracellular Microbial Polysaccharides*; American Chemical Society: Washington, DC, USA, 1977; Volume 45, pp. 40–57.

22. Sutherland, IW. Structure-function relationship in microbial exopolysaccharides. *Biotechnol. Adv* **1994**, *12*, 393–448.

23. Vanhooren, PT; Vandamme, EJ. Biosynthesis, physiological role, use and fermentation process characteristics of bacterial exopolysaccharides. *Rec. Res. Devel. Ferment. Bioeng* **1998**, *1*, 253–299.

24. Samain, E; Milas, M; Bozzi, L; Dubreucq, M; Rinaudo, M. Simultaneous production oft wo different gel-forming exopolysaccharides by an *Alteromonas* strain originating from deep-sea hydrothermal vents. *Carbohydr. Polym* **1997**, *34*, 235–241.

25. Christensen, BE; Kjosbakken, J; Smithrod, O. Partial chemical and physical characterization of two extracellular polysaccharides produced by a marine, periphytic *Pseudomonas* sp. strain NCMB 2021. *Appl. Environ. Microbiol* **1985**, *50*, 837–845.

26. Kumar, AS; Mody, K; Jha, B. Bacterial exopolysaccharides - a perception. *J. Bas. Microbiol* **2007**, *47*, 103–107.

27. Ko, S-H; Lee, H-S; Park, S-H; Lee, H-K. Optimal conditions for the production of exopolysaccharide by marine microorganism *Hahella chejuensis*. *Biotechnol. Bioproc. Eng* **2000**, *5*, 181–185.

28. Lee, HK; Chun, J; Moon, EJ; Ko, SH; Lee, DS; Lee, HS; Bae, KS. *Hahella chejuensis* gen. nov., sp. nov., an extracellular-polysaccharide-producing marine bacterium. *IJSEM* **2001**, *51*, 661–666.

29. Poli, A; Schiano Moriello, V; Esposito, E; Lama, L; Gambacorta, A; Nicolaus, B. Exopolysaccharide production by a new *Halomonas* strain CRSS isolated from saline lake Cape Russell in Antarctica growing on complex and defined media. *Biotechnol. Lett* **2004**, *26*, 1635–1638.

30. Poli, A; Esposito, E; Orlando, P; Lama, L; Giordano, A; de Appolonia, F; Nicolaus, B; Gambacorta, A. *Halomonas alkaliantarctica* sp. nov., isolated from saline lake Cape Russell in Antarctica, an alkalophilic moderately halophilic, exopolysaccharide-producing bacterium. *Syst. Appl. Microbiol* **2007**, *30*, 31–38.

31. Quesada, E; Béjar, V; Del Moral, A; Ferrer, MR; Calvo, C; Llamas, I; Martinez-Checa, F; Arias, S; Ruiz-Garcìa, C; Martinez-Cánovas, J; Páez, R. Ventosa, A, Ed.; Moderately halophilic exopolysaccharide-producing bacteria. In *Halophilic Microorganisms*; Springer-Verlag: Heidelberg, Germany, 2004; pp. 297–314.

32. Farres, J; Caminal, G; Lopez-Santin, J. Influence of phosphate on rhamnose-containing exopolysaccharide rheology and production by *Klebsiella* I-174. *Appl. Microbiol. Biotechnol* **1997**, *48*, 522–527.

33. Gorret, N; Maubois, JL; Engasser, JM; Ghoul, M. Study of the effects of temperature, pH and yeast extract on growth and exopolysaccharide production by *Propionibacterium acidi-propionici* on milk microfiltrate using a response surface methodology. *J. Appl. Microbiol* **2001**, *90*, 788–796.

34. Schiano Moriello, V; Lama, L; Poli, A; Gugliandolo, C; Maugeri, TL; Gambacorta, A; Nicolaus, B. Production of exopolysaccharides from a thermophilic microorganism isolated from a marine hot spring in flegrean areas. *J. Ind. Microbiol. Biotechnol* **2003**, *30*, 95–101.

35. Maugeri, TL; Gugliandolo, C; Caccamo, D; Panico, A; Lama, L; Gambacorta, A; Nicolaus, BA. Halophilic thermotolerant *Bacillus* isolated from a marine hot spring able to produce a new exopolysaccharide. *Biotechnol. Lett* **2002**, *24*, 515–519.

36. Bozal, N; Tudela, E; Rossello-Mora, R; Lalucat, J; Guinea, J. *Pseudoalteromonas antartica* sp. nov., isolated from an Antarctica coastal environment. *Int. J. Syst. Bacteriol* **1997**, *47*, 345–351.

37. Raguénès, G; Pignet, P; Gauthier, G; Peres, A; Christen, R; Rougeaux, H; Barbier, G; Guezennec, J. Description of a new polymer-secreting bacterium from a deep-sea hydrothermal vent, *Alteromonas macleodii* subsp. *fijiensis*, and preliminary characterization of the polymer. *Appl. Environ. Microbiol* **1996**, *62*(1), 67–73.

38. Morin, A. Dumitriu, S, Ed.; Screening of polysaccharide-producing microorganisms, factors influencing the production and recovery of microbial polysaccharides. In *Polysaccharides-Structural Diversity and Functional Versatility*; Marcel Dekker Inc. Publication: New York, NY, USA, 1998; pp. 275–296.

39. Rougeaux, H; Guezennec, J; Carlson, RW; Kervarec, N; Pichon, R; Talaga, P. Structural determination of the exopolysaccharide of *Pseudoalteromonas* strain HYD 721 isolated from a deep-sea hydrothermal vent. *Carbohydr. Res* **1999a**, *315*, 273–285.

40. Guezennec, JG. Deep-sea hydrothermal vents: a new source of innovative bacterial exopolysaccharides of biotechnological interest? *J. Ind. Microbiol. Biotechnol* **2002**, *29*, 204–208.

41. Rougeaux, H; Talaga, P; Carlson, RW; Guezennec, J. Structural studies of an exopolysaccharide produced by *Alteromonas macleodii* subsp. *fijiensis* originating from a deep-sea hydrothermal vent. *Carbohydr. Res* **1998**, *312*, 53–59.

42. Zanchetta, P; Guezennec, J. Surface thermodynamics of osteoblasts: relation between hydrophobicity and bone active biomaterials. *Colloids Surf* **2001**, *22*, 301–307.

43. Colliec Jouault, S; Chevolot, L; Helley, D; Ratiskol, J; Bros, A; Sinquin, C; Roger, O; Fischer, A-M. Characterization, chemical modifications and *in vitro* anticoagulant properties of an exopolysaccharide produced by *Alteromonas infernus*. *BBA* **2001**, *1528*, 141–151.

44. Manca, MC; Lama, L; Improta, R; Esposito, E; Gambacorta, A; Nicolaus, B. Chemical composition of two exopolysaccharides from *Bacillus thermoantarcticus*. *Appl. Environ. Microbiol* **1996**, *62*, 3265–3269.

45. McDowell, EM; Trump, BF. Histologic fixative suitable for diagnostic light and electron microscopy. *Arch. Pathol. Lab. Med* **1976**, *100*, 405–414.

46. Ito, M; Baba, M; Hirabayashi, K; Matsumoto, T; Suzuki, M; Suzuki, S; Shigeta, S; de Clercq, E. *In vitro* activity of mannan sulfate, a novel sulfated polysaccharide, against human immunodeficiency virus type I and other enveloped viruses. *Eur. J. Clin. Microbiol. Infec. Dis* **1989**, *8*, 171–173.

47. Rinker, KD; Kelly, RM. Growth Physiology of the Hyperthermophilic Archaeon *Thermococcus litoralis*: Development of a Sulfur-Free Defined Medium, Characterization of an Exopolysaccharide, and Evidence of Biofilm Formation. *Appl. Environ. Microbiol* **1996**, *12*, 4478–4485.

48. Nicolaus, B; Lama, L; Panico, A; Schiano Moriello, V; Romano, I; Gambacorta, A. Production and characterization of exopolysaccharides excreted by thermophilic bacteria from shallow, marine hydrothermal vents of flegrean areas (Italy). *Syst. Appl. Microbiol* **2002**, *25*, 319–325.

49. Gugliandolo, C; Maugeri, TL. Temporal variations of culturable mesophilic heterotrophic bacteria from a marine shallow hydrothermal vent of the island of Vulcano (Eolian Islands, Italy). *Microb. Ecol* **1998**, *36*, 13–22.

50. Guezennec, JG; Pignet, P; Raguenes, G; Deslandes, E; Lijour, Y; Gentric, E. Preliminary chemical characterization of unusual eubacterial exopolysaccharides of deep-sea origin. *Carbohydr. Polym* **1994**, *24*, 287–294.

51. Baba, M; Snoeck, R; Pauwels, R; de Clercq, E. Sulfated polysaccharides are potent and selective inhibitors of various enveloped viruses, including herpes simplex virus, cytomegalovirus, vesicular stomatitis virus, and human immunodeficiency virus. *Antimicrob. Agents Chemother* **1988**, *32*, 1742–1745.

52. Hayashi, TK; Hayashi, M; Kojima, I. Calcium spirulan, an inhibitor of enveloped virus replication, from a blue-green alga *Spirulina platensis*. *J. Nat. Prod* **1996**, *39*, 83–87.

53. Arena, A; Gugliandolo, C; Stassi, G; Pavone, B; Iannello, D; Bisignano, G; Maugeri, TL. An exopolysaccharide produced by *Geobacillus thermodenitrificans* strain B3-72: Antiviral activity on immunocompetent cells. *Immunol. Lett* **2009**, *123*, 132–137.

54. Arena, A; Maugeri, TL; Pavone, B; Iannello, D; Gugliandolo, C; Bisignano, G. Antiviral and immunoregulatory effect of a novel exopolysaccharide from a marine thermotolerant *Bacillus licheniformis*. *Int. Immunopharmacol* **2006**, *6*, 8–13.

55. Rademacher, TW; Parekh, RB; Dwek, RA. Glycobiology. *Ann. Rev. Biochem* **1988**, *57*, 785–838.

56. Qin, K; Zhu, L; Chen, L; Wang, PG; Zhang, Y. Structural characterization and ecological roles of a novel exopolysaccharide from the deep-sea psychrotolerant bacterium *Pseudoalteromonas* sp. SM9913. *Microbiology* **2007**, *153*, 1566–1572.

57. Mancuso Nichols, CA; Bowman, JP; Guezennec, J. Effects of incubation temperature on growth and production of exopolysaccharides by an Antarctic sea ice bacterium grown in batch culture. *Appl. Environ. Microbiol* **2005c**, *71*, 3519–3523.

58. Li, WW; Zhou, WZ; Zhang, YZ; Wang, J; Zhu, XB. Flocculation behavior and mechanism of an exopolysaccharide from the deep-sea psychrophilic bacterium *Pseudoalteromonas* sp. SM9913. *Bioresour. Technol* **2008**, *99*(15), 6893–6899.

59. Flemming, HC; Wingender, J. Relevance of microbial extracellular polymeric substances (EPSs)—Part I: structural and ecological aspects. *Water Sci. Technol* **2001**, *43*, 1–8.

60. Delille, D; Rosier, C. Seasonal changes of Antarctic marine bacterioplankton and sea ice bacterial assemblages. *Pol. Biol* **1996**, *16*, 27–34.

61. Mancuso Nichols, CA; Garon, S; Bowman, JP; Raguénès, G; Guézennec, J. Production of exopolysaccharides by Antarctic marine bacterial isolates. *J. Appl. Microbiol* **2004**, *96*(5), 1057–1066.

62. Brown, MV; Lester, JN. Role of bacterial extracellular polymers in metal uptake in pure bacterial culture and activated sludge. *Water Res* **1982**, *16*, 1539–1548.

63. Marx, JG; Carpenter, SD; Deming, JW. Production of cryoprotectant extracellular polysaccharide substances (EPS) by the marine psychrophilic bacterium *Colwellia psychrerythraea*strain 34H under extreme conditions. *Can. J. Microbiol* **2009**, *55*(1), 63–72.

64. Martinez-Checa, F; Toledo, FL; Vilchez, R; Quesada, E; Calvo, C. Yield production, chemical composition, and functional properties of emulsifier H28 synthesized by *Halomonas eurihalina* strain H-28 in media containing various hydrocarbons. *Appl. Microbiol. Biotechnol* **2002**, *58*, 358–363.

65. Pepi, M; Cesaro, A; Liut, G; Baldi, F. An antarctic psychrotrophic bacterium *Halomonas* sp. ANT-3b, growing on n-hexadecane, produces a new emulsifying glycolipid. *FEMS Microbiol. Ecol* **2005**, *53*, 157–166.

66. Rosenberg, E. Microbial diversity as a source of useful biopolymers. *J. Ind. Microbiol* **1993**, *11*, 131–137.

67. Weiner, RM. Biopolymers from marine prokaryotes. *Mar. Biotechnol* **1997**, *15*, 390–394.

68. Anton, J; Meseguer, I; Rodriguez-Valera, F. Production of an extracellular polysaccharide by *Haloferax mediterranei*. *Appl. Environ. Microbiol* **1988**, *10*, 2381–2386.

69. Parolis, H; Parolis, LAS; Boán, IF; Rodríguez-Valera, F; Widmalm, G; Manca, Mc; Jansson, P-E; Sutherland, IW. The structure of the exopolysaccharide produced by the halophilic Archaeon *Haloferax mediterranei* strain R4 (ATCC 33500). *Carbohydr. Res* **1996**, *295*, 147–156.

70. Yim, JH; Kim, SJ; Aan, SH; Lee, HK. Physicochemical and rheological properties of a novel emulsifier, EPS-R, produced by a marine bacterium *Hahella chejuensis*. *Biotechnol. Bioproc. Eng* **2004**, *9*, 405–413.

71. Cambon-Bonavita, M-A; Raguénès, G; Jean, J; Vincent, P; Guezennec, J. A novel polymer produced by a bacterium isolated from a deep-sea hydrothermal vent polychaete annelid. *J. Appl. Microbiol* **2002**, *93*, 310–315.

72. Dubreucq, G; Domon, B; Fournet, B. Structure determination of a novel uronic acid residue isolated from the exopolysaccharide produced by a bacterium originating from deep sea hydrothermal vents. *Carbohydr. Res* **1996**, *290*, 175–181.

73. Bozzi, L; Milas, M; Rinaudo, M. Characterization and solution properties of a new exopolysaccharide excreted by the bacterium *Alteromonas* sp. strain 1644. *Int. J. Biol. Macromol* **1996a**, *18*, 9–17.

74. Bozzi, L; Milas, M; Rinaudo, M. Solution and gel rheology of a new exopolysaccharide excreted by the bacterium *Alteromonas* sp. strain 1644. *Int. J. Biol. Macromol* **1996b**, *18*, 83–91.

75. Raguénès, G; Christen, R; Guerzennec, J; Pignet, P; Barbier, G. *Vibrio diabolicus* sp. nov., a new polysaccharide-secreting organism isolated from a deep-sea hydrothermal vent polychaete annelid, *Alvinella pompejana*. *Int. J. Syst. Bacteriol* **1997**, *47*, 989–995.

76. Rougeaux, H; Kervarec, N; Pichon, R; Guerzennec, J. Structure of the exopolysaccharide of *Vibrio diabolicus* isolated from a deep–sea hydrothermal vent. *Carbohydr. Res* **1999b**, *322*, 40–45.

77. Colliec Jouault, S; Boisson-Vidal, C; Jozefonvicz, J. A low molecular weight fucoidan fraction from the brown seaweed *Pelvetia canaliculata*. *J. Phytochem* **1994**, *35*, 697–700.

78. Nardella, A; Chaubet, F; Boisson-Vidal, C; Blondin, C; Durand, P; Jozefonvicz, J. Anticoagulant low molecular weight fucans produced by radical process and ion exchange chromatography of high molecular weight fucans extracted from the brown seaweed *Ascophyllum nodosum*. *Carbohydr. Res* **1996**, *289*, 201–208.

79. Casu, B; Grazioli, G; Razi, N; Guerrini, M; Naggi, A; Torri, G; Oreste, P; Tursi, F; Zoppetti, G; Lindahl, U. Heparin-like compounds prepared by chemical modification of capsular polysaccharide from *E.coli* K5. *Carbohydr. Res* **1994**, *263*, 356–365.

80. Zanchetta, P; Lagarde, N; Guezennec, J. A new bone-healing material: a hyaluronic acid-like bacterial exopolysaccharide. *Calcif. Tissue Int* **2003a**, *72*, 74–79.

81.   Zanchetta, P; Lagarde, N; Guezennec, J. Systematic effects on bone healing of a new hyaluronic acid-like bacterial exopolysaccharide. *Calcif. Tissue Int* **2003b**, *73*, 232–236.

Samples Availability: Available from the authors.

marine drugs

MDPI

Review

Recent Advances in Exopolysaccharides from *Paenibacillus* spp.: Production, Isolation, Structure, and Bioactivities

Tzu-Wen Liang and San-Lang Wang [†,*]

Department of Chemistry/Life Science Development Centre, Tamkang University, New Taipei City 25137, Taiwan; ltw27@ms55.hinet.net

* Author to whom correspondence should be addressed; sabulo@mail.tku.edu.tw; Tel.: +886-2-2621-5656; Fax: +886-2-2620-9924.

† Present address: No. 151, Yingchuan Rd., Tamsui, New Taipei City 25137, Taiwan.

Academic Editor: Paola Laurienzo

Received: 24 February 2015; Accepted: 25 March 2015; Published: 1 April 2015

Abstract: This review provides a comprehensive summary of the most recent developments of various aspects (*i.e.*, production, purification, structure, and bioactivity) of the exopolysaccharides (EPSs) from *Paenibacillus* spp. For the production, in particular, squid pen waste was first utilized successfully to produce a high yield of inexpensive EPSs from *Paenibacillus* sp. TKU023 and *P. macerans* TKU029. In addition, this technology for EPS production is prevailing because it is more environmentally friendly. The *Paenibacillus* spp. EPSs reported from various references constitute a structurally diverse class of biological macromolecules with different applications in the broad fields of pharmacy, cosmetics and bioremediation. The EPS produced by *P. macerans* TKU029 can increase *in vivo* skin hydration and may be a new source of natural moisturizers with potential value in cosmetics. However, the relationships between the structures and activities of these EPSs in many studies are not well established. The contents and data in this review will serve as useful references for further investigation, production, structure and application of *Paenibacillus* spp. EPSs in various fields.

Keywords: *Paenibacillus* spp.; exopolysaccharides; squid pen; production; purification; bioactivity

1. Introduction

Natural polysaccharides are highly susceptible to natural biodegradation and are less harmful than synthetic polymers. Over the past few decades, the number of known polysaccharides produced by microbial fermentation has been gradually increasing. Most of the polysaccharides derived from microorganisms are of the exo-polysaccharide (EPS) type; these EPSs attach to the cell surfaces or are released into the extracellular medium [1,2]. The microbial exo-polysaccharides can be divided into homo-polysaccharides, which are composed of a single monosaccharide unit, and hetero-polysaccharides, in which regular repeat units are formed from two to eight monosaccharides [3]. The repeating units of these EPSs are very regular, branched or unbranched, and interconnected by glycosidic linkages. Microbial EPSs are water-soluble polymers and may be ionic or non-ionic in nature. In addition, they are usually biodegradable, biocompatible, edible and nontoxic toward humans and the environment. Due to their many interesting physico-chemical and rheological properties in that they disperse in water to give a thickening or a viscosity building effect [4]. The shear thickening effect of EPSs is one in which viscosity increases with the rate of shear strain. Microbial EPSs are generally of uniform structure and fairly limited polydispersity, which depend on their molecular weights and structure. For example, the microbial levan similar to bacterial dextran is a polymer of fructose linked by β-(2→6) fructofuranosidic bond. It is known that levans with different molecular weights are needed for different purposes. The low molecular-weight, less branched levan usually provides a low viscosity, and can be used as a tablet binder in immediate-release dosage forms, while levans of medium- and high-viscosity grade are used in controlled-release matrix formulations [5]. Besides, curdlan, a linear glucan interconnected by β-(1→3) glucosidic linkages, is produced by the *Agrobacterium* species or *Alcaligenes faecalis* in initial studies [6]. This biopolymer is a thickener with unique thermal gelling properties; hence it is a good gelling material for the improvement of the textural quality, water-holding capacity and thermal stability of various commercialized products in food industry [6]. Microbial EPSs have found many applications in the food, pharmaceutical and other industries, such as the production of textiles, detergents, adhesives, cosmetics, microbial enhanced oil recovery, controlled drug delivery and wastewater treatment [6,7] because they showed properties of various physiological activities, stabilization, suspension of particulates, control of crystallization, inhibition of synaeresis, encapsulation and formation of film [8–10].

In recent years, significant progress has been made in the search for novel microbial EPSs that possess novel and highly functional properties [5]. The different biopolymers that have been extensively studied and are currently being marketed as commercial products include hyaluronic acid from *Streptococcus equii* and *Streptococcus zooepidemicus* [11]; levan from *Bacillus subtilis* and *Bacillus polymyxa* [12,13]; pullulan from *Aureobasidium pullulans* [14]; dextran from *Leuconostoc mesenteroides* [15]; xanthan from *Xanthomonas campestris* [16]; gellan from *Sphingomonas paucimobilis* [17]; alginate from *Pseudomonas* species and *Azotobacter chrococcum* [18]; cellulose from *Acetobacter xylinium* [19]; curdlan from *Agrobacterium* and *Alcaligenes* species [20]; and succinoglycan from *Rhizobium* [1]. However, the novel EPSs from *Paenibacillus* spp. were only discovered in the past ten years.

The genus *Paenibacillus* consists of over 89 species of facultative anaerobes, endospore-forming, neutrophilic, periflagellated heterotrophic and low G + C Gram-positive bacilli, which were originally included within the genus *Bacillus* and then reclassified as a separate genus in 1993 [21]. Bacteria belonging to this genus have been detected in a variety of environments, such as soil, water, rhizospheres, vegetable matter, forage and insect larvae, as well as clinical samples [7]. *Paenibacillus* spp. produced a wide variety of different EPSs with diverse physiological and biotechnological functions. *Paenibacillus* spp. EPSs have also attracted great interest because of their biotechnological potential in different industrial processes and wastewater treatment [6,7,22–32]. Thorough reviews on the production and applications of *Paenibacillus* spp., EPSs have shown that there is a growing interest in using *Paenibacillus* spp. EPSs as biomaterials; extensive research has been performed, leading to a large number of publications in recent years [6,7,22–32]. The present review is devoted to a survey of the main achievements in the production, isolation and structural composition of *Paenibacillus* spp. EPSs. Furthermore, the applications of *Paenibacillus* spp. EPSs in various fields are extensively reviewed.

2. Production of EPSs from *Paenibacillus* spp.

As mentioned earlier, *Paenibacillus* spp. were discovered to produce various EPS. Some of the production conditions have been investigated to optimize the production for commercial usage. The production of EPS is not species-specific and each strain of the species produces different types of EPSs with different biotechnological properties. In our reviews, very recently, researchers observed that the culture of *Paenibacillus* spp. could produce two different, commercial, well-known EPSs. One of these polymers is a levan-type EPS from *P. polymyxa* EJS-3 [25], and the second is curdlan gum from *P. polymyxa* ATCC 21830 [6]. Levan is a polymer of fructose linked by a β-(2→6) fructofuranosidic bond and is present in many plants and microbial products [33]. The microbial levans are produced from sucrose-based substrate by a transfructosylation reaction of levansucrase (beta-2,6-fructan:D-glucose-fructosyl transferase, EC 2.4.1.10) by microorganisms. Liu *et al.* [23–25,29] reported that the shake flask cultures of the endophytic bacterium *P. polymyxa* EJS-3 grown in a culture medium (sucrose 188.2 g/L, yeast extract 25.8 g/L, K_2HPO_4 5 g/L, $CaCl_2$ 0.34 g/L, and initial pH 8.0) typically yielded 35.26 g/L of a levan-type EPS after cultivation for 60 h at 24 °C. Curdlan is a neutral, bacterial extracellular polysaccharide without branched chains, composed entirely of β-(1→3)-d-glycosidic linkages. Curdlan is a thickener with unique thermal gelling properties and is widely used in the food industry. Curdlan is biodegradable, nontoxic toward humans and the environment, edible and it has growing capacity in the pharmaceutical industry because of its potent biological activities [6]. Modified curdlan has pharmaceutical applications, such as controlled drug delivery [34].

To improve the efficiency and productivity of *Paenibacillus* spp. EPSs fermentation processes, many investigators have studied the effects of various process factors on the maximal production of EPSs to optimize the fermentation conditions, such as medium composition, temperature, pH, and culture vessel [6,7,22–32]. Table 1 provides a summary of the strains, culture conditions and EPS yields of *Paenibacillus* spp. that have been reported in the literature. The EPS yields varied over a wide range from 3.44 to 41.25 g/L with the bacterial species and culture conditions. In particular, our group previously demonstrated that EPSs were induced from a squid pen powder (SPP)-containing medium by *Paenibacillus* sp. TKU023 [31] and *P. macerans* TKU029 [32]. The production of inexpensive EPS is an important factor in the utilization of fishery waste products. The discovery of an inexpensive EPS not only solves environmental problems but promotes the economic value of marine waste. Until now, there have only been some reports on the culture conditions for the production of EPSs from *Paenibacillus* spp. Thus, there is a need to enhance the EPS productivity from *Paenibacillus* spp. through effective strategies of process intensification in the future.

2.1. Influence of Carbon/Nitrogen (C/N) Sources, pH and Temperature

Studies of EPS have indicated that the medium composition plays a critical role in EPS production [23,35]. The most commonly used carbon sources for EPS production are sugars, namely glucose and sucrose [6,23,27–30,36]. As shown in Table 1, sucrose is often employed as the most suitable carbon source for EPS production. It has been reported recently that the use of sucrose in the medium results in high yields of levan-type EPSs [23,29]. This is consistent with the findings of many other earlier investigators [33,37]. It has been reported that levansucrase with strong sucrose hydrolysing activity is involved in many *P. polymyxa* strains, which may be responsible for the high yield of EPS with sucrose as a carbon source [33,35,37]. However, the high cost of these carbon sources has a direct impact on production costs, which limits the market potential of these biopolymers. To decrease the production costs, it is important to look for less expensive carbon sources, such as wastes or industrial by-products [9]. Fishery by-products contain a large amount of chitin, which is a long-chain polysaccharide of a *N*-acetylglucosamine. Therefore, to lower the production cost of EPSs and efficiently reutilize the chitin-containing fishery by-products, we screened and isolated EPS-producing bacteria using squid pen powder (SPP) as the sole C/N source. We first produced and characterized the EPSs from the SPP culture supernatant of two bacteria, *Paenibacillus* sp. TKU023 [31] and *P. macerans* TKU029 [32]. *Paenibacillus* sp. TKU023 and *P. macerans* TKU029 have a chitin hydrolysing mechanism because there is EPS production by TKU023 and TKU029 with SPP as the C/N source without sucrose (Table 1). For media containing 1.5% (w/v) and 2% (w/v) SPP, the EPS yields from *Paenibacillus* sp. TKU023 and *P. macerans* TKU029 were 4.55 g/L and 3.46 g/L, respectively. However, in other literature, using high concentrations of sucrose in the medium resulted in EPS yields increasing up to 25.63 g/L [27] and 35.26 g/L [23,29]. This inspired us to increase the concentration of SPP to obtain higher yields of EPSs. The results demonstrated that when media contained 10% (w/v) SPP, the EPS yields from *Paenibacillus* sp. TKU023 and *P. macerans* TKU029 were 41.25 g/L and 35.75 g/L, respectively. Additionally, during fermentation, due to liquefaction of protein and chitin, bioactive material rich liquor is formed, including enzymes, biosurfactants, peptides, and chitooligosaccharides [31,32,38]. Furthermore, the fermented SPP in the culture broth could be recovered for biological applications in dye removal [39]. These advantages of using SPP as the sole C/N source to produce EPS are different from using sucrose.

Table 1. EPS production by fermentation of *Paenibacillus* spp.

Bacteria Source	Fermentation Conditions					EPS Yield (g/L)	References
	Medium Composition	Temperature (°C)	pH	Culture Vessel	Period (Days)		
Paenibacillus sp. TKU023	1.5% SPP, 0.1% K_2HPO_4, and 0.05% $MgSO_4 \cdot 7H_2O$	37	7.23	50 mL in a 250 mL flask at 150 rpm	5	4.55	[31]
	10% SPP, 0.1% K_2HPO_4, and 0.05% $MgSO_4 \cdot 7H_2O$	37	7.23	50 mL in a 250 mL flask at 150 rpm	5	41.25	
P. macerans TKU029	2% SPP, 0.1% K_2HPO_4, and 0.05% $MgSO_4 \cdot 7H_2O$	30	7.21	100 mL in a 250 mL flask at 150 rpm	4	3.46	[32]
	10% SPP, 0.1% K_2HPO_4, and 0.05% $MgSO_4 \cdot 7H_2O$	30	7.21	100 mL in a 250 mL flask at 150 rpm	4	35.75	
P. polymyxa SQR-21	Galactose 48.5 g/L, Fe^{3+} 242 μM and Ca^{2+} 441 μM	30	6.5	250 mL in a 1 L flask	4	3.44	[7]
P. polymyxa DSM 365	5 g/L yeast extract	30–40		1 L in 2 L Jar fermenter: agitation speed, 500 rpm			[22]
P. polymyxa EJS-3	Sucrose 188.2 g/L, yeast extract 25.8 g/L, K_2HPO_4 5 g/L, $CaCl_2$ 0.34 g/L	24	8	200 mL in a 1 L flask	2.5	35.26	[23–25,29]
P. jamilae CP-38	80% olive mill wastewaters (OMW)	30	7	2 L bioreactor at 150 rpm	3	4.2	[26]
P. elgii B69	Sucrose 51.35 g/L, peptone 6.78 g/L and yeast extract 0.47 g/L	30	7.2	100 mL in a 250 mL flask at 220 rpm	4	25.63	[27]
P. polymyxa ATCC 21830	Glucose 100 g/L, yeast extract 3 g/L	50	7	400 mL in a 1 L flask at 150 rpm	4	6.89	[6]
P. polymyxa	Sucrose 20 g/L, yeast extract 0.2 g/L, K_2HPO_4 0.25 g/L, $MgSO_4 \cdot 7H_2O$ 0.1 g/L, NaCl 0.05 g/L, agar 15 g/L	30		Petri dishes of 90 mm of diameter	5		[28]
P. polymyxa JB115	MSM broth containing 10% sucrose	30		1 L medium at 180 rpm	3	10	[30]

Many results have also reported that the production of EPS is cell growth-associated [7,31,32]. In addition to the C/N source, the initial liquid culture pH and culture temperature are both important and may affect the cell growth, the uptake of different nutrients and EPS production. An optimal value of pH for the EPS production from *Paenibacillus* spp. lies in the range between 6.5 and 7.2 (Table 1) [6,7,22,26,27,31,32]. Rafigh *et al.* found that when the initial pH of the fermentation broth was increased from 5.5 to 7.0, there was an increase in curdlan gum and biomass production, approximately 39.3% and 4.8%, respectively. However, higher values (pH 8.5) caused a decrease in their production [6]. These EPSs from *Paenibacillus* spp. were in agreement with previous study that for EPS production by *P. polymyxa* KCTC 8648P, the optimum pH value of 7.0 was reported [37]. However, there were only a few reports where the highest EPS content was achieved by cultivating the microorganism at slightly alkaline pH [23–25,29]. In the experiments with the strain *P. polymyxa* EJS-3, Liu *et al.* [23–25,29] found that the specific growth conditions of *P. polymyxa* EJS-3 might be due to its living in the tissues of plants.

Incubation temperature is another critical factor for EPS biosynthesis [23]. Rafigh *et al.* [6] found that when the temperature varied from 30 to 40 °C, the yield of curdlan production increased rapidly and then slightly increased as the fermentation proceeded from 40 to 50 °C. Batch fermentation processes were conducted at 25 °C separately, but the yield of curdlan production was very low, only a few μg/L. In addition, above 50 °C, curdlan may be curdled. Therefore, a higher temperature ratio was not employed in the production of curdlan [6]. However, Liu *et al.* reported that the optimal temperatures of *P. polymyxa* EJS-3 for cell growth and EPS production were 27 °C and 24 °C, respectively [23]. *P. polymyxa* EJS-3 favoured lower temperature for EPS production compared with others.

2.2. Fermentation Techniques

The influence of aeration on the vital activity of *Paenibacillus* spp. producing EPS was studied [6, 31,32]. Under anaerobic conditions, the cell population neither grows nor produces EPS. An intense aeration during fermentation leads to a significant increase of the EPS concentration. Previous studies also argued that agitation could be beneficial to the growth and performance of the microbial cells through improving the characteristics of mass transfer with respect to substrates, products and oxygen [40]. The findings by Rafigh *et al.* [6] also indicated that both biomass and curdlan production from *P. polymyxa* ATCC 21830 underwent significant enhancements as the agitation speed increased from 120 to 150 rpm. The highest yield of curdlan gum and biomass production was at an agitation speed of 150 rpm. The curdlan gum and biomass production were low at 120 rpm, which can be attributed to the limitation of oxygen transfer. However, at 180 rpm, lower levels of curdlan gum and biomass were observed, which can be ascribed to bacterial fragmentation mediated by several shearing mechanisms. However, for some other species of strains, higher agitation speed (600 rpm) has been used for curdlan production, e.g., from *Agrobacterium* sp. [20], but not from *Paenibacillus* sp.

3. Isolation and Purification of EPS

In our previous studies, the EPS sample after SPP fermentation was immediately autoclaved for 20 min to reduce the ropy condition of the culture and was centrifuged (12,000 *g* for 20 min) to remove the remaining SPP and biomass. The supernatant was filtered through a 0.45-μm membrane filter, mixed with two volumes of methanol, stirred vigorously and kept overnight at 4 °C. The precipitate from the ethanol dispersion was collected by centrifugation at 12,000 *g* for 15 min, washed three times with sterilized distilled water, and then lyophilized to yield the crude EPS. This is the most common and convenient method for isolating EPS from culture supernatant [7,23,27,30–32]. Figure 1 summarizes the isolation procedures of EPS from *Paenibacillus* spp.

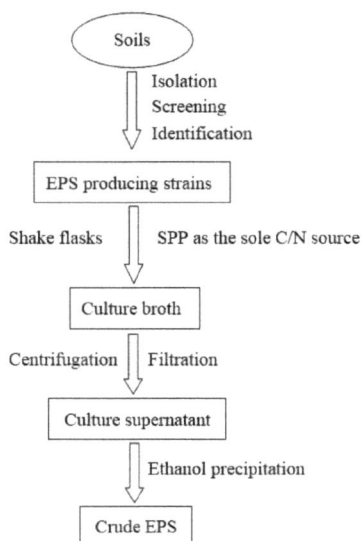

Figure 1. Schematic diagram for the isolation of EPS from *Paenibacillus* spp.

The crude EPS was re-dissolved in distilled water and stirred vigorously at 80 °C for 30 min, mixed with four volumes of anhydrous ethanol, stirred vigorously and kept overnight at 4 °C. The precipitate from the ethanol dispersion was collected by centrifugation at 12,000× g for 15 min, re-dissolved in distilled water and deproteinized with 1/5 volume of Sevag reagent (CHCl$_3$-BuOH, v/v = 5/1) seven times [41]. The deproteinized solution was then dialyzed against distilled water, concentrated and lyophilized to yield deproteinized EPS.

The deproteinized EPS was purified sequentially through column chromatography, including ion-exchange chromatography, gel filtration chromatography and affinity chromatography [7,23,27,30–32]. Elution was conducted with an appropriate running buffer, followed by collection, concentration, dialysis, and lyophilization, and the carbohydrates were determined by the phenol-sulfuric acid method using glucose as the standard [42]. In addition, based on the different solubility of the EPSs in ethanol, isopropanol, and other solvents, the EPSs were simply and effectively fractionated. Huang *et al.* recently isolated EPS fractions from a fermentation medium by gradient ethanol precipitation [43]. Their results suggest that the method is simple and workable for the initial fractionation of EPSs, proteins, and their complexes with different molecular sizes and for further identification of bioactive components.

4. Physicochemical Characterization

Paenibacillus spp. EPSs with different monosaccharide constituents and chemical structures have been isolated from their culture supernatant. Many research groups have elucidated the chemical structures of purified EPSs using infrared spectroscopy, liquid-state nuclear magnetic resonance (NMR) (one and two dimensions), solid-state NMR, gas chromatography (GC), GC-mass spectroscopy (GC-MS), high-performance liquid chromatography (HPLC), acid hydrolysis, methylation analysis, periodate-oxidation, and Smith degradation [7,23,27,30,44]. The different structural characteristics of EPS from *Paenibacillus* spp. showing various bioactivities have been investigated. The sources and chemical compositions of the EPSs are summarized in Table 2.

4.1. Molecular-Weight Distribution of EPS from Paenibacillus spp.

Various techniques, such as viscometry, osmometry, sedimentation, and HPLC, have been used to determine the average polymer molecular weight (MW) and polydispersity index. Among them, high-performance gel permeation chromatography (HPGPC) is a common method for determining the MW of polysaccharides and has also been used by many researchers for the MW of EPSs. Size-exclusion chromatography with multi-angle laser light scatter detection is also an efficient method for the evaluation of the absolute MW of polysaccharides and provides greater resolution than traditional GPC [6,7,23,27,30,45,46]. The average molecular weight of the EPS from *Paenibacillus* spp. varies in very broad ranges, from hundreds to thousands of kDa, depending on the culture strain, pH, cultivation techniques and C/N sources used. It has been reported that on a sucrose-containing medium, the bacterium produces high-molecular-weight EPS [23]. The molecular weights of EPS-1 and EPS-2 from *P. polymyxa* EJS-3 were estimated to be 1220 and 869 kDa, respectively [23]. The average molecular weight of curdlan produced using glucose as the carbon source by *P. polymyxa* ATCC 21830 was 170 kDa [6]. The molecular weight of EPS from *Paenibacillus* spp. varies significantly and is affected by variables such as temperature, pH and the initial carbon source concentration [11].

4.2. Monosaccharide Composition

The monosaccharide composition analysis usually involves cleavage of glycosidic linkages by acid hydrolysis, derivatization, and detection and quantification by GC. In addition, high-performance anion-exchange chromatography with pulsed amperometric detection has been developed to supplement traditional methods because it does not require derivatization of the monosaccharide with high resolution [47]. Recently, a 1-phenyl-3-methy-5-pyrazolone pre-column derivatization method has been used to determine the monosaccharide composition [48].

Many different EPSs from *Paenibacillus* spp. have been obtained; the monosaccharide composition is usually glucose, mannose, galactose and glucuronic acid in various mole ratios (Table 2). There is a large variety of EPS produced by *Paenibacillus* spp. depending on the type of *Paenibacillus* sp. strain, culture conditions and medium composition. The characterization of the EPS from *Paenibacillus* sp. TKU023 demonstrated that it was mainly composed of glucose and maltose [31]. Madden *et al.* reported that EPS produced by *P. polymyxa* NCIB 11429 was composed of glucose, mannose, galactose, glucuronic acid and pyruvate [49]. Han and Clarke found that EPS from *P. polymyxa* NRRL B-18475 was β-(2-6) linked fructan [33]. The EPS produced by *P. polymyxa* KCTC 8648P was composed of glucose, galactose, mannose, fucose and glucuronic acid [37]. Liu *et al.* reported that mannose, fructose and glucose were the constituents of the EPS from *P. polymyxa* EJS-3 [23]. The EPS from *P. polymyxa* SQR-21 was composed of mannose, glucose, fructose and glucuronic acid [7]. In particular, Li *et al.* found that the EPS produced by *P. elgii* B69 was composed of glucose, glucuronic acid, xylose, and mannose [27]. This finding differed from the EPS produced by other *Paenibacillus* strains. Xylose is rarely described as a main component of other bacterial polysaccharides. These dissimilarities reflect the species-specific production and biotechnological potential of EPS. Increasing numbers of microbial strains have been screened to produce bioactive EPS to replace synthetic chemicals, which have many side effects.

Table 2. Chemical structures of EPSs from *Paenibacillus* spp.

Microorganisms	Chemical Composition of EPSs	References
Paenibacillus sp. TKU023	glucose and maltose	[31]
P. polymyxa KCTC 8648P	Glucose, galactose, mannose, fucose and glucuronic acid	[37]
P. polymyxa NCIB 11429	Glucose, mannose, galactose, glucuronic acid and pyruvate	[49]
P. polymyxa NRRL B-18475	β-(2-6) linked fructan	[33]
P. elgii B69	Glucose:glucuronic acid:xylose:mannose = 1:0.53:1.15:0.46	[27]
P. polymyxa ATCC 21830	linear glucan interconnected by β-(1→3) glucosidic linkages	[6]
P. polymyxa JB115	glucan having β-(1,3) and β-(1,6) linkages	[30]
P. polymyxa EJS-3	Mannose, fructose and glucose	[23]
P. polymyxa SQR-21	Mannose, glucose, fructose and glucuronic acid	[7]

5. Bioactivities and Application of *Paenibacillus* spp. EPSs

In recent years, microbial EPSs and their derivatives have found many applications in the food, pharmaceutical and other industries because they have different physiological activities from natural gums and synthetic polymers [9,10]. Moreover, they are highly susceptible to natural biodegradation and are less harmful than synthetic polymers. Their applications are diverse, ranging from the laboratory through clinical to tableting; they have found applications in such diverse bio-medical fields as ophthalmology, orthopedic surgery, tissue engineering, implantation of medical devices and artificial organs, prostheses, dentistry, bone repair and many other medical fields [5]. In addition, they have therapeutic and pharmaceutical uses in that they enable the controlled, slow-release of drugs into the body. They also make possible targeting of drugs into sites of inflammation or tumors for disease treatment, and they can be used for skin rejuvenation and wound healing [5,32]. Several of these microbial polysaccharides are commercial industrial products, whereas others are in various stages of developments. The multiple bioactivities and environmental benefits of EPSs from *Paenibacillus* spp. are summarized in Table 3 and are compared in detail below.

Table 3. Potential applications of EPSs from *Paenibacillus* spp.

Microorganisms	EPSs	Applications	References
Paenibacillus sp. TKU023		Antioxidant	[31]
P. macerans TKU029		Improvement of human skin hydration	[32]
P. polymyxa EJS-3	Levan and its derivatives	Antioxidant / Antitumor	[23–25]
P. jamilae CP-38		Reduction in the toxicity of olive mill wastewaters / Heavy metal biosorption capacity	[26]
P. elgii B69		Bioflocculant	[27]
P. polymyxa ATCC 21830	Curdlan	Drug-delivery carriers for the sustained release of drugs and a support matrix for immobilization of enzymes	[6]
P. polymyxa		Removal of cadmium	[28]
P. polymyxa SQR-21		Antioxidant / Bioflocculant / Metal chelating capacity	[7]
P. polymyxa JB115	Glucan	Animal feed additive for the purpose of enhancing immunity	[30]

5.1. Antioxidant and Antitumor Activity

Free radicals are harmful to living organisms [50]. To reduce the damage caused by free radicals, both synthetic and natural antioxidants are used. However, synthetic antioxidants are thought to cause liver damage and carcinogenesis [23]. Therefore, it is essential to develop natural nontoxic antioxidants to protect humans from free radicals. Novel natural antioxidants have gained importance in science and medicine in recent decades. The antioxidant properties of EPSs have been reported from many types of EPSs derived from filamentous fungi [51], such as *Cordyceps militaris* SU5-08 [51], *Fusarium solani* SD5 [50], *Pleurotus sajor-caju* [52], *Fomes fomentarius* [53], *Tremella fuciformis* [54], *Agrocybe cylindracea* [55], *Collybia maculate* [56], *Cordyceps jiangxiensis* [57] and *Tremella mesenterica* [58]. However, relative to fungal EPS, reports concerning the antioxidant activities of bacterial polysaccharides [59–61] are more rare, such as *Paenibacillus* spp. EPSs.

Antioxidant activities have been attributed to various reactions and mechanisms. The *in vitro* antioxidant capacities of EPS were evaluated using various assay methods and activity indices [7,23–25, 29,31]. In our previous studies, *Paenibacillus* sp. TKU023 could produce EPS and antioxidant by using SPP as the sole C/N source. The culture supernatant incubated for four days using a baffled base flask showed strong 2,2-diphenyl-1-picrylhydrazyl (DPPH) radical scavenging activity, reducing powers, ferrous ion chelating activity and high total phenolic content, but maximum EPS production was found at the fifth day when using a flat base flask. The production of two invaluable environmentally friendly biomaterials (EPS and antioxidant) is unprecedented. In addition, the use of SPP (waste) is green, which made the whole process more valuable and attractive [31]. Raza *et al.* reported that *P. polymyxa* SQR-21 produced one type of EPS using yeast extract and galactose as the best N and C sources, respectively. Their EPS showed good superoxide scavenging and moderate inhibition of lipid peroxidation and reducing activities [7]. Furthermore, Liu *et al.* studied the *in vitro* and *in vivo* antioxidant activity of the levan-type EPSs from endophytic bacterium *P. polymyxa* EJS-3. In antioxidant assays *in vitro*, both crude EPS and its purified fractions (EPS-1 and EPS-2) were found to have moderate DPPH radical scavenging activity, hydrogen peroxide scavenging activity, lipid peroxidation inhibition effects, and strong ferrous ion chelating activity. In antioxidant assays *in vivo*, mice were subcutaneously injected with D-galactose for six weeks and administered EPS-1 via gavage. As a result, administration of EPS-1 significantly increased the thymus and spleen indices of D-galactose in aging mice. Moreover, EPS-1 administration significantly enhanced the activities of antioxidant enzymes and the total antioxidant capacity and decreased the levels of malondialdehyde in both serums and livers of aging mice [23,24]. Very recently, Liu *et al.* successfully acetylated, phosphorylated and benzylated the levan-type EPSs from *P. polymyxa* EJS-3 to obtain the derivatives of acetylated levan (AL), phosphorylated levan (PL) and benzylated levan (BL). For the antioxidant and antitumor activities *in vitro* of the natural polysaccharide and its derivatives, AL, BL and PL all exhibited higher reducing power, scavenging activity against superoxide radicals and scavenging activity of hydroxyl radicals compared to the natural polysaccharide, EPS-1. In addition, AL, BL and PL also exhibited higher antiproliferative activity against human gastric cancer BGC-823 cells *in vitro* than EPS-1. The enhanced activities of the derivatives were probably due to the introduction of acetyl, benzyl, or phosphoryl groups into the EPS-1 molecules, which increased the electron-donating ability and affinity with the receptors on immune cells. The results suggested that the derivatives could be explored as promising antioxidant and antitumor agents [25].

5.2. Improvement of Skin Hydration

Polysaccharides of bacterial origin are very important in the cosmetic and pharmaceutical industries [62]. One important criterion for the evaluation of cosmetic products is their effect on skin hydration. We found that EPSs from *P. macerans* TKU029 can significantly increase skin hydration [32]. The same amount of TKU029 EPSs (5%, w/v), propylene glycol/butylene glycol/water (1.5:1.5:1; v/v), and hyaluronic acid were applied to the skin of fifteen female volunteers (average age of 21 years). Skin hydration was measured in the test and control areas of each volunteer at each time point. The changes in skin capacitance 180 min after the application of TKU029 EPSs and hyaluronic acid were measured. TKU029 EPSs increased skin hydration significantly more than hyaluronic acid. During the application period, TKU029 EPSs increased skin hydration from 37.3% to 44.3%. Application of hyaluronic acid led to a slight increase at the beginning of the period, but then skin hydration decreased from 41.0% (at 30 min) to 37.0% (at 180 min). The untreated control remained nearly unchanged, *i.e.*, 37.3% at 0 min and 36.2% at 180 min. The untreated control revealed an increase in skin hydration after 180 min of 8.1% for TKU029 EPSs and of approximately 0.8% for hyaluronic acid [32]. The EPSs produced by *P. macerans* TKU029 can increase *in vivo* skin hydration and may be a new source of natural moisturizers with potential value in cosmetics.

5.3. Bioremediation of Wastewater

Chemical contamination of water from a wide range of toxic compounds, particularly dyeing pigments (from the textile industry), heavy metal ions, and other toxic suspended particles, remains a serious environmental problem causing serious risk to public health. Various physical and chemical processes have been developed for removing pollutants from wastewater. One of the most popular methods is flocculation due to its economic advantages and potency. Bioflocculants are more eco-friendly and biodegradable and are less harmful to the environment than inorganic flocculants and organic synthetic flocculants [63]. Many research groups have evaluated the flocculating activity of EPSs from *Paenibacillus* spp. Bioflocculation of high-ash Indian coals using *P. polymyxa* showed a 60% decrease in ash, suggesting that selective flocculation of coal is possible [64]. *P. polymyxa* P13 was reported as an EPS producer that exhibited significant biosorption capacity of Cu^{2+} produced in several industries [65]. The EPS from *P. polymyxa* SQR-21 showed a high flocculating activity towards activated carbon [7]. *P. jamilae* CP-38 was able to grow and produce EPS using olive mill wastewaters as the sole nutrient and energy source, with a concomitant reduction in the toxicity of the waste [26]. The EPS produced by *P. polymyxa* had cadmium sorption capacity in aqueous solution [28]. In particular, Li *et al.* discovered a new EPS-based broad-spectrum bioflocculant produced by a newly isolated strain, *P. elgii* B69 [27]. This bioflocculant had high activities towards all tested pollutants, including kaolin clay, dyeing pigment, heavy metal ion, and real wastewater. The multiple-pollutant-removal performance of *P. elgii* B69 is a significant advantage [27].

5.4. Other Bioactivities

As aforementioned, *Paenibacillus* spp. could produce different types of EPSs with different biotechnological properties. Many other important bioactivities of *Paenibacillus* spp. EPSs have also been investigated. *P. polymyxa* JB115 was isolated from Korean soil as a glucan producer for the development of animal feed additives showing activities as a biological response modifier, natural immuno-modulator and a potential anti-tumor agent for livestock [30]. Rafigh *et al.* demonstrated that *P. polymyxa* ATCC 21830 is capable of producing curdlan gum [6]. Curdlan has potential applications in the manufacture of food products, and it is also known as a drug-delivery carrier for the sustained release of drugs and as a support matrix for the immobilization of enzymes [66–68]. In addition, curdlan has been used together with activated carbon adsorbents for heavy metal removal [69].

6. Conclusions and Perspectives

This review contains the most recent information on the production of various *Paenibacillus* spp. EPSs with applications in bioactivity and bioremediation. Economical, environmentally friendly and high quantity production of *Paenibacillus* spp. EPSs is necessary for various applications. Squid pen waste was utilized to produce high yield, inexpensive EPSs. The discovery of inexpensive EPSs not only solves environmental problems but also promotes the economic value of marine wastes. Furthermore, this EPS produced by *P. macerans* TKU029 using SPP can increase *in vivo* skin hydration and may be a new source of natural moisturizers with potential value in cosmetics. Novel uses of *Paenibacillus* spp. EPSs due to its inexpensive production and its bulk will be developed. In addition, the movement toward "greener" products and technologies that are more environmentally friendly is prevailing. Production methods of EPSs are eagerly anticipated in the future, and many have already reached the market. However, the relationship between the structural features, solution behavior, space conformation, and the bioactivity of *Paenibacillus* spp. EPSs is unclear due to the structural diversity and complexity of polysaccharide molecules. In addition, alteration of the chemical properties of the original EPSs will also greatly enhance their values and extend their range of applications. The elaboration of either biotechnological or technical procedures for the production of EPSs of diverse structures (e.g., varied stereochemical composition or molecular sizes) and ultimate product functions (e.g., varied water-solubility and physical activity) to meet the special demands of practical application are being launched and will soon provide a broad spectrum of new EPSs. These will be useful to understand the chemical structures, chain conformations and the biological activities for applications in various fields.

Acknowledgments: This work was supported in part by a grant from the Ministry of Science and Technology, Taiwan (NSC 102-2313-B-032-001-MY3, NSC 102-2621-M-032-005, and NSC 102-2811-B-032-001).

Author Contributions: All authors contributed to the analysis of data and preparation of the manuscript.

Conflicts of Interest: The authors declare no conflict of interest.

References

1. Sutherland, I.W. Novel and established applications of microbial polysaccharides. *Trends Biotechnol.* **1998**, *16*, 41–46. [CrossRef] [PubMed]
2. Roller, S.; Dea, I.C.M. Biotechnology in the production and modification of biopolymers for foods. *Crit. Rev. Biotechnol.* **1992**, *12*, 261–277. [CrossRef]
3. Byrom, D. Polymer synthesis by micro-organisms: Technology and economics. *Trends Biotechnol.* **1987**, *5*, 246–250. [CrossRef]
4. Johns, M.R.; Noor, E. Recovery and purification of polysaccharides from microbial broth. *Aust. J. Biotechnol.* **1991**, *5*, 73–77. [PubMed]
5. Shih, I.L. Microbial exo-polysaccharides for biomedical applications. *Mini Rev. Med. Chem.* **2010**, *10*, 1345–1355. [CrossRef] [PubMed]
6. Rafigh, S.M.; Yazdi, A.V.; Vossoughi, M.; Safekordi, A.A.; Ardjmand, M. Optimization of culture medium and modeling of curdlan production from *Paenibacillus polymyxa* by RSM and ANN. *Int. J. Biol. Macromol.* **2014**, *70*, 463–473. [CrossRef] [PubMed]
7. Raza, W.; Makeen, K.; Wang, Y.; Xu, Y.; Qirong, S. Optimization, purification, characterization and antioxidant activity of an extracellular polysaccharide produced by *Paenibacillus polymyxa* SQR-21. *Bioresour. Technol.* **2011**, *102*, 6095–6103. [CrossRef] [PubMed]
8. Dziezak, J.D. Special report: A focus on gums. *Food Technol.* **1991**, *45*, 116–132.
9. Kumar, A.S.; Mody, K.; Jha, B. Bacterial exopolysaccharides–A perception. *J. Basic Microbiol.* **2007**, *47*, 103–117. [CrossRef] [PubMed]
10. Freitas, F.; Alves, V.D.; Pais, J.; Costa, N.; Oliveira, C.; Mafra, L.; Hilliou, L.; Oliveir, R.; Reis, M.A.M. Characterization of an extracellular polysaccharide produced by a *Pseudomonas* strain grown on glycerol. *Bioresour. Technol.* **2009**, *100*, 859–865. [CrossRef] [PubMed]
11. Armstrong, D.C.; Johns, M.R. Culture conditions affect the molecular weight properties of hyaluronic acid produced by *Streptococcus zooepidemicus*. *Appl. Environ. Microbiol.* **1997**, *63*, 2759–2764. [PubMed]
12. Shih, I.L.; Yu, Y.T.; Shieh, C.J.; Hsieh, C.Y. Selective production and characterization of levan by *Bacillus subtilis* (natto) Takahashi. *J. Agric. Food Chem.* **2005**, *53*, 8211–8215. [CrossRef] [PubMed]
13. Han, Y.W. Levan production by *Bacillus polymyxa*. *J. Ind. Microbiol.* **1989**, *4*, 447–452. [CrossRef]
14. Youssef, F.; Roukas, T.; Biliaderis, C.G. Pullulan production by a non-pigmented strain of *Aureobasidium pullulans* using batch and fed-batch culture. *Process Biochem.* **1999**, *34*, 355–366. [CrossRef]
15. Qader, S.A.U.; Iqbal, L.; Aman, F.; Shireen, E.; Azhar, A. Production of dextran by newly isolated strains of *Leuconostoc mesenteroides* PCSIR-3 and PCSIR-9. *Turk. J. Biochem.* **2005**, *31*, 21–26.
16. Jana, A.K.; Ghosh, P. Stimulation of xanthan production by *Xanthomonas campestris* using citric acid. *World J. Microbiol. Biotechnol.* **1997**, *13*, 261–264. [CrossRef]
17. Manna, B.; Gambhir, A.; Ghosh, P. Production and rheological characteristics of the microbial polysaccharide gellan. *Lett. Appl. Microbiol.* **1996**, *23*, 141–145. [CrossRef]
18. Espevik, T.; Skjåk-Bræk, G. Application of alginate gels in biotechnology and biomedicine. *Carbohydr. Eur.* **1996**, *14*, 19–25.
19. Krystynowicz, A.; Czaja, W.; Wiktorowska-Jezierska, A.; Goncalves-Miśkiewicz, M.; Turkiewicz, M.; Bielecki, S. Factors affecting the yield and properties of bacterial cellulose. *J. Ind. Microbiol. Biotechnol.* **2002**, *29*, 189–195. [CrossRef]
20. Lee, I.Y.; Kim, M.K.; Lee, J.H.; Seo, W.T.; Jung, J.K.; Lee, H.W.; Park, Y.H. Influence of agitation speed on production of curdlan by *Agrobacterium* species. *Bioprocess Eng.* **1999**, *20*, 283–287.

21. Ash, C.; Priest, F.G.; Collins, M.D. Molecular identification of rRNA group 3 *bacilli* (Ash, Farrow, Wallbanks and Collins) using a PCR probe test. Proposal for the creation of a new genus *Paenibacillus*. *Antonie Van Leeuwenhoek* **1993**, *64*, 253–260. [CrossRef] [PubMed]

22. Häßler, T.; Schieder, D.; Pfaller, R.; Faulstich, M.; Sieber, V. Enhanced fed-batch fermentation of 2,3-butanediol by *Paenibacillus polymyxa* DSM 365. *Bioresour. Technol.* **2012**, *124*, 237–244. [CrossRef] [PubMed]

23. Liu, J.; Luo, J.; Ye, H.; Sun, Y.; Lu, Z.; Zeng, X. Production, characterization and antioxidant activities *in vitro* of exopolysaccharides from endophytic bacterium *Paenibacillus polymyxa* EJS-3. *Carbohydr. Polym.* **2009**, *78*, 275–281. [CrossRef]

24. Liu, J.; Luo, J.; Ye, H.; Sun, Y.; Lu, Z.; Zeng, X. *In vitro* and *in vivo* antioxidant activity of exopolysaccharides from endophytic bacterium *Paenibacillus polymyxa* EJS-3. *Carbohydr. Polym.* **2010**, *82*, 1278–1283. [CrossRef]

25. Liu, J.; Luo, J.; Ye, H.; Zeng, X. Preparation, antioxidant and antitumor activities *in vitro* of different derivatives of levan from endophytic bacterium *Paenibacillus polymyxa* EJS-3. *Food Chem. Toxicol.* **2012**, *50*, 767–772. [CrossRef] [PubMed]

26. Aguilera, M.; Quesada, M.T.; del Águila, V.G.; Morillo, J.A.; Rivadeneyra, M.A.; Ramos-Cormenzana, A.; Monteoliva-Sánchez, M. Characterisation of *Paenibacillus jamilae* strains that produce exopolysaccharide during growth on and detoxification of olive mill wastewaters. *Bioresour. Technol.* **2008**, *99*, 5640–5644. [CrossRef] [PubMed]

27. Li, O.; Lu, C.; Liu, A.; Zhu, L.; Wang, P.M.; Qian, C.D.; Jiang, X.H.; Wu, X.C. Optimization and characterization of polysaccharide-based bioflocculant produced by *Paenibacillus elgii* B69 and its application in wastewater treatment. *Bioresour. Technol.* **2013**, *134*, 87–93. [CrossRef] [PubMed]

28. Mokaddem, H.; Sadaoui, Z.; Boukhelata, N.; Azouaou, N.; Kaci, Y. Removal of Cadmium from aqueous solution by polysaccharide produced from *Paenibacillus polymyxa*. *J. Hazard. Mater.* **2009**, *172*, 1150–1155. [CrossRef] [PubMed]

29. Liu, J.; Luo, J.; Ye, H.; Sun, Y.; Lu, Z.; Zeng, X. Medium optimization and structural characterization of exopolysaccharides from endophytic bacterium *Paenibacillus polymyxa* EJS-3. *Carbohydr. Polym.* **2010**, *79*, 206–213. [CrossRef]

30. Jung, H.K.; Hong, J.H.; Park, S.C.; Park, B.K.; Nam, D.H.; Kim, S.D. Production and physicochemical characterization of β-glucan produced by *Paenibacillus polymyxa* JB115. *Biotechnol. Bioprocess Eng.* **2007**, *12*, 713–719. [CrossRef]

31. Wang, C.L.; Huang, T.H.; Liang, T.W.; Fang, C.Y.; Wang, S.L. Production and characterization of exopolysaccharides and antioxidant from *Paenibacillus* sp.TKU023. *N. Biotechnol.* **2011**, *28*, 559–565. [CrossRef] [PubMed]

32. Liang, T.W.; Wu, C.C.; Cheng, W.T.; Chen, Y.C.; Wang, C.L.; Wang, I.L.; Wang, S.L. Exopolysaccharides and antimicrobial biosurfactants produced by *Paenibacillus macerans* TKU029. *Appl. Biochem. Biotechnol.* **2014**, *172*, 933–950. [CrossRef] [PubMed]

33. Han, Y.W.; Clarke, M.A. Production and characterization of microbial levan. *J. Agric. Food Chem.* **1990**, *38*, 393–396. [CrossRef]

34. Jiang, L.F. Effect of nitrogen source on curdlan production by *Alcaligenes faecalis* ATCC 31749. *Int. J. Biol. Macromol.* **2013**, *52*, 218–220. [CrossRef] [PubMed]

35. Park, N.H.; Choi, H.J.; Oh, D.K. Lactosucrose production by various microorganisms harboring levansucrase activity. *Biotechnol. Lett.* **2005**, *27*, 495–497. [CrossRef] [PubMed]

36. Gauri, S.S.; Mandal, S.M.; Mondal, K.C.; Dey, S.; Pati, B.R. Enhanced production and partial characterization of an extracellular polysaccharide from newly isolated *Azotobacter* sp. SSB81. *Bioresour. Technol.* **2009**, *100*, 4240–4243. [CrossRef] [PubMed]

37. Lee, I.Y.; Seo, W.T.; Kim, G.J.; Kim, M.K.; Ahn, S.G.; Kwon, G.S.; Park, Y.H. Optimization of fermentation conditions for production of exopolysaccharide by *Bacillus polymyxa*. *Bioprocess Eng.* **1997**, *16*, 71–75. [CrossRef]

38. Wang, S.L.; Lin, C.L.; Liang, T.W.; Liu, K.C.; Kuo, Y.H. Conversion of squid pen by *Serratia ureilytica* for the production of enzymes and antioxidants. *Bioresour. Technol.* **2009**, *100*, 316–323. [CrossRef] [PubMed]

39. Liang, T.W.; Huang, C.T.; Nguyen, A.D.; Wang, S.L. Squid pen chitin chitooligomers as food colorants absorbers. *Mar. Drugs* **2015**, *13*, 681–696. [CrossRef] [PubMed]

40. García-Ochoa, F.; Santos, V.E.; Casas, J.A.; Gómez, E. Xanthan gum: Production, recovery, and properties. *Biotechnol. Adv.* **2000**, *18*, 549–579. [CrossRef]

41. Staub, A.M. Removal of protein-Sevag method. *Methods Carbohyd. Chem.* **1965**, *5*, 5–6.

42. Dubois, M.; Gilles, K.A.; Hamilton, J.K.; Rebers, P.A.; Smith, F. Colorimetric method for determination of sugars and related substances. *Anal. Chem.* **1956**, *28*, 350–356. [CrossRef]

43. Huang, Q.L.; Siu, K.C.; Wang, W.Q.; Cheung, Y.C.; Wu, J.Y. Fractionation, characterization and antioxidant activity of exopolysaccharides from fermentation broth of a *Cordyceps sinensis* fungus. *Process Biochem.* **2013**, *48*, 380–386. [CrossRef]

44. Yan, J.K.; Wang, W.Q.; Li, L.; Wu, J.Y. Physiochemical properties and antitumour activities of two a-glucans isolated from hot water and alkaline extracts of *Cordyceps* (Cs-HK1) fungal mycelia. *Carbohydr. Polym.* **2011**, *85*, 753–758. [CrossRef]

45. Boukari, I.; Putaux, J.L.; Cathala, B.; Barakat, A.; Saake, B.; Rémond, C.; O'Donohue, M.; Chabbert, B. *In vitro* model assemblies to study the impact of lignin-carbohydrate interactions on the enzymatic conversion of xylan. *Biomacromolecules* **2009**, *10*, 2489–2498. [CrossRef] [PubMed]

46. Hilliou, L.; Freitas, F.; Oliveira, R.; Reis, M.A.M.; Lespineux, D.; Grandfils, C.; Alves, V.D. Solution properties of an exopolysaccharide from a *Pseudomonas* strain obtained using glycerol as sole carbon source. *Carbohydr. Polym.* **2009**, *78*, 526–532. [CrossRef]

47. Panagiotopoulos, C.; Sempéré, R.; Lafont, R.; Kerhervé, P. Sub-ambient temperature effects on the separation of monosaccharide by high-performance anion-exchange chromatography with pulse amperometric detection: Application to marine chemistry. *J. Chromatogr. A* **2001**, *920*, 13–22. [CrossRef] [PubMed]

48. Chen, S.; Siu, K.C.; Wang, W.Q.; Liu, X.X.; Wu, J.Y. Structure and antioxidant activity of a novel poly-*N*-acetylhexosamine produced by a medicinal fungus. *Carbohydr. Polym.* **2013**, *94*, 332–338. [CrossRef] [PubMed]

49. Madden, J.K.; Dea, I.C.M.; Steer, D.C. Structural and rheological properties of the extracellular polysaccharides from *Bacillus polymyxa*. *Carbohydr. Polym.* **1986**, *6*, 51–73. [CrossRef]

50. Mahapatra, S.; Banerjee, D. Evaluation of *in vitro* antioxidant potency of exopolysaccharide from endophytic *Fusarium solani* SD5. *Int. J. Biol. Macromol.* **2013**, *53*, 62–66. [CrossRef] [PubMed]

51. Lin, R.; Liu, H.; Wu, S.; Pang, L.; Jia, M.; Fan, K.; Jia, S.; Jia, L. Production and *in vitro* antioxidant activity of exopolysaccharide by a mutant, *Cordyceps militaris* SU5-08. *Int. J. Biol. Macromol.* **2012**, *51*, 153–157. [CrossRef] [PubMed]

52. Confortin, F.G.; Marchetto, R.; Bettin, F.; Camassola, M.; Salvado, M.; Dillon, A.J. Production of *Pleurotus sajor-caju* strain PS-2001 biomass in submerged culture. *J. Ind. Microbiol. Biotechnol.* **2008**, *35*, 1149–1155. [CrossRef] [PubMed]

53. Chen, W.; Zhao, Z.; Chen, S.F.; Li, Y.Q. Optimization for the production of exopolysaccharide from *Fomes fomentarius* in submerged culture and its antitumor effect *in vitro*. *Bioresour. Technol.* **2008**, *99*, 3187–3189. [CrossRef] [PubMed]

54. Cho, E.J.; Oh, J.Y.; Chang, H.Y.; Yun, J.W. Production of exopolysaccharidesby submerged mycelial culture of a mushroom *Tremella fuciformis*. *J. Biotechnol.* **2006**, *127*, 129–140. [CrossRef] [PubMed]

55. Kim, H.O.; Lim, J.M.; Joo, J.H.; Kim, S.W.; Hwang, H.J.; Choi, J.W.; Yun, J.W. Optimization of submerged culture condition for the production of mycelial biomass and exopolysaccharides by *Agrocybe cylindracea*. *Bioresour. Technol.* **2005**, *96*, 1175–1182. [CrossRef] [PubMed]

56. Lim, J.M.; Kim, S.W.; Hwang, H.J.; Joo, J.H.; Kim, H.O.; Choi, J.W.; Yun, J.W. Optimization of medium by orthogonal matrix method for submerged mycelial culture and exopolysaccharide production in *Collybia maculate*. *Appl. Biochem. Biotechnol.* **2004**, *119*, 159–170. [CrossRef] [PubMed]

57. Xiao, J.H.; Chen, D.X.; Liu, J.W.; Liu, Z.L.; Wan, W.H.; Fang, N.; Xiao, Y.; Qi, Y.; Liang, Z.Q. Optimization of submerged culture requirements for the production of mycelial growth and exopolysaccharide by *Cordyceps jiangxiensis* JXPJ 0109. *J. Appl. Microbiol.* **2004**, *96*, 1105–1116. [CrossRef]

58. De Baets, S.; Du Laing, S.; Francois, C.; Vandamme, E.J. Optimization of exopolysaccharide production by *Tremella mesenterica* NRRL Y-6158 through implementation of fed-batch fermentation. *J. Ind. Microbiol. Biotechnol.* **2002**, *29*, 181–184. [CrossRef] [PubMed]

59. Raveendran, S.; Palaninathan, V.; Chauhan, N.; Sakamoto, Y.; Yoshida, Y.; Maekawa, T.; Mohanan, P.V.; Kumar, D.S. *In vitro* evaluation of antioxidant defense mechanism and hemocompatibility of mauran. *Carbohydr. Polym.* **2013**, *98*, 108–115. [CrossRef] [PubMed]

60. Li, J.Y.; Jin, M.M.; Meng, J.; Gao, S.M.; Lu, R.R. Exopolysaccharide from *Lactobacillus planterum* LP6: Antioxidation and the effect on oxidative stress. *Carbohydr. Polym.* **2013**, *98*, 1147–1152. [CrossRef] [PubMed]

61. Kanmani, P.; Satish kumar, R.; Yuvaraj, N.; Paari, K.A.; Pattukumar, V.; Arul, V. Production and purification of a novel exopolysaccharide from lactic acid bacterium *Streptococcus phocae* PI80 and its functional characteristics activity *in vitro*. *Bioresour. Technol.* **2011**, *102*, 4827–4833. [CrossRef] [PubMed]

62. Satpute, S.K.; Banat, I.M.; Dhakephalkar, P.K.; Banpurkar, A.G.; Chopade, B.A. Biosurfactants, bioemulsifiers and exopolysaccharides from marine microorganisms. *Biotechnol. Adv.* **2010**, *28*, 436–450. [CrossRef] [PubMed]

63. Shih, I.; van, Y.; Yeh, L.; Lin, H.; Chang, Y. Production of a biopolymer flocculant from *Bacillus licheniformis* and its flocculation properties. *Bioresour. Technol.* **2001**, *78*, 267–272. [CrossRef] [PubMed]

64. Vijayalakshmi, S.P.; Raichur, A.M. Bioflocculation of high-ash Indian coals using *Paenibacillus polymyxa*. *Int. J. Miner. Process* **2002**, *67*, 199–210. [CrossRef]

65. Acosta, M.P.; Valdman, E.; Leite, S.G.F.; Battaglini, F.; Ruzal, S.M. Biosorption of copper by *Paenibacillus polymyxa* cells and their exopolysaccharide. *World J. Microbiol. Biotechnol.* **2005**, *21*, 1157–1163. [CrossRef]

66. Takeda-Hirokawa, N.; Neoh, L.P.; Akimoto, H.; Kaneko, H.; Hishikawa, T.; Sekigawa, I.; Hashimoto, H.; Hirose, S.I.; Murakami, T.; Yamamoto, N.; *et al.* Effect of curdlan sulfate on the production of β-chemokines and cytokines. *Microbiol. Immunol.* **1997**, *41*, 741–745. [CrossRef] [PubMed]

67. Kim, B.; Jung, I.; Kim, J.; Lee, J.H.; Lee, I.; Lee, K. Curdlan gels as protein drug delivery vehicles. *Biotechnol. Lett.* **2000**, *22*, 1127–1130. [CrossRef]

68. Saudagar, P.S.; Singhal, R.S. Curdlan as a support matrix for immobilization of enzyme. *Carbohydr. Polym.* **2004**, *56*, 483–488. [CrossRef]

69. Moon, C.J.; Lee, J.H. Use of curdlan and activated carbon composed adsorbents for heavy metal removal. *Process Biochem.* **2005**, *40*, 1279–1283. [CrossRef]

marine drugs

MDPI

Article

Characterization of the Exopolysaccharide Produced by *Salipiger mucosus* A3ᵀ, a Halophilic Species Belonging to the *Alphaproteobacteria*, Isolated on the Spanish Mediterranean Seaboard

Inmaculada Llamas [1], Juan Antonio Mata [1], Richard Tallon [2], Philippe Bressollier [2,3], María C. Urdaci [3], Emilia Quesada [1] and Victoria Béjar [1,*]

[1] Microbial Exopolysaccharide Research Group, Department of Microbiology, Faculty of Pharmacy, Cartuja Campus, University of Granada, 18071 Granada, Spain; illamas@ugr.es (I.L.); jonanmg@gmail.com (J.A.M.); equesada@ugr.es (E.Q.)

[2] IUT, Département Génie Biologique, allée André Maurois, 87065 Limoges, France; richard_tallon@yahoo.fr (R.T.); philippe.bressollier@unilim.fr (P.B.)

[3] Laboratorie de Microbiologie et Biochimie Appliquée, ENITA de Bordeaux, 1 Cours du Général de Gaulle, BP 201, 33175 Gradignan, France; m-urdaci@enitab.fr

* *Author to whom correspondence should be addressed; vbejar@ugr.es; Tel. +34-958-241705; Fax. +34-958-246235.

Received: 8 June 2010; in revised form: 16 July 2010; Accepted: 28 July 2010; Published: 30 July 2010

Abstract: We have studied the exopolysaccharide produced by the type strain of *Salipiger mucosus*, a species of halophilic, EPS-producing (exopolysaccharide-producing) bacterium belonging to the *Alphaproteobacteria*. The strain, isolated on the Mediterranean seaboard, produced a polysaccharide, mainly during its exponential growth phase but also to a lesser extent during the stationary phase. Culture parameters influenced bacterial growth and EPS production. Yield was always directly related to the quantity of biomass in the culture. The polymer is a heteropolysaccharide with a molecular mass of 250 kDa and its components are glucose (19.7%, w/w), mannose (34%, w/w), galactose (32.9%, w/w) and fucose (13.4%, w/w). Fucose and fucose-rich oligosaccharides have applications in the fields of medicine and cosmetics. The chemical or enzymatic hydrolysis of fucose-rich polysaccharides offers a new efficient way to process fucose. The exopolysaccharide in question produces a solution of very low viscosity that shows pseudoplastic behavior and emulsifying activity on several hydrophobic substrates. It also has a high capacity for binding cations and incorporating considerable quantities of sulfates, this latter feature being very unusual in bacterial polysaccharides.

Keywords: exopolysaccharides; Salipiger mucosus; halophilic bacteria; fucose; sulfates

1. Introduction

Microbial exopolysaccharides (EPSs) are polymers that consist principally of carbohydrates and are excreted by some bacteria and fungi onto the outside of their cell walls. Their composition and structure is very varied: they may be either homo- or heteropolysaccharides and may also contain a number of different organic and inorganic constituents [1]. Microbial EPSs occur naturally in many habitats. They are important in the formation of biofilm, a structure involved in the adherence to surfaces and in the protection of bacteria against noxious influences of the environment. They can be readily prepared in the laboratory by fermentation. Increasing attention is being paid to these molecules because of their bioactive role and their extensive range of potential applications in pharmaceuticals as antiangiogenic [2] or antiviral agents [3], and in agriculture and various other industrial areas [4].

Mar. Drugs **2010**, *8*, 2240–2251

The advantages of microbial polysaccharides over plant or marine macroalgal polymers are their novel functionality, easily reproducible chemical and physical properties and stable cost and supply [5]. During the past 50 years a considerable number of EPSs have been described, but few have achieved great commercial success, due either to their being unable to offer better properties than those already on the market or to difficulties in finding new practical applications [4].

A new approach to encountering EPSs with novel properties might entail investigating different environments such as hypersaline habitats, since most of the strains used so far for the industrial production of EPSs belong to a small number of taxa, which are generally non-halophilic, soil-dwelling diazotrophs and often plant-associated. Because of the extreme nature of hypersaline environments they may feasibly harbor unusual microorganisms of biotechnological interest and so for the last few years we have been carrying out a wide research program, looking at microorganisms living in habitats such as these in an attempt to find new EPSs with different characteristics. This has resulted in our describing the EPSs synthesized by several moderately halophilic species published by our group, belonging to the γ-*Proteobacteria*: *Halomonas eurihalina* [6], *H. maura* [7], *H. ventosae* [8], *H. anticariensis* [9], *Alteromonas hispanica* [10] and *Idiomarina ramblicola* and *I. fontislapidosi* [11]. In general, the polymers produced by these bacteria show potential interest as viscosifying, jellifying and emulsifying agents and as metal-binding compounds [12,14]. The most interesting polymers obtained so far from halophilic bacteria are mauran, produced by *H. maura*, which has a viscosifying activity similar to that of xanthan gum [15,16], and polymers from *H. eurihalina* species, which have emulsifying activity and jellifying properties at acidic pH [17,23].

We describe here an EPS produced by the type strain (A3^T) of *Salipiger mucosus* [24], a halophilic bacterium belonging to the *Alphaproteobacteria* class. To our knowledge, this is the first time a halophilic *Alphaproteobacteria* has been found to produce an EPS. To understand better the characteristics of EPSs and thus be able to apply them successfully to biotechnological ends it is essential to optimize their production and to identify their components and physical-chemical properties.

2. Results

2.1. Kinetics of Growth and EPS Production

To study the synthesis of the EPS as a function of the growth phase, *S. mucosus* strain A3^T was grown in MY medium containing 7.5% (w/v) total salts and an initial glucose concentration of 10 g/L for eight days at 32 °C and 100 rpm without control over pH, as shown in Figure 1. Glucose metabolism led to an increase in biomass (final OD$_{520}$ of 2.6) and an EPS accumulation of 1.35 g/L (0.1 g of EPS per gram of dry cell weight). The kinetics of EPS production by *S. mucosus* strain A3^T showed that it was excreted mainly during the exponential growth phase but continued to a lesser extent during the stationary phase. Production ceased when residual glucose was about 1.5 g/L. The quantity of EPS in the culture decreased considerably after 72 hours of incubation.

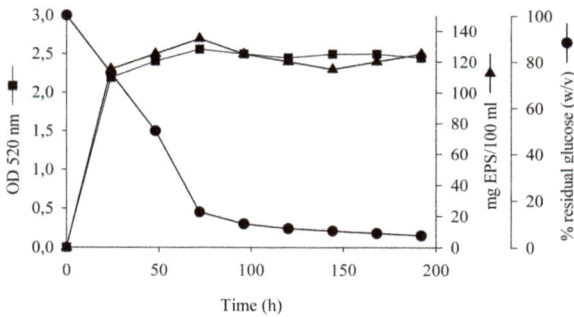

Figure 1. Growth profile and EPS production by *S. mucosus* strain A3^T in MY medium at 7.5% total salts *versus* consumption of glucose. 100% of residual glucose corresponds to 10 g/L of glucose.

2.2. Influence of Culture Parameters on Growth and EPS Synthesis

We studied the influence of different cultural parameters in order to be able to improve EPS production by *S. mucosus* strain A3^T. Yield was always directly related to the quantity of biomass in the culture. A sea-salt concentration of 2.5% (w/v) proved best in terms of production. One noteworthy result was that the bacterium grew and produced EPS with all the carbon sources assayed. Growth remained unimpeded when glucose was increased to above 1% (w/v) in the culture but without any significant increase in synthesis. As far as incubation temperature is concerned, yields were substantially lower at both 22 °C and 42 °C than at 32 °C, concomitant with a significant inhibition in bacterial growth. Similar results were obtained when the pH was lower or higher than 7.0. In the same way, both static incubation and a stirring rate of more than 100 rpm gave similarly lower growth and lower EPS yields.

2.3. Electron Microscopy

Figure 2 shows the appearance of strain A3^T after 24 hours incubation. It is noteworthy that during the early stage of incubation the EPS was in close contact with the cell, but afterwards it was diffused fairly evenly throughout the medium as an amorphous slime with no indication of any higher concentrations near the cell wall (data no shown). This phenomenon explains why we were able to extract higher quantities of EPS from the third day onwards.

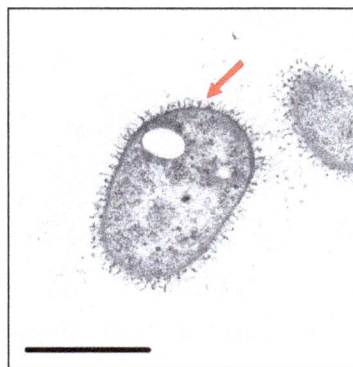

Figure 2. Transmission electronic microscopy photograph of *S. mucosus* strain A3^T stained with the specific stain for polysaccharide ruthenium red. Bar: 1 μm. Arrow indicates the EPS of the strain.

2.4. Chemical Composition

When collected under optimum conditions the EPS produced by *S. mucosus* strain A3[T] was composed of 53.1% (w/w) carbohydrates, 1.6% (w/w) proteins, 0.9% (w/w) acetyls, 0.9% (w/w) sulfates, 0.1% (w/w) phosphates, 1.9% (w/w) hexosamines and 6% (w/w) pyruvic acid; each value representing the average of three measurements. The EPS collected under other conditions assayed had the same basic composition but the carbohydrate content was always lower (data not shown). The samples obtained under optimum conditions were purified by anion exchange chromatography (AEC). A single major peak of sugar-bearing material was eluted from the column with NaCl concentrations of about 1 M, indicating a single acidic EPS. The components of the carbohydrate fraction of the EPS from *S. mucosus* strain A3[T] were (%, w/w): 19.7 glucose, 34 mannose, 32.9 galactose and 13.4 fucose; each value representing the average of three measurements.

2.5. Physical Properties

The average molecular mass of the EPS produced by *S. mucosus* strain A3[T] extracted after three days from MY medium with 2.5% (w/v) salts was 250 kDa. To evaluate the rheological properties of the EPS we measured the viscosity of an aqueous solution with 0.5% (w/v) of EPS over a range of different shear rates. The flow curve in Figure 3 shows the pseudoplastic character of a solution of the polymer, its viscosity decreasing concomitantly with shear rate. The viscosity of the solution was quite low.

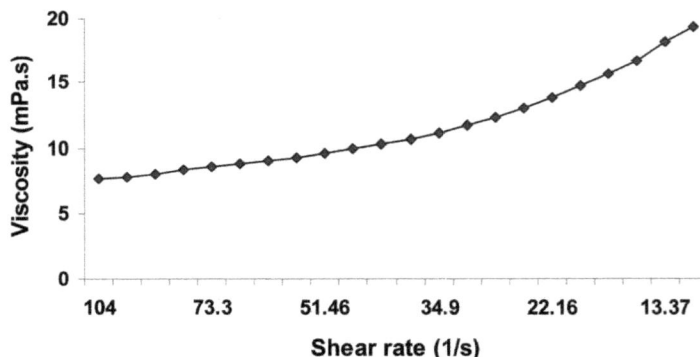

Figure 3. Viscosity of a 0.5% (w/v) solution of EPSs produced by *S. mucosus* A3[T] in MY medium at 2.5% total salts.

The emulsifying activity of the EPS is shown in Table 1. It was capable of stabilizing different mixtures of oil and water in which the hydrophobic phase was either a hydrocarbon or a vegetable or mineral oil; the polymer's activity was more efficient than the chemical surfactants (Triton X-100 and Tween 80) used as controls when tetradecane, octane, kerosene and xylene were used as substrates. The highest emulsifying activity was obtained with crude oil (higher than with xanthan). As far as metal chelation is concerned, it chelated 15.7, 43.5 and 8.7 mg of copper, lead and cobalt, respectively, per gram of EPS.

Table 1. Emulsifying activities of the EPS produced by *S. mucosus* strain A3^T *.

	S. mucosus A3^T	Triton X-100	Tween 80	Xanthan gum
Sunflower oil	70 ± 1.41	62.5 ± 1.82	62 ± 1.67	88.95 ± 1.60
Mineral oil	71 ± 1.60	67.5 ± 1.41	70 ± 1.15	90.3 ± 1.43
Olive oil	60.3 ± 1.52	60 ± 1.52	62.5 ± 1.64	100 ± 1.94
Tetradecane	75 ± 2.08	62.5 ± 1.82	62.5 ± 1.10	90.3 ± 0.98
Octane	70 ± 1.51	60 ± 1.52	60 ± 0.70	93.3 ± 2.18
Kerosene	70 ± 0.51	62.1 ± 2.12	60 ± 1.40	89 ± 1.02
Isopropyl myristate	67.5 ± 0.70	67.5 ± 0.70	67.5 ± 1.02	100 ± 1.34
Petrol	20 ± 3.01	70 ± 1.52	32.5 ± 2.60	92.5 ± 2.45
Diesel	37.5 ± 0.70	62.5 ± 2.12	65 ± 1.70	87.5 ± 3.01
Crude oil	95 ± 0.57	60 ± 0.70	60 ± 0.60	89.75 ± 1.32
Xylene	47.5 ± 0.70	12.5 ± 1.52	12.5 ± 1.30	86.5 ± 1.78
Toluene	52.6 ± 3.53	12.5 ± 0.70	60 ± 1.80	100 ± 1.75
Vaseline oil	52.6 ± 2.50	65 ± 0.57	60 ± 2.55	100 ± 1.67
Hexane	50 ± 2.51	50 ± 1.52	42 ± 0.66	100 ± 2.25

* Expressed as the percentage of the total height occupied by the oil-water emulsion after 24 h; 0.5% (w/v) EPS or chemical surfactant was used as emulsifier. Each value represents the average of three measurements.

3. Discussion

We report here, the first isolation and characterization of an exopolysaccharide produced by a halophilic bacterium belonging to the *Alphaproteobacteria* class. *Salipiger mucosus* A3^T excretes significant quantities of EPS when cultivated under optimum growth conditions, that is to say, the conditions that result in the highest speed of cell division and ultimately the greatest quantity of bacterial growth. These results do not agree with those of some other authors, who maintain that cell growth and EPS formation usually have different nutritional requirements [25,26]. The different culture conditions assayed did not change meaningfully the chemical composition of the EPS, although production decreased significantly when they were unfavorable (data not shown). EPS production by *S. mucosus* strain A3^T exhibited a fermentation kinetic similar to that of mauran [15] and those produced by *H. ventosae* and *H. anticariensis* [13] and *Alteromonas hispanica* and *Idiomarina fontislapidosi* [14]. It began early, during the exponential growth phase, then increased concomitantly with the rise in number of viable cells, only ceasing when the glucose substrate was almost completely consumed. The highest quantity of EPS was obtained after a period of three days. After the optimum incubation time there was a decrease in the quantity of EPS in the culture, which may well be due to enzymatic degradation, as has also been reported with other EPSs [27].

With regard to the chemical composition of our EPS, its sulfate content, together with the presence of phosphates and fucose, are especially interesting. Sulfates are not commonly found in microbial EPSs, although they are present in all the EPSs produced by the halophilic bacteria described by our group [see for example [13]–[15],[18]] and also in many marine bacteria and in cell wall polysaccharides from red and brown macroalgae. Sulfated EPSs are of great potential interest in medicine since they have a number of bioactive properties: anticoagulant, antiangiogenic, antiproliferative, antiviral, *etc*. [2,3,28]. Phosphate groups, which have also been observed in other EPSs [13,29], could confer important properties on them because they are essential to the activation of lymphocytes [30] and in some antitumoral processes [31]. Fucose and fucose-rich oligosaccharides can be used in biocosmetics, in the field of medicine and in the food industry [32]. The polymer from *S. mucosus* strain A3^T may prove to be a simple source of fucose, as reported for the EPSs excreted by *Klebsiella pneumoniae* and *Clavibacter michiganensis* [32] since chemical synthesis or extraction from algae is laborious and expensive and they are often in short supply.

One important feature of the emulsions produced by the EPS described here is that they are very stable and are composed of small, uniform droplets, resulting in a fine, smooth consistency. Other known polysaccharides, such as xanthan, are capable of producing stable emulsions but they tend to be thicker and more viscous, which is not very desirable for some of the uses for which emulsifiers such as these are intended [33]. Proteins play an important role in the emulsifying capacity of some exopolysaccharides [13,34]. The EPS from our halophilic strain contains protein concentrations of 1.6% (w/w). In the same way, emulsan, a biosurfactant obtained from *Acinetobacter calcoaceticus*, is a complex of polysaccharides and proteins that emulsifies mixtures of water and aliphatic, aromatic or cyclic hydrocarbons but cannot emulsify pure hydrocarbons [35]. *Salipiger mucosus* strain A3^T, on the other hand, produces an EPS capable of emulsifying higher percentages of pure hydrocarbons (tetradecane, octane, kerosene, xylene and crude oil) than the chemical surfactants used in comparison, and where crude oil is concerned, higher even than xanthan. Apart from this, the presence of acetyl groups renders the EPS somewhat hydrophobic, which might contribute to its emulsifying capacity, as Ashtaputre *et al.* [36] have described for the EPS produced by *S. paucimobilis*.

EPSs with high concentrations of charged components often form gels in the presence of metal ions and have great potential for removing toxic metals from polluted environments and wastewater as an alternative to other more aggressive methods [37]. EPS solutions from *S. mucosus* strain A3^T did not produce stable gels in the presence of the salts we tested, but they did chelate several metals with considerable efficiency. Anionic EPSs generally prefer to bind cations with large ionic radii [38], which agree with our findings, since our polysaccharide showed a high capacity to bind lead. It may well be that acetyls bring more electron-donating groups into the vicinity of the binding site, thus allowing the larger Pb ions to bind more firmly [38]. Although we are not yet sure of the actual mechanism involved when ions bind to these polymers, this type of chelation could be classified as biosorption, as mentioned by Valls and de Lorenzo [39]. Whatever the case may be, the strong chelating property of this polymer offers the possibility of its being used as a biosorbent in the treatment of polluted water and other such environments.

4. Experimental

4.1. Bacterial Strain

We used *Salipiger mucosus* strain A3^T (=CECT 5855^T), described by our group after an extensive program of isolating halophilic EPS producers from 19 hypersaline environments in Spain and Morocco [40]. *S. mucosus* strain A3^T was isolated from a hypersaline soil located in a solar saltern in Calbanche (Murcia) on the Spanish Mediterranean seaboard.

4.2. Optimization and Analysis of the Kinetics of EPS Production

The bacterial strain was cultivated in MY medium [41] supplemented with 7.5% (w/v) salts [42].

The EPS was isolated using the method described by Quesada *et al.* [23]. Briefly, the culture was centrifuged and the supernatant precipitated with cold ethanol before being ultracentrifuged, dialyzed against distilled water and lyophilized [43].

To establish which conditions lead to optimum EPS production, we assayed the following variables: incubation time (1–8 days), incubation temperature (22, 32 and 42 °C), sea-salt concentration (1, 2.5, 5, 7.5, 10, 15, and 20% (w/v)), carbon source (glucose, sucrose, mannose, galactose), glucose concentration (0, 1, 2, 5, 7 and 10% (w/v)) and incubation either in a rotating shaker (100 and 200 rpm) or static conditions.

Microbial growth and EPS production were monitored in batch cultures in 500-mL Erlenmeyer flasks containing 100 mL medium (three replicate flasks per experiment). Bacterial growth was determined by measuring optical density at 520 nm. Any residual carbon source was calculated using the glucose-oxidase technique [44].

4.3. Electron Microscopy

Ultrathin sections of bacterial cells were negatively stained as described elsewhere [7].

4.4. Chemical Analysis

Total carbohydrates [45], proteins [46], acetyls [47], pyruvate [48], hexosamines [49], sulfates and phosphates [13] were analyzed.

The EPS was purified and its negative net charge analyzed by AEC on a 1.5 m × 20 cm quaternary methyl ammonium Accel Plus column (Waters) eluted at a flow rate of 2 mL/min with 0.05 M NH_4HCO_3 (pH 8.0), followed by a linear gradient of 0.05 to 2 M NaCl in the same buffer. It was also monitored by UV detection at 210 nm. Five milliliters fractions were collected to determine their sugar composition and molecular mass.

Sugar composition was determined as described by Chaplin [50]. The purified EPS (100 μg) was subject to methanolysis with methanolic HCl (0.9 M) for 16 h at 80 °C. The resulting mixture of methylglycosides was dried under nitrogen at room temperature and re-N-acetylated by the addition of 50 μL dry methanol, 5 μL pyridine and 5 μL acetic anhydride, dried under nitrogen and then derivated with 15 μL trimethylsilylimidazole (Alltech) at room temperature for 30 min. The re-N-acetylated trimethylsilylated glycosides were analyzed on a BP1 fused-silica capillary column (12 m × 0.32 mm, SGE) with a Peri 2000 GLC-FID chromatograph (Perichrom) using a temperature program of 140–240 °C at 6.1 °C/min followed by isothermal elution.

4.5. Physical Properties

Apparent molecular mass was determined by high-performance size-exclusion chromatography (HPSEC) done on a 600 E system (Waters) equipped with a PL aquagel-OH 60, 8 μm column (30 cm × 7.5 mm) (Polymer Laboratories), eluted with a 0.2 M sodium-acetate buffer (pH 5.1) at a flow rate of 0.8 mL/min. The sample volume was 20 μL, containing 25 μg of EPS. Compounds were detected using refractive-index monitoring (Model 475, Kontron Instruments) and dextrans (7×10^4 Da to 4.9×10^6 Da, Sigma) were used as standards.

For rheological analysis the EPS was dissolved in distilled water (0.5%, w/v) and measurements were made at 25 °C in a controlled-stress Bohlin CSR10 rheometer.

The emulsifying activity of the EPS was determined by a modified version of the procedure described by Cooper and Goldenberg [51]. Equal volumes of the different EPS solutions (0.5%, w/v) in distilled water and various hydrophobic substrates were added to 105 × 15 mm glass tubes. The mixtures were shaken vigorously using a vortex and allowed to stand for 24 h at 4 °C. Emulsifying activity was expressed as the percentage of the total height occupied by the emulsion. The hydrophobic substrates were sunflower and olive oils (commercial brands), mineral oil, tetradecane, octane, kerosene, isopropyl myristate, xylene, toluene, vaseline oil, hexadecane (all from Sigma), and petrol, diesel and crude oil. Tween 80 and Triton X-100 (from Sigma) and xanthan were used as control surfactants. The emulsions were observed under a light microscope to determine their size and the uniformity of the drops in the oil phase and whether or not there was creaming or flocculation.

Metal-binding analyses were made as described by Geddie and Sutherland [38]. The EPS was applied to an Amberlite IR 120H+ cation exchange column (Avocado) buffered with doubly distilled water to convert it into the acidic form. 5 mL polysaccharide solutions (0.5%, w/v) were put into dialysis tubing in flasks containing 200 mL of each appropriate metal-salt solution and shaken at 100 rpm for 24 h at 30 °C. The quantity of metal removed from the solution, *i.e.*, bound to the polymer, was calculated by measuring the ions in solution at 0 h and those remaining after 24 h by atomic absorption spectrophotometry. Controls were made by placing 5 mL distilled water in dialysis tubing with the various metal-salt solutions. The metal salts used were cupric sulfate, cobalt chloride and lead acetate (Sigma).

5. Conclusions

Few EPSs with properties of interest to biotechnology have been described during the last decade. Among these few, however, we include the EPS produced by *Salipiger mucosus* strain A3^T, the first halophilic, EPS-producing bacterium discovered belonging to the *Alphaproteobacteria*. The most remarkable property of the EPS in question is its unusual composition—with a high fucose content—and therefore it offers possible novel applications as a biological agent. We are currently studying its antitumoral activity and also its use as a source of fucose and fucose-oligosaccharides.

Acknowledgments: This research was supported by grants from the Dirección General de Investigación Científica y Técnica (BOS2003-00498; CGL 2005-05947) and from the Plan Andaluz de Investigación, Spain. We thank our colleague J. Trout for revising our English text.

References

1. Sutherland, IW. *Biotechnology of Microbial Exopolysaccharides*; Cambridge University: New York, NY, USA, 1990.
2. Matou, S; Colliec-Jouault, S; Galy-Fauroux, I; Ratiskol, J; Sinquin, C; Guezennec, J; Fischer, AM; Helley, D. Effect of an oversulfated exopolysaccharide on angiogenesis induced by fibroblast growth factor-2 or vascular endothelial growth factor *in vitro*. *Biochem. Pharmacol* **2005**, *69*, 751–759.
3. Arena, A; Gugliandolo, C; Stassi, G; Pavone, B; Iannello, D; Bisignano, G; Maugeri, TL. An exopolysaccharide produced by *Geobacillus thermodenitrificans* strain B3-72: antiviral activity on immunocompetent cells. *Immunol. Lett* **2009**, *123*, 132–137.
4. Sutherland, IW. Vandamme, EJ, De Baets, S, Steinbüchel, A, Eds.; Polysaccharides from Prokaryotes. In *Biopolymers: Polysaccharides I*; Wiley-VCH: Weinheim, Germany, 2002; pp. 1–19.
5. MacCormick, CA; Harris, JE; Jay, AJ; Ridout, EJ; Colquhoun, IJ; Morris, VJ. Isolation and characterization of new extracellular polysaccharide from an *Acetobacter* species. *J. Appl. Bacteriol* **1996**, *81*, 419–424.
6. Quesada, E; Valderrama, MJ; Béjar, V; Ventosa, A; Gutiérrez, MC; Ruiz-Berraquero, F; Ramos-Cormenzana, A. *Volcaniella eurihalina* gen. nov., sp. nov., a moderately halophilic non motile gram-negative rod. *Int. J. Syst. Evol. Microbiol* **1990**, *40*, 261–267.
7. Bouchotroch, S; Quesada, E; del Moral, A; Llamas, I; Béjar, V. *Halomonas maura* sp. nov., a novel moderately halophilic, exopolysaccharide-producing bacteria. *Int. J. Syst. Evol. Microbiol* **2001**, *51*, 1625–1632.
8. Martínez-Cánovas, MJ; Quesada, E; Llamas, I; Béjar, V. *Halomonas ventosae* a new moderately halophilic, denitrifying, exopolysaccharide-producing bacterium. *Int. J. Syst. Evol. Microbiol* **2004**, *54*, 733–734.
9. Martínez-Cánovas, MJ; Béjar, V; Martínez-Checa, F; Quesada, E. *Halomonas anticariensis* sp. nov., from Fuente de Piedra, a saline-wetland wild-fowl reserve in Malaga, southern Spain. *Int. J. Syst. Evol. Microbiol* **2004**, *54*, 1329–1332.
10. Martínez-Checa, F; Béjar, V; Llamas, I; del Moral, A; Quesada, E. *Alteromonas hispanica* sp. nov., a polyunsaturated-fatty-acid-producing, halophilic bacterium isolated from Fuente de Piedra, south-east Spain. *Int. J. Syst. Evol. Microbiol* **2005**, *55*, 2385–2390.
11. Martínez-Cánovas, MJ; Béjar, V; Martínez-Checa, F; Páez, R; Quesada, E. *Idiomarina fontislapidosi* sp. nov. and *Idiomarina ramblicola* sp. nov., isolated from inland hypersaline habitats in Spain. *Int. J. Syst. Evol. Microbiol* **2004**, *54*, 1793–1797.
12. Quesada, E; Béjar, V; Ferrer, MR; Calvo, C; Llamas, I; Martínez-Checa, F; Arias, S; Ruíz-, García C; Páez, R; Martínez-Cánovas, MJ; del Moral, A. Ventosa, A, Ed.; Moderately halophilic exopolysaccharide-producing bacteria. In *Halophilic Microorganisms*; Springer: Berlin, Germany, 2004; pp. 295–314.
13. Mata, JA; Béjar, V; Llamas, I; Arias, S; Bressollier, P; Tallon, R; Urdaci, MC; Quesada, E. Exopolysaccharides produced by the recently described bacteria *Halomonas ventosae* and *Halomonas anticariensis*. *Res. Microbiol* **2006**, *157*, 827–835.
14. Mata, JA; Béjar, V; Bressollier, P; Tallon, R; Urdaci, MC; Quesada, E; Llamas, I. Characterization of exopolysaccharides produced by thres moderately halophilic bacteria belonging to the family *Alteromonadaceae*. *J. Appl. Microbiol* **2008**, *105*, 521–528.
15. Arias, S; del Moral, A; Ferrer, MR; Tallon, R; Quesada, E; Béjar, V. Mauran, an exopolysaccharide produced by the halophilic bacterium *Halomonas maura*, with a novel composition and interesting properties for biotechnology. *Extremophiles* **2003**, *7*, 319–326.
16. Bouchotroch, S; Quesada, E; Izquierdo, I; Rodríguez, M; Béjar, V. Bacterial exopolysaccharides produced by newly discovered bacteria belonging to the genus *Halomonas*, isolated from hypersaline habitats in Morocco. *J. Ind. Microbiol. Biotechnol* **2000**, *24*, 374–378.

17.   Béjar, V; Calvo, C; Moliz, J; Díaz-Martínez, F; Quesada, E. Effect of growth conditions on the rheological properties and chemical composition of *Volcaniella eurihalina* exopolisaccharide. *Appl. Biochem. Biotechnol* **1996**, *59*, 77–86.

18.   Béjar, B; Llamas, I; Calvo, C; Quesada, E. Characterization of exopolysaccharides produced by 19 halophilic strains of the species *Halomonas eurihalina*. *J. Biotechnol* **1998**, *61*, 135–141.

19.   Calvo, C; Ferrer, MR; Martínez-Checa, F; Béjar, V; Quesada, E. Some rheological properties of the extracellular polysaccharide produced by *Volcaniella eurihalina* F2-7. *Appl. Biochem. Biotechnol* **1995**, *55*, 45–54.

20.   Calvo, C; Martínez-Checa, F; Mota, A; Béjar, V; Quesada, E. Effect of cations, pH and sulphate content on the viscosity and emulsifying activity of the *Halomonas eurihalina* exopolysaccharide. *J. Ind. Microbiol. Biotechnol* **1998**, *20*, 205–209.

21.   Calvo, C; Martinez-Checa, F; Toledo, FL; Porcel, J; Quesada, E. Characteristics of bioemulsifiers synthesised in crude oil media by *Halomonas eurihalina* and their effectiveness in the isolation of bacteria able to grow in the presence of hydrocarbons. *Appl. Microbiol. Biotechnol* **2002**, *60*, 347–351.

22.   Martínez-Checa, F; Calvo, C; Caba, MA; Ferrer, MR; Béjar, V; Quesada, E. Efecto de las condiciones nutricionales sobre la viscosidad y la capacidad emulgente del biopolímero V2-7 de *Volcaniella eurihalina*. *Microbiología SEM* **1996**, *12*, 55–60.

23.   Quesada, E; Béjar, V; Calvo, C. Exopolysaccharide production by *Volcaniella eurihalina*. *Experientia* **1993**, *49*, 1037–1041.

24.   Martínez-Cánovas, MJ; Quesada, E; Martínez-Checa, F; del Moral, A; Béjar, V. *Salipiger mucosus* gen. nov., sp. nov., a moderately halophilic, exopolysaccharide-producing bacterium isolated from hypersaline soil, belonging to the α-*Proteobacteria*. *Int. J. Syst. Evol. Microbiol* **2004**, *54*, 1735–1740.

25.   Cheirslip, B; Shimizu, H; Shioya, S. Modelling and optimization of environmental conditions for kefiran production by *Lactobacillus kefiranofaciens*. *Appl. Microbiol. Biotechnol* **2001**, *57*, 639–643.

26.   Gorret, AU; Maubois, N; Engasser, JL; Ghoul, JM. Study of the effects of temperature, pH and yeast extract on growth and exopolysaccharide production by *Propionibacterium acidipropionici* on milk microfiltrate using a response surface methodology. *J. Appl. Microbiol* **2001**, *90*, 788–796.

27.   Pham, PL; Dupont, I; Roy, D; Lapointe, G; Cerning, J. Production of exopolysaccharide by *Lactobacillus rhamnosus* and analysis of its enzymatic degradation during prolonged fermentation. *Appl. Environ. Microbiol* **2000**, *66*, 2302–2310.

28.   Yim, JH; Son, E; Pyo, S; Lee, HK. Novel sulphated polysaccharide derived from red-tide microalga *Gyrodinium impudicum* strain KG03 with immunostimulating activity *in vivo*. *Mar. Biotechnol* **2005**, *7*, 331–338.

29.   Kitazawa, H; Ishii, Y; Uemura, J; Kawai, Y; Saito, T; Kaneko, T; Noda, K; Itoh, T. Augmentation of macrophage functions by an extracellular phosphopolysaccharide from *Lactobacillus delbrueckii* ssp. *bulgaricus*. *Food Microbiol* **2000**, *17*, 109–118.

30.   Nishimura-Uemura, J; Kitazawa, H; Kawai, Y; Itoh, T; Oda, M; Saito, T. Functional alteration of murine macrophages stimulated with extracellular polysaccharides from *Lactobacillus delbrueckii* ssp. *bulgaricus* OLL1073R-1. *Food Microbiol* **2003**, *20*, 267–273.

31.   Ebina, T; Ogata, N; Murata, K. Antitumor effect of *Lactobacillus bulgaricus* 878R. *Biotherapy* **1995**, *9*, 65–70.

32.   Vanhooren, PT; Vandamme, EJ. L-fucose: occurrence, physiological role, chemical, enzymatic and microbial synthesis. *J. Chem. Technol. Biotechnol* **1999**, *74*, 479–497.

33.   Desai, JD; Banat, IM. Microbial production of surfactants and their commercial potential. *Microbiol. Mol. Biol. Rev* **1997**, *61*, 47–64.

34.   Navon-Venezia, S; Banin, E; Ron, EZ; Rosenberg, E. The bioemulsifier alasan: role of protein in maintaining structure and activity. *Appl. Microbiol. Biotechnol* **1998**, *49*, 382–384.

35.   Rosenberg, E; Zuckerberg, A; Rubinovitz, C; Gutnick, DL. Emulsifier of *Arthrobacter* RAG-1: isolation and emulsifying properties. *Appl. Environ. Microbiol* **1979**, *37*, 402–408.

36.   Ashtaputre, AA; Shah, AK. Emulsifying property of a viscous exopolysaccharide from *Sphingomonas paucimobilis*. *World J. Microbiol. Biotechnol* **1995**, *11*, 219–222.

37.   Sutherland, IW. Structure-function relationships in microbial exopolysaccharides. *Biotechnol. Adv* **1994**, *12*, 393–448.

38.   Geddie, JL; Sutherland, IW. Uptake of metals by bacterial polysaccharides. *J. Appl. Bacteriol* **1993**, *74*, 467–472.

39.   Valls, M; de Lorenzo, V. Exploiting the genetic and biochemical capacities of bacteria for the remediation of heavy metal pollution. *FEMS Microbiol. Rev* **2002**, *26*, 327–338.

40. Martínez-Cánovas, MJ; Quesada, E; Martínez-Checa, F; Béjar, V. A taxonomic study to establish the relationship between exopolysaccharide-producing bacterial strains living in diverse hypersaline habitats. *Curr. Microbiol* **2004**, *48*, 348–353.

41. Moraine, RA; Rogovin, P. Kinetics of polysaccharide B-1459 fermentation. *Biotechnol. Bioeng* **1996**, *8*, 511–524.

42. Rodríguez-Valera, F; Ruíz-Berraquero, F; Ramos-Comenzana, A. Characteristics of the heterotrophic bacterial populations in hypersaline environments of different salt concentrations. *Microbiol. Ecol* **1981**, *7*, 235–243.

43. Quesada, E; del Moral, A; Béjar, V. Comparative methods for isolation of *Volcaniella eurihalina* exopolysaccharide. *Biotechnol. Tech* **1994**, *8*, 701–706.

44. Bergmeyer, HV; Bent, E. *Method of Enzymatic Analysis*; Academic: New York, NY, USA, 1965; pp. 123–130.

45. Dubois, M; Gilles, KA; Hamilton, JK; Rebers, PA; Smith, F. Colorimetric method for determination of sugars and related substances. *Anal. Chem* **1956**, *28*, 350–356.

46. Bradford, MM. A rapid and sensitive method for the quantification of microgram quantities of protein utilizing the principle protein-dye binding. *Anal. Biochem* **1976**, *72*, 248–254.

47. McComb, EA; McCready, RM. Determination of acetyl in pectin and in acetylated carbohydrate polymers. *Anal. Chem* **1957**, *29*, 819–821.

48. Sloneker, JH; Orentas, DG. Quantitative determination of pyruvic acid. *Nature* **1962**, *194*, 478.

49. Johnson, AR. Improved method of hexosamine determination. *Anal. Biochem* **1971**, *44*, 628–635.

50. Chaplin, MF. A rapid and sensitive method for the analysis of carbohydrate components in glycoproteins using gas-liquid chromatography. *Anal. Biochem* **1982**, *123*, 336–341.

51. Cooper, D; Goldenberg, G. Surface active agents from two *Bacillus* species. *Appl. Environ. Microbiol* **1987**, *54*, 224–229.

Samples Availability: Available from the authors.

MDPI AG
St. Alban-Anlage 66
4052 Basel, Switzerland
Tel. +41 61 683 77 34
Fax +41 61 302 89 18
http://www.mdpi.com

Marine Drugs Editorial Office
E-mail: marinedrugs@mdpi.com
http://www.mdpi.com/journal/marinedrugs

www.ingramcontent.com/pod-product-compliance
Lightning Source LLC
Chambersburg PA
CBHW051838210326
41597CB00033B/5700